TAYLOR & FRANCIS *MONOGRAPHS ON PHYSICS*

EDITOR

B. R. COLES, D.PHIL., B.SC.
Professor of Solid State Physics, Imperial College of Science and Technology, London

CONSULTANT EDITOR

SIR NEVILL MOTT, F.R.S.
Cavendish Professor of Experimental Physics, University of Cambridge

THE FERMI SURFACES OF METALS

THE FERMI SURFACES OF METALS

A DESCRIPTION OF THE FERMI SURFACES OF THE METALLIC ELEMENTS

ARTHUR P. CRACKNELL

Carnegie Laboratory of Physics, University of Dundee

TAYLOR AND FRANCIS LTD
LONDON

BARNES & NOBLE
NEW YORK

1971

First published 1971 *by Taylor & Francis Ltd., London and Barnes & Noble, New York.*

Reprinted from ADVANCES IN PHYSICS, *Volume* 18, *No.* 76, *November* 1969 *and*
Volume 20, *No.* 83, *January* 1971

Printed and bound in Great Britain by Taylor & Francis Ltd., 10–14 *Macklin Street,*
London WC2B 5NF

Taylor & Francis ISBN 0 85066 048 3
Barnes & Noble ISBN 389 04459 8

Preface

In 1960 there was held a very significant conference on the Fermi surface (see " *The Fermi surface.* Proceedings of an International Conference held at Cooperstown, New York, on 22–24 August 1960 ", edited by W. A. Harrison and M. B. Webb (New York: Wiley)). By that time it had become clear not only that the concept of the Fermi surface could be extremely useful in understanding many of the properties of a metal, but also that the detailed shape of the Fermi surface of any given metal could be determined by a number of rather direct methods and that these methods had been applied successfully to a few of the more common metallic elements. During the decade that followed there was a substantial effort devoted to the determination of the shapes of the Fermi surfaces of nearly all the metallic elements.

There are now a number of review articles and books that describe in considerable detail both the principal methods that are available for the determination of the shape of the Fermi surface of a given metal and also the use of the Fermi surface in understanding many different physical properties of a metal. What did appear to be missing from the literature was a systematic account of the shapes of the Fermi surfaces of the various metals and this monograph is an attempt to fill that gap. This monograph, which comprises a review that was originally published in two parts in *Advances in Physics*, seeks to summarize the state of knowledge of the shapes of the Fermi surfaces of the metallic elements at about the end of the 1960's; it is aimed at post-graduate students and research workers with some previous knowledge of the physics of electrons in metals. In part I after a brief discussion of Brillouin zones, Fermi surfaces and the free-electron approximation, the state of knowledge of the details of the topology and dimensions of the Fermi surfaces of the s-block and p-block metals and semi-metals is described in detail. It is shown that the free-electron model gives a good first approximation to the Fermi surfaces of many of these metals and the implication of this fact in the use of pseudopotential methods in describing the properties of these materials is briefly discussed. In part II there is a similar description of the Fermi surfaces of the d-block and f-block metals; these include the transition metals, the rare-earth metals (or lanthanide metals) and the actinide metals. In an encyclopaedic work of this nature it is almost inevitable that some references have been omitted; any such omission is accidental and I should be pleased to hear from any author who feels that his work has been ignored or misrepresented. The two articles have been reprinted with only one or two very small alterations; a subject index has been added but, because of the very large number of references (~ 2000), many of which are in multiple authorship, it was felt that the compilation of an author index would not be worth the effort involved. It should be noted that the References to Parts I and II are listed separately.

I am grateful to Professor B. R. Coles and Dr. L. Mackinnon for their encouragement at various stages in the preparation of this work, to the various authors, editors and publishers who so readily gave their permission for the reprinting of diagrams and tables, the sources of which are indicated appropriately in the text, and to Mr. M. K. Kettle for re-drawing all the diagrams.

February 1971 A. P. CRACKNELL

Contents

PART II

THE FERMI SURFACE

PART I

s- block and p- block metals

§ 1. Introduction

1.1. *Scope of Review*

The present state of our knowledge of the topology of the Fermi surfaces of the various metals is a delicate synthesis of both theoretical and experimental results. The present review and its successor (on d-block and f-block metals) are concerned with the collection and evaluation of what is known at present about the shape of the Fermi surface of each of the metallic elements in the periodic table, or, as it has sometimes been called, the ' Fermiology ' of the metals. The reader is assumed to be familiar already, in outline at least, with the main theoretical and experimental methods for determining the shape of the Fermi surface of a metal. There have been many reviews of the methods used in the calculation of band structures and of the various interpolation schemes which are often used to find the constant-energy surfaces, the density of states, the Fermi energy and hence the shape of the Fermi surface of a given metal (see, for example, Reitz (1955), Pincherle (1960), Callaway (1964), Heine (1964, 1969), Slater (1965), and Cornwell (1969)). There have also been several discussions of the various experimental methods that can be used in the determination of the shape of the Fermi surface of a metal (see, for example, Chambers (1956), Harrison and Webb (1960), Pippard (1960, 1965),

Rosenberg (1963), Shoenberg (1964), Simon (1965), Mackinnon (1966), Mercouroff (1967), Cochran and Haering (1968), and several of the articles in the book edited by Ziman (1969)).

In the present review and in part II we shall concentrate almost entirely on collecting together the theoretical and experimental results which exist for each metal ; the author hopes to be forgiven if the account seems rather like a stamp catalogue or an old-fashioned textbook of inorganic chemistry ; the author also hopes to be forgiven for any omissions of references to relevant work on these metals. The present state of our knowledge of the Fermi surfaces of some of these metals has recently been described by Shoenberg (1969) (namely the alkali metals, the noble metals, Be, Mg, Zn, Cd, Al, Pb, Bi and, briefly, the transition metals) ; slightly older reviews also exist of the electronic structure of graphite (Haering and Mrozowski 1960) and of Bi (Boyle and Smith 1963). An extensive bibliography, up to 1967, will be found in the book by Slater (1967).

1.2. *The Periodic Table*

The periodic table provides the most sensible classification for the study of the Fermi surfaces of the various metallic elements, because this classification is itself based on the electronic structures of the atoms of these elements. In table 1 a convenient version of the periodic table is given with the division into s-block, p-block, d-block, and f-block elements.

It is most useful for our present purposes to consider together those metals which are in the same group of the periodic table because such elements have very similar properties, whereas neighbours in horizontal periods exhibit gradual changes in their chemical, physical and electronic properties. In an arbitrary division we consider s-block and p-block elements now and reserve the consideration of d-block and f-block elements for part II. This division between part I and part II is then, roughly, that part I deals mostly with ‘ simple ’ metals for which the free-electron model gives a good first approximation to the shape of the Fermi surface, whereas part II deals mostly with ‘ non-simple ’ metals for which it is less clear that the free-electron model gives even a first approximation to the shape of the Fermi surface. Part II includes the transition metals and rare-earth metals. A clear and useful account of the periodic table and its relationship to the underlying electronic structures of the elements is given by Cooper (1968). For the sake of those readers not familiar with current chemical usage we mention that the terms ‘ s-block ’, ‘ p-block ’, etc., refer to the angular momentum quantum number of the subshell of electrons in an atom which either is incomplete or, for the inert gases, was the last subshell to be completed as the atomic number Z increased up to the Z of the element in question. The examples of electronic structure in table 2 should clarify this.

Table 1. The periodic table

	s-block		d-block										p-block					
Group	IA	IIA											IIIB	IVB	VB	VIB	VIIB	0
	Li	Be											B	C	N	O	F	Ne
	Na	Mg	IIIA	IVA	VA	VIA	VIIA	—	VIII	—	IB	IIB	Al	Si	P	S	Cl	Ar
	K	Ca	Sc	Ti	V	Cr	Mn	Fe	Co	Ni	Cu	Zn	Ga	Ge	As	Se	Br	Kr
	Rb	Sr	Y	Zr	Nb	Mo	Tc	Ru	Rh	Pd	Ag	Cd	In	Sn	Sb	Te	I	Xe
	Cs	Ba	La[1]	Hf	Ta	W	Re	Os	Ir	Pt	Au	Hg	Tl	Pb	Bi	Po	At	Rn
	Fr	Ra	Ac[2]															
			f-block															
			[1]Ce	Pr	Nd	Pm	Sm	Eu	Gd	Tb	Dy	Ho	Er	Tm	Yb	Lu		
			[2]Th	Pa	U	Np	Pu	Am	Cm	Bk	Cf	Es	Fm	Md	No	Lw		

Cooper (1968).

Table 2. Examples of electronic structure of elements

Atomic No.	Element	Electronic structure	s, p, d or f-block
3	Li	$1s^2 . 2s$	s
4	Be	$1s^2 . 2s^2$	
5	B	$1s^2 . 2s^2 . 2p$	p
10	Ne	$1s^2 . 2s^2 . 2p^6$	
11	Na	$\ldots 3s$	s
12	Mg	$\ldots 3s^2$	
13	Al	$\ldots 3s^2 . 3p$	p
18	A	$\ldots 3s^2 . 3p^6$	
19	K	$\ldots 3p^6 . 4s$	s
20	Ca	$\ldots 3p^6 . 4s^2$	
21	Sc	$\ldots 3p^6 . 3d . 4s^2$	d
22	Ti	$\ldots 3p^6 . 3d^2 . 4s^2$	
59	Pr	$\ldots 4d^{10} . 4f^3 . 5s^2 . 5p^6 . 6s^2$	f
60	Nd	$\ldots 4d^{10} . 4f^4 . 5s^2 . 5p^6 . 6s^2$	

In this article we discuss, in turn, each of the groups of metals in the s-block and p-block of the periodic table. We shall have to be a little bit arbitrary in drawing the line about which elements to include and which ones to exclude. Thus Si, Ge and grey Sn are regarded as semi-conductors and are excluded although their band structures and constant-energy surfaces have been studied extensively. Se and Te are also excluded on the grounds of being semiconductors rather than metals. On the other hand, graphite, though perhaps not usually regarded as a metal, does have a small overlap between the conduction and valence bands and therefore is included in our discussion ; it has the lowest conductivity of the semi-metals that we shall include. Fr and Ra are excluded on the ground that, for fairly obvious reasons, hardly anything is known about their electronic structures. The metals Cu, Ag and Au of group IB of the periodic table and Zn, Cd and Hg of group IIB are left until part II on the rather arbitrary grounds that they are d-block elements. In any other less rigid scheme of classification into ' simple ' and ' non-simple ' metals there are bound to be similar difficulties associated with determining the dividing line.

At the start of each section we give the electronic structure of each element in that group, for example,

Ga Gallium 2.8.18.(2, 1),

Fe Iron 2.8.(2, 6, 6).2,

where a completely full shell (as defined by the principal quantum number n) is indicated by a number without parentheses, but for an incompletely full shell the subshell structure is enclosed in parentheses (this is only done, in fact, when the incomplete shell is close to the conduction band or forms part of the conduction band). Thus Ga has the incomplete shell $4s^2 4p$ of electrons and Fe has the incomplete shell $3s^2 3p^6 3d^6$. One word of warning, however, should be added : the concept of orbits and shells that has been developed for describing the electronic structure of atoms will only be even approximately true for those electrons in a metal that are localized on one atom or ion core. In general, those electrons that are deep inside the ion core are localized and belong to well-defined and approximately **k**-independent levels. However, it is possible that electrons which one might, at first sight, have thought to be localized are in fact spread out into bands with a definite band structure. This is illustrated nicely in the band structure calculated by Mattheiss (1964) for the non-metal solid argon, see fig. 1. So long as such bands remain

Fig. 1

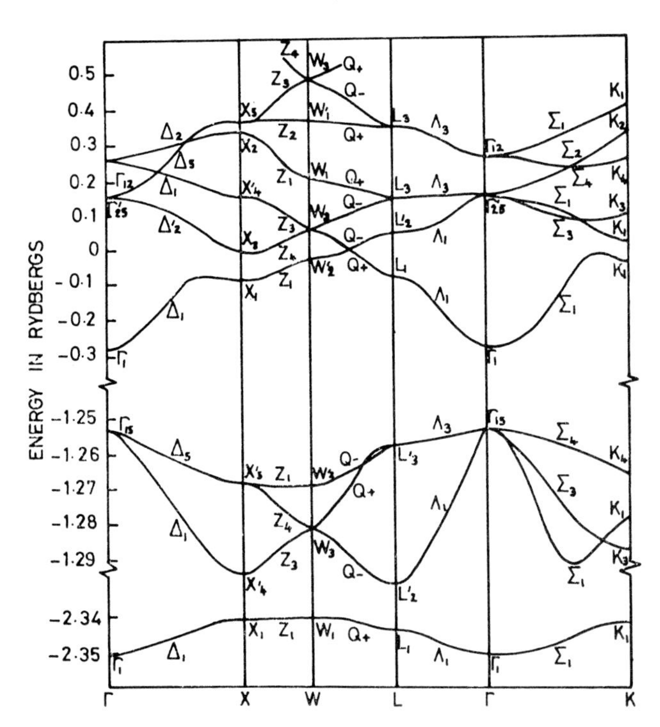

Energy band structure for f.c.c. solid argon (Mattheiss 1964).

separated from the conduction band by a large energy gap E_0 ($E_0 \gg kT$) this broadening is not very important.. However, the point to be emphasized now is that there is not necessarily always a clear distinction in a metal between the electrons in the ion core and the conduction electrons.

§ 2. Brillouin Zones, Energy Bands and Fermi Surfaces

2.1. *Brillouin Zones and the Free-electron Approximation*

Nearly every metal has either the hexagonal close-packed structure (h.c.p., P6$_3$/mmc, D_{6h}^4, 194) or the face-centred cubic structure (f.c.c., Fm3m, O_h^5, 225) or the body-centred cubic structure (b.c.c., Im3m, O_h^9, 229). The basis vectors of these three lattices can be written, referred to orthogonal Cartesian axes, as:

h.c.p.:

$$\left.\begin{aligned} \mathbf{t}_1 &= (0, a, 0) \\ \mathbf{t}_2 &= \tfrac{1}{2}(\sqrt{3}a, a, 0) \\ \mathbf{t}_3 &= (0, 0, c) \end{aligned}\right\}, \qquad (2.1.1)$$

f.c.c.:

$$\left.\begin{aligned} \mathbf{t}_1 &= \tfrac{1}{2}(0, a, a) \\ \mathbf{t}_2 &= \tfrac{1}{2}(a, 0, a) \\ \mathbf{t}_3 &= \tfrac{1}{2}(a, a, 0) \end{aligned}\right\}, \qquad (2.1.2)$$

and b.c.c.:

$$\left.\begin{aligned} \mathbf{t}_1 &= \tfrac{1}{2}(-a, a, a) \\ \mathbf{t}_2 &= \tfrac{1}{2}(a, -a, a) \\ \mathbf{t}_3 &= \tfrac{1}{2}(a, a, -a) \end{aligned}\right\}, \qquad (2.1.3)$$

where a and c have their usual meaning as the lengths of the sides of the conventional unit cell (Henry and Lonsdale 1965). For each structure the reciprocal lattice vectors $\mathbf{g}_1$, $\mathbf{g}_2$ and $\mathbf{g}_3$ can be determined using $\mathbf{g}_i \cdot \mathbf{t}_j = 2\pi\delta_{ij}$ and so the appropriate Brillouin zone can be constructed (Bouckaert *et al.* 1936, Herring 1942, Koster 1957).

The energy bands of the electrons in a metal can be determined in the individual-particle approximation by solving Schrödinger's equation:

$$-\frac{\hbar^2}{2m}\nabla^2\psi(\mathbf{r}) + V(\mathbf{r})\psi(\mathbf{r}) = E\psi(\mathbf{r}), \qquad (2.1.4)$$

where, in a perfect crystal, $V(\mathbf{r})$ possesses the periodic properties of the Bravais lattice of the metal. The solutions of this equation are known to take the form $\psi_{\mathbf{k}}(\mathbf{r}) = \exp(-i\mathbf{k}\cdot\mathbf{r})u_{\mathbf{k}}(\mathbf{r})$, where $u_{\mathbf{k}}(\mathbf{r})$ also possesses the periodicity of the Bravais lattice and $\mathbf{k}$ is the wave vector of the electron. In any real metal it is very difficult to determine theoretically a realistic expression for $V(\mathbf{r})$, the potential that is experienced by an individual electron in the conduction band of the metal. However, by first exploiting the periodic properties of $V(\mathbf{r})$ and subsequently setting $V(\mathbf{r})$ equal to zero everywhere, exact solutions of the one-electron Schrödinger equation can be obtained; this is called the *free-electron approximation* (Shockley 1937). It is a valuable approach because it illustrates many of the general features of energy band structures and their use in determining the shape of the Fermi surface. A fairly detailed treatment of the free-electron band structure is given by Jones (1960). Indeed, as we shall see in the course of this review, for many of the simpler non-transition and non-rare-earth metals the free-electron model gives a very good first approximation to the actual band structure and Fermi surface of the metal. This means that the net scattering of a conduction electron by the atoms in one of these metals is weak. This does not, however, imply that the potential $V(\mathbf{r})$ itself is weak, which is obviously not true near the centres of the atoms in the metal. It is this fact that the net scattering of the electrons is weak that underlies the considerable success that has attended the use of pseudopotential methods (see, for example, Ziman (1964), Harrison (1966), Heine (1969)). We shall return to a brief discussion of pseudopotentials in § 8.1.

The solutions of eqn. (2.1.4), where $\psi_{\mathbf{k}}(\mathbf{r}) = \exp(-i\mathbf{k}\,.\,\mathbf{r})u_{\mathbf{k}}(\mathbf{r})$ and when $V(\mathbf{r})$ is then set equal to zero, actually take the form:

$$\psi_{\mathbf{k}}(\mathbf{r}) = \exp\{-i(\mathbf{k} + n_1\mathbf{g}_1 + n_2\mathbf{g}_2 + n_3\mathbf{g}_3)\,.\,\mathbf{r}\}, \quad . \quad . \quad . \quad (2.1.5)$$

where n_1, n_2, and n_3 are integers and the corresponding energy eigenvalues are given by:

$$E_{\mathbf{k}} = \frac{\hbar^2}{2m}|\mathbf{k} + n_1\mathbf{g}_1 + n_2\mathbf{g}_2 + n_3\mathbf{g}_3|^2. \quad . \quad . \quad . \quad . \quad . \quad (2.1.6)$$

If we use the extended zone scheme $n_1 = n_2 = n_3 = 0$ and $\mathbf{k}$ is allowed to terminate anywhere in reciprocal space and is not required to be confined to the appropriate first Brillouin zone. In the extended zone scheme eqn. (2.1.6) reduces to:

$$E_{\mathbf{k}} = \frac{\hbar^2\mathbf{k}^2}{2m}, \quad . \quad . \quad . \quad . \quad . \quad . \quad . \quad . \quad . \quad (2.1.7)$$

so that the constant-energy surfaces are spheres. Having determined the eigenstates of the electrons either in the free-electron approximation or for a more realistic non-zero potential $V(\mathbf{r})$ we now have to consider the occupancy of these states. Which of these energy levels are actually occupied will be determined by the Fermi–Dirac distribution function $1/\{\exp[(E - E_F)/kT] + 1\}$. The constant-energy surface for which $E_{\mathbf{k}} = E_F$ is said to be the *Fermi surface*. At absolute zero the Fermi surface marks the boundary, in $\mathbf{k}$ space, between occupied and unoccupied states. At temperatures other than absolute zero this is still approximately true since typical values of kT ($\sim 1/40$ ev at room temperature) are much smaller than values of E_F (~ 5 ev for a typical metal). In the free-electron approximation the Fermi surface in the extended zone scheme will therefore be a sphere. In practice one seldom uses the extended-zone scheme, so that it is necessary to investigate what happens to those parts of this spherical

Fermi surface that have **k** vectors which terminate outside the Brillouin zone. Any piece of the sphere which lies outside the first Brillouin zone can be moved through a suitable reciprocal lattice vector $(n_1\mathbf{g}_1+n_2\mathbf{g}_2+n_3\mathbf{g}_3)$, that is chosen so that these occupied states come within the Brillouin zone. The details of the problem of decomposing the free-electron spherical Fermi surface into equivalent pieces that are all in the first Brillouin zone depend on the geometry of the lattice and on the number of conduction electrons contributed by the atoms in each unit cell of the metal.

If k_F is the radius of the Fermi surface the volume contained within the sphere of the free-electron Fermi surface is:

$$V_F = \tfrac{4}{3}\pi k_F{}^3. \qquad (2.1.8)$$

If a crystal contains N fundamental unit cells each of volume V, the volume of the first Brillouin zone is $(8\pi^3/V)$ and it contains N allowed wave vectors. Allowing for spin this means that in the Brillouin zone each band contains $2N$ states that may be occupied by electrons. If the atoms in each fundamental unit cell of the metal contribute between them x electrons to the conduction band of the metal the volume contained within the Fermi surface must be $\frac{1}{2}x$ times the volume of the Brillouin zone, so that

$$\tfrac{1}{2}x\left(\frac{8\pi^3}{V}\right) = \tfrac{4}{3}\pi k_F{}^3. \qquad (2.1.9)$$

Therefore

$$k_F = \left(\frac{2\pi}{a}\right)\left(\frac{3xa^3}{8\pi V}\right)^{1/3}, \qquad (2.1.10)$$

it being convenient to express k_F in terms of $(2\pi/a)$. The values of V for the lattices in question are:

f.c.c.: $\frac{1}{4}a^3$,

b.c.c.: $\frac{1}{2}a^3$,

h.c.p.: $\dfrac{\sqrt{3}}{2}\rho a^3$,

where $\rho = c/a$ for a hexagonal close-packed metal and may depart from the ideal value of $\sqrt{(8/3)}$ (McClure 1955, Heine and Weaire 1966, Weaire 1968). The value of k_F given by eqn. (2.1.10) can then be used, together with the known values of $\mathbf{g}_1$, $\mathbf{g}_2$ and $\mathbf{g}_3$ and the shape of the Brillouin zone, to determine the shape of the free-electron Fermi surface for metals of each structure and of various valence. This has been done for f.c.c. and b.c.c. metals with valence 1, 2, 3 and 4 and for h.c.p. metals with valence 1, 2 and 3 by Harrison (1960 b), see fig. 2.

Fig. 2

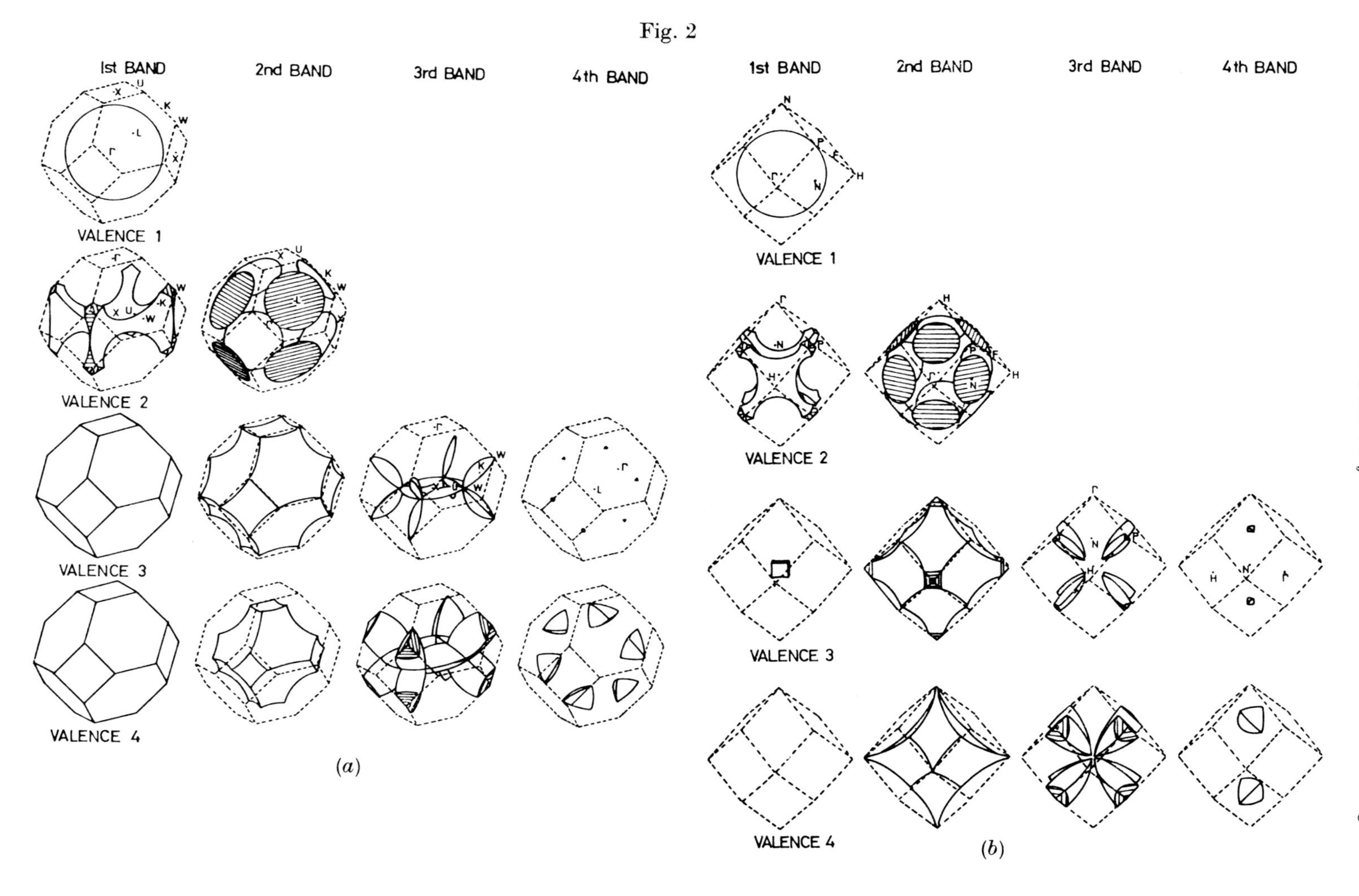
1st BAND
2nd BAND
3rd BAND
4th BAND
VALENCE 1
VALENCE 2
VALENCE 3
VALENCE 4
(a)
1st BAND
2nd BAND
3rd BAND
4th BAND
VALENCE 1
VALENCE 2
VALENCE 3
VALENCE 4
(b)

Fig. 2 (*continued*)

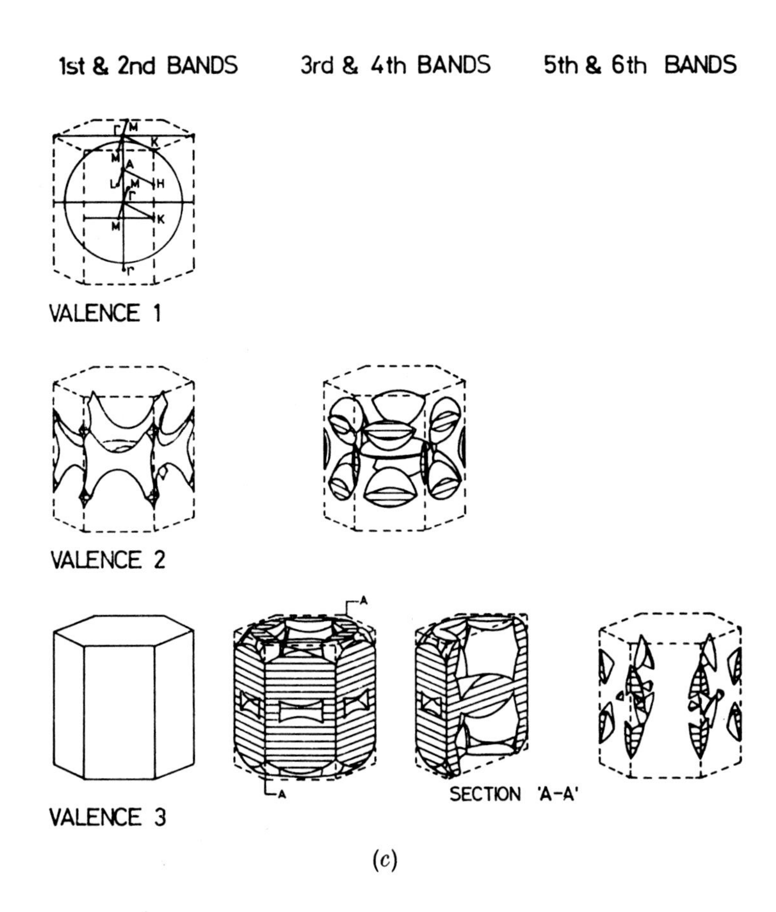

Free-electron Fermi surfaces for (*a*) f.c.c., (*b*) b.c.c., and (*c*) h.c.p. metals of various valence (Harrison 1960 b).

2.2. *Spin Effects*

In the past band structure calculations have mostly been performed with no spin-dependent terms in the Hamiltonian used in the calculation. In such approximations it is as though electron spin did not exist and the transformation properties of the wave functions and the degeneracies of the energy bands are described by the single-valued space-group representations (Bouckaert *et al.* 1936, Herring 1942). One particularly important result is that there is a double-degeneracy, due to time-reversal symmetry, all over the large top hexagonal face of the Brillouin zone of the h.c.p. structure (Herring 1937). For this reason a double Brillouin zone is often used for h.c.p. metals. The introduction of spin–orbit coupling effects and, more recently, of full-scale relativistic band structure calculations necessitates the replacement of the single-valued space-group representations by the double-valued space-group representations (Elliott 1954). This will lead, for some points and lines of symmetry in the Brillouin zone, to a lifting of some of the degeneracies. In particular, one finds that the double-degeneracy at a general point on the top face AHL of the Brillouin zone of an h.c.p. metal is lifted; it is also lifted at the point H (see fig. 2 (*c*)) but it still survives along the line AL (Cohen and Falicov 1960). Numerical estimates of this splitting at H for Mg were made by Cohen and Falicov (1960) using the results of a six-O.P.W. band structure calculation (Falicov 1962) and it was found to be very small, see § 4.3.

Fig. 3

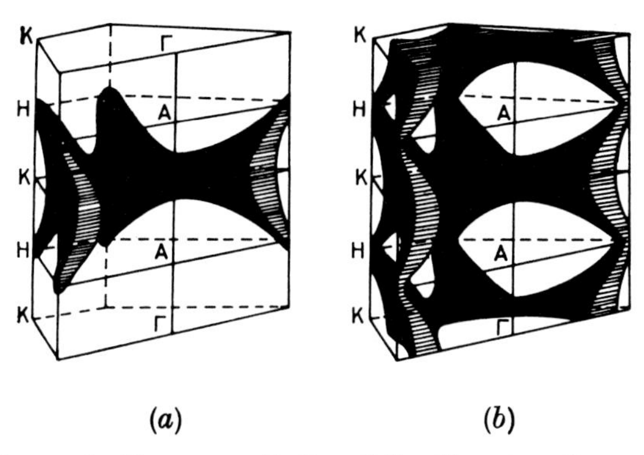

(*a*) (*b*)

The change in the connectivity of the Fermi surface of Mg due to spin–orbit coupling (Cohen and Falicov 1960).

Although the actual splitting due to the introduction of spin-dependent terms in the Hamiltonian may be small, the topological features of the Fermi surface of a pure metal may be quite drastically changed. For example, without spin–orbit coupling, that piece of the Fermi surface corresponding to holes in the first and second bands in Mg (Falicov 1962) is multiply-connected with infinite extent normal to the *c* axis but not parallel to it, see fig. 3 (*a*). With such a Fermi surface the observed lack of saturation of the magnetoresistance for fields normal to the *c* axis (Alekseevskiĭ and Gaĭdukov 1960, Lifshitz and Peschanskii 1958) cannot be understood. The spin–orbit splitting, however, causes the Fermi surface in the second band to extend infinitely along the *c* axis, see fig. 3 (*b*). This explains the observed magnetoresistance, because now open orbits can exist for all directions of magnetic field normal to the *c* axis. In practice it is sometimes difficult to distinguish between a Fermi surface that has open orbits for zero (or nearly zero) magnetic fields and situations in which open orbits arise as a result of magnetic breakdown between nearly connected pieces of Fermi surface.

§ 3. s-block : Group IA : the Alkali Metals

Li Lithium 2.1

Na Sodium 2.8.1

K Potassium 2.8.8.1

Rb Rubidium 2.8.18.8.1

Cs Caesium 2.8.18.18.8.1

3.1. *Structures*

The first point to establish in the discussion of these metals is their structure. According to Smithells (1967) Li, Na and K have the b.c.c. structure at room temperature and Rb and Cs have the f.c.c. structure at room temperature. Li and Na partially transform to the h.c.p. structure below 78°K and 35°K, respectively (Barrett 1956). Similar transformations were not found for K, Rb and Cs down to 1·2°K (Barrett 1955). Schneider and Stoll (1967 b) calculated the total energy at 0°K for each of the five alkali metals and for each of the three important common possible structures b.c.c., f.c.c., and h.c.p. Their calculated energies using a model pseudopotential determined from the measured phonon dispersion curves for Na and K and from the elastic constants for Li, Rb and Cs are shown in table 3.

Table 3. Relative energies of b.c.c., f.c.c. and h.c.p. structures of alkali metals at 0°K

	Li	Na	K	Rb	Cs
b.c.c.	20·47	13·88	0·76	0	0
f.c.c.	1·84	1·14	1·60	1·76	5·03
h.c.p.	0	0	0	0·12	3·37

Energies are in units of 10^{-16} erg per ion. (Schneider and Stoll (1967 b).)

These calculations are in agreement with the calculations of Harrison (1964) on Na and with the experimental results we have just described (Li, Na h.c.p. at very low temperatures, others b.c.c.) with one exception. Schneider and Stoll (1967 b) found that for K at 0°K the h.c.p. structure has the lowest energy and they therefore predict a transition in K to the h.c.p. structure at very low temperature, below that used in previous experimental structure determinations, i.e. below 1·2°K. Since Fermi surface determinations are usually performed above this temperature the Fermi surfaces that have been found for K do apply to the high temperature form.

It is clearly important in studying the topology of Fermi surfaces to ensure that the correct structure is being used; this is mostly because the shape of the Brillouin zone depends on the crystal structure of the metal, see § 2.1. For various experimental reasons it is nearly always necessary to use low temperatures in experiments for measuring the shape of the Fermi surface; the actual conditions which have to be satisfied in the various experimental methods are summarized in a convenient table by Mercouroff (1967, page 82). However, the importance of the Fermi surface lies in helping to explain the various properties of a metal not only at low temperatures but very often at high temperatures also, that is, so long as $kT \ll E_F$. If the metal undergoes a phase transformation, then it may be impossible to determine directly by experiment the shape of the Fermi surface for the high temperature phase. This has proved to be a serious difficulty in the way of determining the shape of the Fermi surface of Li which has a transition temperature of 78°K. The possibility of a direct determination of the Fermi surface of b.c.c. Li above 78°K is unlikely (see Mercouroff (1967), page 82), except possibly by positron annihilation experiments, and there would seem to be little chance of freezing in the b.c.c. structure at a sufficiently low temperature either. Even if one felt that a determination of the shape of the Fermi surface of the h.c.p. phase of Li would be useful, in order to help in explaining the low temperature properties of Li, this would still be a difficult experimental problem. This is because even if one obtains a single crystal of b.c.c. Li at room temperature it is very unlikely that it will turn into a single crystal of h.c.p. Li as it is cooled below the transition temperature unless very special annealing conditions are used. Moreover, if one did succeed in growing such a crystal of h.c.p. Li it would always have to be kept at a temperature below 78°K. However, for Na the transition temperature is only 35°K and, therefore, there is a good chance that as a single crystal b.c.c. sample of Na is cooled below 35°K the b.c.c. structure will be frozen in. The Fermi surface determination would then automatically be a determination of the Fermi surface of the high temperature form in a metastable state below the transition temperature (Lee 1966). It is, of course, important in such experiments to check that the sample still has the b.c.c. structure, either by X-ray work at the low temperature or by checking the rotational symmetry of the effect that is being used to determine the shape of the Fermi surface. For K any possible transition to the h.c.p. structure is unimportant because it must be at a temperature below that normally used in Fermi surface studies.

In the alkali metals, since the Fermi surface is only very slightly distorted from the free-electron sphere, which was described in § 2.1 and is completely contained within the first Brillouin zone, the effect of the periodic lattice on the conduction electrons is very small. Therefore, even when a phase transformation occurs in these metals there will be very little change in the contours of the energy surfaces or in the shape of the Fermi surface.

3.2. *Band Structure Calculations*

Some of the earliest calculations of the energy levels of electrons in metals were performed on the alkali metals (Wigner and Seitz 1933, 1934; Herring and Hill 1940; von der Lage and Bethe 1947; Howarth and Jones

1952). More recently these metals have continued to receive a large proportion of the time and energy of those theoreticians engaged in band structure calculations; this is doubtless because of the relative simplicity of the electronic structures of these metals. It is not our intention in this article to include any lengthy discussion of band theory because, as mentioned in § 1.1, several reviews already exist; we shall only mention a few of these calculations for the alkali metals in this section.

The cellular method has been applied to Li (Callaway and Kohn 1962), Na (Howarth and Jones 1952, Callaway 1961 a), K (Callaway 1956, 1960) and Cs (Callaway and Haase 1957, Callaway 1958 b). The tight-binding method was used for Li (Lafon and Lin 1966) and plane wave methods were used for Li (Brown and Krumhansl 1958, Glasser and Callaway 1958, Harrison 1963 c, Schlosser and Marcus 1963), Na (Callaway 1958 a, Harrison 1963 c, Hughes and Callaway 1964), K (Callaway 1956, Harrison 1963 c) and Cs (Callaway and Haase 1957, Mahanti and Das 1969). Energy band calculations were performed for Li, Na, K, Rb and Cs by Ham (1962 a) using the quantum defect method and the Green's function method, while the so-called GI method (Goddard 1967 a, b) has been applied to Li (O'Keefe and Goddard 1969 a, b). Definite trends in the band structure through the alkali metal series and as a function of lattice constant were noted by Ham. An interpolation scheme was then used (Ham 1962 b) to determine the shape of the Fermi surface of each metal. Li and Cs were found to be the most distorted from the spherical free-electron Fermi surface and exhibited bulges in the [110] directions, while those of Na and K were found to be nearly spherical. According to a band structure calculation by Callaway (1961 b) the Fermi surface for b.c.c. Li is distorted from the spherical free-electron Fermi surface by having bulges of the order of 5% along the [110] directions. By performing his band structure calculations for various values of the lattice constant, corresponding to known possible changes in the external pressure Ham (1962 b) predicted that the Fermi surface of Li should not touch the Brillouin zone boundary even under substantial pressure, while contact should occur for Cs at a slight compression. Various other parameters, such as the density of states, the Fermi energy, the cyclotron and optical effective masses were also calculated, see table 4. The band structures of both the b.c.c. and the h.c.p. phases of Na were calculated, using a pseudopotential method, by Hughes and Callaway (1964). They deduced that the Fermi surface is nearly spherical in both phases with a maximum distortion of the order of $\frac{1}{4}$%. Within the accuracy of the calculations the Fermi energy and effective mass were found to be the same in the two phases

In many of the earlier band structure calculations it was not possible to determine the energy eigenvalues for electrons with a general wave vector $\mathbf{k}$ but only for those wave vectors that terminate on special points of symmetry or lines of symmetry in the Brillouin zone. Various interpolation schemes were then devised to estimate the eigenvalues for general wave vectors, $\mathbf{k}$, and thence to find the density of states, the Fermi energy

Table 4. Calculated parameters of the Fermi surfaces of the alkali metals

	Li	Na	K	Rb	Cs
a (A.U.)	6·651	8·109	10·049	10·742	11·458
m_t/m_0	1·64	1·00	1·07	1·18	1·75
$(m_t/m_0)_s$	1·32	1·00	1·01	0·99	1·06
m_a/m_0	1·45	1·00	1·02	1·06	1·29
m_t/m_a	1·13	1·00	1·06	1·11	1·36
$m_{c[110]}/m_0$	1·48	1·00	1·035	1·07	1·46
$m_{c[100]}/m_0$	1·65	1·00	1·063	1·16	1·92
$m_{c[111]}/m_0$	1·82	1·00	1·092	1·25	2·38
$A_{[110]}/A_0$	0·976	1·00	0·995	0·979	0·94
$A_{[100]}/A_0$	0·993	1·00	1·001	0·996	0·99
$A_{[111]}/A_0$	1·011	1·00	1·007	1·013	1·04
S/S_0	1·06	1·00	1·03	1·06	1·12
$(S/S_0)^2$	1·11	1·00	1·06	1·13	1·25
$k_{[110]}/k_F$	1·023	1·00	1·007	1·018	1·08
$k_{[100]}/k_F$	0·973	1·00	0·994	0·980	0·94
$k_{[111]}/k_F$	0·983	1·00	0·994	0·980	0·94

a is the lattice constant, m_t thermal mass, m_a optical mass, m_c cyclotron mass, A extremal area of cross section of the Fermi surface, S surface area of the Fermi surface, k radius of Fermi surface. S_0 and A_0 denote the corresponding areas of a spherical Fermi surface and k_F its radius. (Adapted from Ham (1962 b).)

and the shape of the Fermi surface. In one such method an l-dependent pseudopotential interpolation scheme, using only a very small number of plane waves in the expansion of the valence electron wave functions, was applied to the alkali metals (Cornwell and Wohlfarth 1960, Cornwell 1961). The general trend suggested by Cohen and Heine (1958), see § 3.3, was reproduced for Li, Na and K in this calculation; however, not enough raw data were available at that time for these calculations to be performed for Rb and Cs.

3.3. *Macroscopic Properties*

While the observation and measurement of macroscopic properties are unlikely, in general, to be particularly sensitive to the fine details of the shape of the Fermi surface of a metal it is, nevertheless, possible to use the macroscopic properties to obtain some general indications of trends in the topology of the Fermi surfaces of the metals in a given group of the periodic table.

A review of the information that could be obtained about the electronic band structures and the topology of the Fermi surfaces of the alkali metals by studying their macroscopic properties was given by Cohen and Heine (1958). These properties include the soft X-ray emission spectrum, the Knight shift, the ratio of the low temperature electrical and thermal resistivities, the thermoelectric power, the variation of the electrical conductivity with pressure and the magnetoresistance. Their general conclusions, which have largely been borne out by later experiments, were that the Fermi surface of Na is nearly spherical, the Fermi surface of K is somewhat distorted, the Fermi surfaces of Rb and Cs are rather more distorted, while the Fermi surface of Li is also considerably distorted and might even touch the Brillouin zone boundary. There is thus a steady trend of increasing distortion from sphericity as the atomic number is increased, with the exception that the behaviour of Li is somewhat anomalous. Subsequent work has shown these conclusions to be substantially correct although, according to Ham (1962 a, b) the Fermi surface of Li does not touch the Brillouin zone boundaries at atmospheric pressure but this contact does occur at a slightly higher pressure (see § 3.2 above). The somewhat anomalous behaviour of Li arises because the Li atom has a very small ion core, Li^+ ($1s^2$), so that the valence electron is, in general, closer to the nucleus than in the other alkali metals. Its wave function is therefore more distorted from a free-electron wave function and, consequently, the energy bands and the Fermi surface itself can be expected to be more distorted from the free-electron model than occurs in the other alkali metals.

One interesting property is that the ratio of the low temperature electrical and thermal resistivities is sensitive to the shape of the Fermi surface (Klemens 1954 a, b, c). For the alkali metals the quantity

$$D' = \frac{64{\cdot}0}{497{\cdot}6}\frac{\Theta_D^2}{L}\frac{\rho_i(T)}{W_i(T)} \quad . \quad . \quad . \quad . \quad . \quad . \quad . \quad (3.3.1)$$

should, qualitatively, increase with increasing distortion of the Fermi surface from sphericity, where Θ_D is the Debye temperature, L is the Lorentz number, $\rho_i(T)$ is the ideal electrical resistivity and $W_i(T)$ is the ideal thermal resistivity in the temperature range where they are proportional to T^5 and T^2, respectively. The values of D' calculated by Cohen and Heine (1958) are:

Li	Na	K	Rb	Cs
19·3	3·1	6·8	14·2	15·7

which illustrate quite clearly the trend which we have already mentioned. The same general trend in the departure from sphericity of the Fermi surface in the alkali metals was also deduced from magnetoresistance data by García-Moliner (1958, 1959) except that the order of K and Rb was reversed. However, the later work of Shoenberg and Stiles (1964) confirmed that the Fermi surface of Rb is more anisotropic than that of K.

There have been a large number of experimental determinations of γ, the temperature coefficient of the electronic contribution to the specific heat. The results were collected and reviewed by Martin (1965) who gives the following values:

$$\begin{aligned} \text{Li}: &\quad (1{\cdot}63 \pm 0{\cdot}02)\ \text{mJ mole}^{-1}\,\text{deg}^{-2} \\ \text{Na}: &\quad (1{\cdot}38 \pm 0{\cdot}02)\ \text{mJ mole}^{-1}\,\text{deg}^{-2} \\ \text{K}: &\quad (2{\cdot}08 \pm 0{\cdot}08)\ \text{mJ mole}^{-1}\,\text{deg}^{-2} \\ \text{Rb}: &\quad (2{\cdot}41\,^{+0{\cdot}29}_{-0{\cdot}16})\ \text{mJ mole}^{-1}\,\text{deg}^{-2} \\ \text{Cs}: &\quad (3{\cdot}20 \pm 1{\cdot}05)\ \text{mJ mole}^{-1}\,\text{deg}^{-2} \end{aligned}$$

(see Martin (1965) for references). Substantially similar values of γ for K, Rb and Cs were obtained by Lien and Phillips (1964). If the effect of electron–phonon interactions is neglected γ is simply related to the density of states at the Fermi level which can be predicted from band structure calculations. The difference between the experimental values of γ and values calculated in this simple way for any given metal gives a good measure of the importance or otherwise of electron–phonon enhancement of γ in that metal, see § 8.1. Therefore in this article we shall quote values of γ for each metal because of current interest in electron–phonon interactions, rather than in terms of providing a test of various band structure calculations. The influence of structural phase changes in the alkali metals on γ has been studied by a number of authors (for example, Martin (1961 b), Stern (1961), and Bhattacharya and Stern (1964)). The discrepancies between the observed values of χ_p, the Pauli spin paramagnetic susceptibility, and the values calculated from band structure calculations of the density of states at the Fermi level are usually quite large (see, for example, the values collected by Lomer (1962)); this is because of the difficulty of either isolating experimentally the spin contribution from other contributions to the susceptibility or of calculating these other contributions theoretically. We shall therefore not bother in this article to quote values of χ_p.

The effect of pressure to over 500 kbar on the electrical resistance of Li, Na, K and Rb was measured at 296°K and 77°K by Stager and Drickamer (1963 b). They generally observed an increase in electrical resistance with increasing pressure; several discontinuities in the resistance or in its slope were observed and interpreted in terms of changes in the crystalline structures of the metals (see also Drickamer (1965) and Stocks and Young (1969)). The effect of pressures up to 15 kbar on the Hall constants of the alkali metals has been measured by Deutsch *et al.* (1961).

Experiments by Mayer and El Naby (1963) on the optical spectra of the alkali metals from the near infra-red to the near ultra-violet, performed under conditions of high surface purity and in high vacuum, have demonstrated the existence of resonance behaviour in the absorptivity at energies below the band-theoretically predicted onset of interband absorption. This resonance is sufficiently large and broad in K and Cs to mask completely the interband threshold. In Na one can clearly distinguish the Drude absorption in the infra-red region, the resonance near 1·7 ev, and the interband absorption region beginning near the value 2·0 ev, predicted by theory, but at low temperatures (-183°C) the resonance disappears leaving only Drude and interband contributions to the absorptivity. The intrinsic absorbing power due to interband transitions in Na has been calculated by Appelbaum (1966), using orthogonalized plane waves to evaluate the matrix elements appearing in the oscillator strengths and a realistic band structure, obtained from a pseudopotential interpolation scheme, to perform the sum over $\mathbf{k}$ space. The ultra-violet values of the conductivity of Li show a weak absorption edge near $\hbar\omega = 3{\cdot}2$ ev (Hodgson 1965) which is near to the minimum energy for interband transitions calculated by Ham (1962 b) to be 3·6 ev. Hodgson (1965) also obtained a value for the optical effective mass of Li of $m^*/m = 1{\cdot}57 \pm 0{\cdot}02$ from the dielectric constant at long wavelengths; this is to be compared with the calculated value due to Ham (1962 b) of 1·45 for the optical effective mass of Li. Several interband transitions were identified in the thermoreflectance measurements of Matatagui and Cardona (1968) on K, Rb and Cs. Although X-ray studies of electronic band structure have been made for several decades it is only recently that they have achieved a high enough degree of accuracy to give accurate detailed measurements of density of states curves (see, for example, the work of Phillips and Weiss (1968) on Li and Na using Compton scattering line shapes).

Measurements of the magnetoresistance of the three phases of Li, b.c.c., f.c.c. and h.c.p., were performed by Gugan and Jones (1963). It was not possible to deduce any firm conclusion as to whether the Fermi surface actually touches the Brillouin zone boundary in Li, but in the h.c.p. phase, the least distorted of the three, the results indicated that the transverse magnetoresistance was approaching saturation, thereby indicating the existence of closed orbits. The effect of strain on the magnetoresistance of Na and K was measured by Jones (1969).

While discussing the observation of macroscopic properties of the alkali metals it is perhaps appropriate to mention calculations and experimental observations of the Knight shift, that is, of the difference between the n.m.r. frequency of a free atom and of the same element in metallic form. The shift can be regarded as due to the difference between the effective magnetic field, at the nucleus, arising from the electrons in the atom or metal. The magnitude of the shift can be shown to be related to the average, over the whole Fermi surface, of the percentage of s-like contribution to the electronic wave function in the metal (Townes *et al.* 1950). Results have been collected and tabulated by Knight *et al.* (1959). More recently, calculations of the Knight shift have been performed by Etienne-Amberg (1966) and further experiments have been performed on Li by Schumacher and VanderVen (1966), on K by Milford and Gager (1961) and van der Lugt and Knol (1967) and on Cs by Holcomb *et al.* (1966). Calculations of the pressure dependence of the Knight shift in alkali metals have recently been made by Micah *et al.* (1969 a, b). Knight shift calculations and experimental results clearly only give very indirect information on the shape of the Fermi surface of a metal.

3.4. *The de Haas–van Alphen Effect*

The de Haas–van Alphen effect, that is, the observation of oscillations in the diamagnetic susceptibility of a metal as a function of applied magnetic field, **B**, gives a direct measurement of the extremal area of cross section of the Fermi surface normal to **B**. The effect requires low temperatures and high purity of the sample (see Mercouroff (1967), page 82). We have already mentioned the difficulties that arise in the experimental investigation of the Fermi surface of Li as a result of the phase transformation which occurs at 78°K.

Early measurements of the de Haas–van Alphen effect in K were performed by Thorsen and Berlincourt (1961 a) and gave an extremal cross-sectional area of the Fermi surface of $1{\cdot}66 \times 10^{16}\,\mathrm{cm}^{-2}$ ($\pm 1\%$) while the corresponding free-electron value is $1{\cdot}68 \times 10^{16}\,\mathrm{cm}^{-2}$. They estimated the effective mass to be $0{\cdot}90 \pm 0{\cdot}09$. The de Haas–van Alphen effect was also observed in Rb by Thorsen and Berlincourt (1961 b) and in Rb and Cs by Okumura and Templeton (1962, 1963). These early measurements of the de Haas–van Alphen effect showed that the Fermi surfaces of the alkali metals are very nearly spherical. It appeared that the Fermi surface of K was spherical to within a few tenths of 1% (Thorsen and Berlincourt 1961 a) while the departures from sphericity in Rb were of order $\frac{1}{2}\%$ (Okumura and Templeton 1962) and in Cs 1 or 2% (Okumura and Templeton 1963). A special field-modulated technique was developed by Shoenberg and Stiles (1963) which is capable of detecting and measuring very small departures from a spherical Fermi surface. This was applied to K and Rb by Shoenberg and Stiles (1964) and again to K by Lee and Falicov (1968) and from the results a rather detailed determination was made of how the Fermi surface departs from a sphere. Using this method Shoenberg and Stiles (1964) were able to observe the de Haas–van Alphen effect in Na for the first time; they also observed the effect in Cs but because the anisotropy is appreciably greater than in Rb resolution of the oscillations obtained by the field-modulated technique was more difficult.

The de Haas–van Alphen effect in Na has been studied by Lee (1966) using the field modulation technique of Shoenberg and Stiles (1963). Reasons were discussed for relating the experimental results to the Fermi surface of the high temperature, b.c.c., phase of the metal. The Fermi surface of Na was shown to be pulled out in the $\langle 110 \rangle$ directions, towards the faces of the first Brillouin zone, as would be expected for nearly-free electrons. The extreme radial distortion of the Fermi surface from a free-electron sphere was found to be rather less than $0{\cdot}1\%$. The inversion of the experimental data to determine the shape of the Fermi surface was done by using a pseudopotential interpolation scheme due to Ashcroft (1965); contours showing the distortion from the free-electron sphere are shown in fig. 4.

Lee and Falicov (1968) showed that their de Haas–van Alphen data for K could not be explained in detail in terms of a local pseudopotential but they obtained a good description of the Fermi surface of K, consistent with

Fig. 4

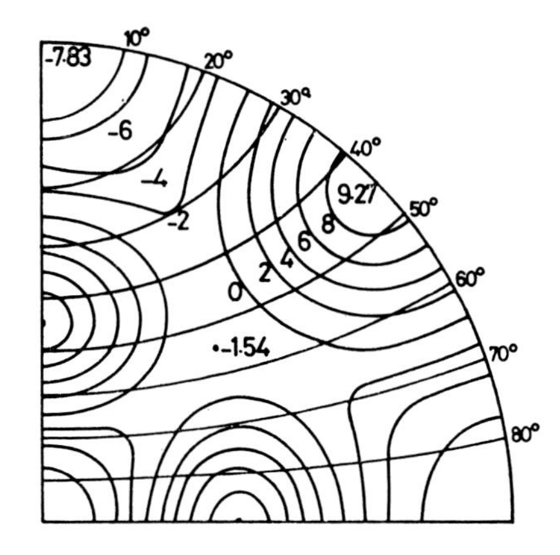

Contours of $\Delta r/r$ of the Fermi surface of Na; units are parts in 10^4 (Lee 1966).

their de Haas–van Alphen data, by fitting the coefficients of a non-local pseudopotential. Contours of $\Delta r/r$ for K and Rb are shown in figs. 5 and 6. The experimental values of $\Delta r/r$ are compared with the results of various band structure calculations in table 5. The effective masses determined by Shoenberg and Stiles (1964) were 1·18 and 1·25 for K and 1·28 for Rb. The value obtained in the cyclotron resonance work of Grimes and Kip (1963) for K was 1·21.

Fig. 5

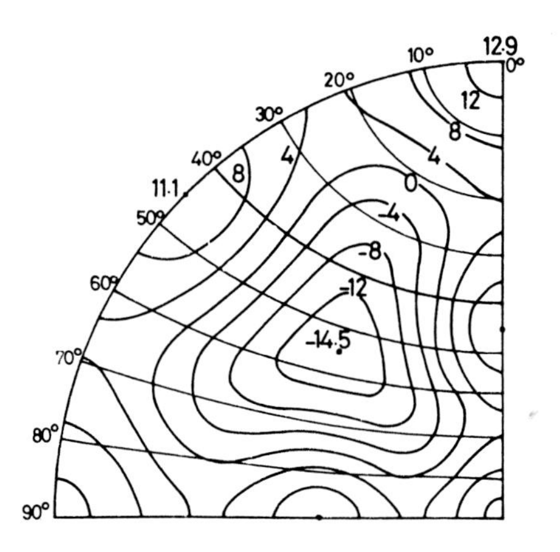

Contours of $\Delta r/r$ of the Fermi surface of K; units are parts in 10^4 (Shoenberg and Stiles 1964).

Fig. 6

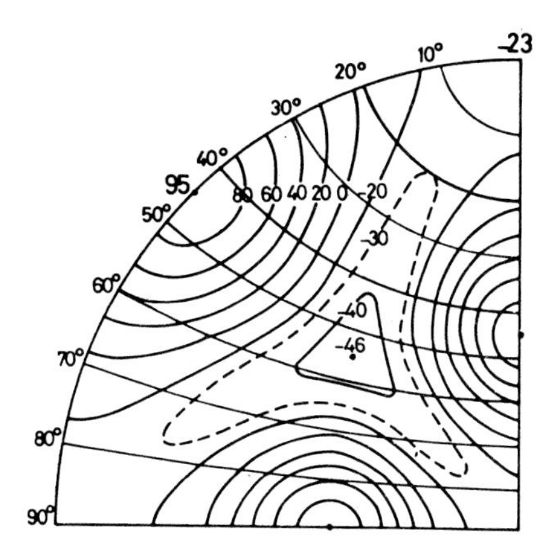

Contours of $\Delta r/r$ of the Fermi surface of Rb; units are parts in 10^4 (Shoenberg and Stiles 1964).

Table 5. Anisotropy of Fermi surfaces of Na, K and Rb

	Shoenberg and Stiles (1964)		Ham (1962 a, b)	Heine and Abarenkov (1964)
Na		<5	0	10 to 45
K	[100]	$12\cdot9 \pm 1\cdot4$	−60	0 to 18
	[110]	$11\cdot1 \pm 0\cdot8$	70	
	[111]	$-14\cdot5 \pm 0\cdot9$	−60	
Rb	[100]	-23 ± 14	−200	25 to 75
	[110]	95 ± 5	180	
	[111]	-46 ± 10	−200	

The anisotropy is given as $\Delta r/r$ (parts in 10^4). (Shoenberg and Stiles (1964).)

A study of the de Haas–van Alphen effect in Cs by the pulsed field method was carried out by Okumura and Templeton (1965). A computer analysis of the results in terms of a cubic harmonic series indicated a probable radial distortion of some $+3\cdot3\%$ in the [110] direction, $-0\cdot9\%$ in the [100] direction and $-1\cdot4\%$ in the [111] direction. The pulsed field method yields results of considerably greater accuracy in Cs than it does in K and Rb where the distortions are so small. In the [110] direction Okumura and Templeton (1965) observed bumps in the Fermi surface and contours of $\Delta r/r$ in Cs as plotted on a stereogram in fig. 7, from which it will be seen that the distortion is considerably larger than in Rb.

Fig. 7

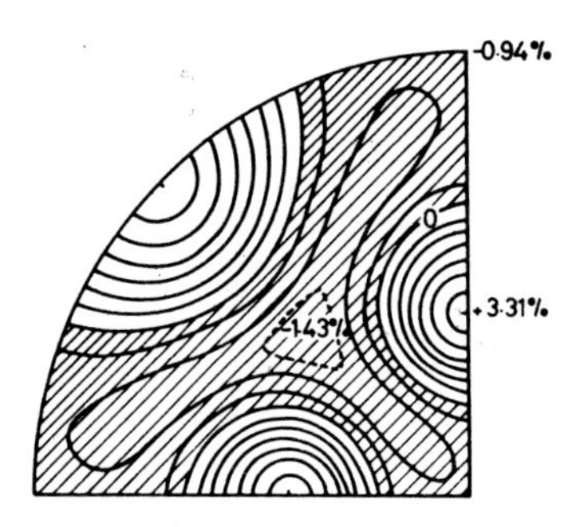

Contours of $\Delta r/r$ of the Fermi surface of Cs (Okumura and Templeton 1965).

3.5. *Positron Annihilation*

Anisotropies in the momentum distribution of positron annihilation γ-radiation have been observed in several single-crystal metals; a review has been given by Wallace (1960). The details of the interpretation of the angular correlation measurements on the γ-rays in terms of the momentum distribution of the electrons in the solid and on the shape of the Fermi surface are complicated. When a positron, which is assumed to be thermalized, annihilates with an electron in a solid, the momentum of the centre of mass of the two particles appears as a small deviation from collinearity of the two γ-rays emitted. Since the positron is assumed to have been thermalized before the annihilation occurs, this momentum is almost entirely that of the electron. The available momenta of the conduction electrons are less than or equal to the Fermi momentum, so that one would expect a sharp drop in the number of γ pairs emitted with a deviation from collinearity corresponding to a momentum larger than the Fermi momentum. One would therefore expect it to be possible, with a single-crystal specimen, to be able to determine the radius of the Fermi surface in any given direction. However, the interpretation of the angular correlation curves in terms of Fermi surface topology is difficult (Berko 1962) because it also involves some knowledge of the details of the departure of the conduction-electron wave function from a plane wave. Experiments on positron annihilation in Na and Li by Donaghy *et al.* (1964) showed that the Fermi surface of Na was spherical within the experimental error (1·5%) but that the Fermi surface of Li was distorted with $|\mathbf{k}_{110}|$ being about 5–6% greater than $|\mathbf{k}_{100}|$ (see also Donaghy and Stewart 1967 a, b). Detailed theoretical predictions of the angular correlation curves for Li were made by Melngailis and De Benedetti (1966) using the O.P.W. method to determine the band structure and the conduction electron wave functions. By including both the Fermi surface topology and the effect of the wave functions, but not just the Fermi surface topology on its own, good agreement with the experimentally observed anisotropies was obtained.

Quantitative agreement with experiment was obtained when, in addition to all other effects, the enhancement factor due to electron–positron attraction calculated by Kahana was included. Previous positron annihilation experiments on Li and Li–Mg alloys by Stewart (1964) were unable to determine the anisotropy of the Fermi surface of Li. This was because the mean penetration of positrons into this light metal was about 0·025 in. or half the slit width; the geometrical resolution of the instrument therefore could not be sufficiently well defined. Stewart (1961) also performed a positron annihilation experiment on the Fermi surface of Na which indicated a substantially larger distortion from sphericity than was later observed in the de Haas–van Alphen results of Lee (1966); the discrepancy is probably due to the difficulties mentioned above in the interpretation of positron annihilation experiments.

3.6. *Ultrasonic Attenuation: Magnetoacoustic Oscillations*

Information about the shape of the Fermi surface of a metal can be obtained from the observation of ultrasonic attenuation in the metal in the presence of an applied magnetic field. It is possible to observe geometrical resonances which are a size effect in which there exists some simple relationship between the wavelength of the ultrasound and the extremal diameter of the orbit of the electron in the magnetic field. Because there is a geometrical similarity between the orbits in real space and in **k** space due to the applied magnetic field, these geometric resonances, which may occur in a variety of relative orientations of crystallographic axes, magnetic field **B**, wave vector **q** of the ultrasound and displacement vector ξ of the ultrasound, enable extremal diameters of the Fermi surface to be measured. Alternatively, quantum oscillations in the ultrasonic attenuation may be observed; these magnetoacoustic oscillations are analogous to the de Haas–van Alphen oscillations and, similarly, enable an extremal cross-sectional area of the Fermi surface, normal to **B**, to be determined.

The magnetic field variation of the ultrasonic attenuation of longitudinal waves of frequencies 10 to 50 MHz in K was observed by Trivisonno *et al.* (1966). The Fermi momentum was found to be $7{\cdot}6 \times 10^{-20}$ g cm sec^{-1}, which is fairly close to the free-electron value. No shifts in the positions of the extremal values of the attenuation were observed as the magnetic field was rotated in the (100) or (110) plane, indicating that the Fermi surface is spherical to better than 1%. Magnetoacoustic attenuation of longitudinal ultrasonic waves in K by Foster *et al.* (1965) showed that the Fermi surface in K is only slightly distorted relative to its free-electron value, bulging by about 0·3% in the ⟨110⟩ direction and pushed in by about 0·4% and 0·5% respectively in the ⟨100⟩ and ⟨111⟩ directions; this is in quite good agreement with the de Haas–van Alphen results of Shoenberg and Stiles (1964). Further magnetoacoustic work on K has been done by Peverley (1968) and several other references are noted by Blaney (1968) in connection with the investigation of the possibility of the existence of a spin-density wave (or charge-density wave) ground state in K.

3.7. *Cyclotron Resonance*

Azbel'–Kaner cyclotron resonance (Azbel' and Kaner 1957) was observed by Grimes and Kip (1963) in oriented single crystals of Na and K for all directions of the magnetic field parallel to the surface of the specimens. The data for both metals was isotropic within experimental error and gave cyclotron masses of $(1{\cdot}21 \pm 0{\cdot}02)$ for K and $(1{\cdot}24 \pm 0{\cdot}02)$ for Na. The values of the cyclotron masses calculated by Ham (1962 b) were about 14% lower than the experimental values. The isotropy of the cyclotron masses for K indicates that its Fermi surface is very nearly spherical with the anisotropy in k_F probably appreciably less than 1%. The interpretation of the data for Na is complicated by the existence of the low-temperature phase transformation which we have mentioned before. From this data and the de Haas–van Alphen data (Thorsen and Berlincourt 1961 a) Grimes and Kip (1963) calculated the radius of the Fermi surface

$$k_F = (7{\cdot}44 \pm 0{\cdot}07) \times 10^7 \text{ cm}^{-1}$$

and the Fermi velocity

$$v_F = (7{\cdot}1 \pm 0{\cdot}2) \times 10^7 \text{ cm sec}^{-1}.$$

In addition to 'ordinary' cyclotron resonance in a metal, the electrons can even undergo cyclotron resonance when the wave frequency is very much smaller than the cyclotron frequency, so long as the mean free path of the electrons is larger than the wavelength. This can occur because an electron at the Fermi surface travels through the wave very rapidly and, as a consequence of the Doppler effect, it experiences an apparent frequency very much larger than the actual frequency of the wave. This was predicted by Kjeldaas (1959) for the case of ultrasonic shear waves propagating in a metal parallel to a magnetic field; later Stern (1963) pointed out that this should also occur for helicon waves. The largest magnetic field at which Doppler-shifted cyclotron resonance is possible has been called the *Kjeldaas edge*. The value of the magnetic field at which the Kjeldaas edge occurs is such that the angular cyclotron frequency is given by:

$$\omega_c = \frac{\omega v_{max}}{s}, \qquad (3.7.1)$$

where ω and s are the angular frequency and the velocity of the ultrasonic waves, and v_{max} is the maximum value of the component of the electron velocity along the direction of the magnetic field. A measurement of the magnetic field at which the absorption edge occurs therefore gives $m^* v_{max}$. The position of the Kjeldaas edge is therefore related to the Gaussian radius of curvature of the Fermi surface at the point where the electrons have the largest component of velocity parallel to the magnetic field. For a spherical Fermi surface this means that the position of the Kjeldaas edge is determined by the radius of the sphere in **k** space. The value obtained by Taylor (1964) for k_F in Na, by observing this Doppler-shifted cyclotron resonance with helicon waves, was $k_F = (0{\cdot}92 \pm 0{\cdot}02) \times 10^8 \text{ cm}^{-1}$, which is to be compared with the theoretical value of $0{\cdot}93 \times 10^8\,cm^{-1}$ on the free-electron theory assuming a lattice constant of 4·225 Å. Although it was only used to obtain a value of k_F it should be possible to use helicon waves with frequencies ~10 000 times smaller than the electron cyclotron frequency to lead to accurate measurements of the curvature at points on the Fermi surface of a metal. The method has also been applied to K by Penz and Bowers (1967).

One of the important consequences of the fact that the Fermi surfaces of the alkali metals are nearly spherical is that the free-electron model is a good approximation to the behaviour of the electrons, For instance, the observed oscillations in the attenuation of sound waves above the Kjeldaas edge in metallic K can be explained well in terms of helicon–phonon interaction using the free-electron model (Blaney 1967).

3.8. *Radio-frequency Size Effect*

Koch and Wagner (1966) made a detailed study of the radio-frequency (r.f.) size effect (or Gantmakher effect (Gantmakher 1962 a)) in thin parallel plates of K. The effect consists of the observation of anomalies in the r.f. surface impedance at that value of the magnetic field where the diameter of the electron orbits, in real space, is equal to the sample thickness. Well-defined signals are observed whenever there is a large group of electrons with nearly the same orbit diameter, that is, for extremal orbits. Measurement of this effect in K provides a test for the conical distortion of the Fermi surface of K predicted by Overhauser (1964), see § 3.9. The data of Koch and Wagner (1966) on the r.f. size effect in the tilted **H** field showed no evidence for this distortion. The experimental results agreed completely with those expected on the basis of a spherical Fermi surface; that is, the distortions from sphericity are too small (Shoenberg and Stiles 1964) to be detected by this method. The radio-frequency size effect has also been used by Tsoi and Gantmakher (1969) to measure the electronic mean free path in K and the line shapes have been investigated experimentally by Wagner and Koch (1968). Samples of K in the form of cylinders rather than the more usual plates were used in the radio-frequency size effect measurements of Blaney (1969).

3.9. *Spin-density Waves (Charge-density Waves)*

Overhauser has suggested that the electronic ground state of certain metals consists of a spin-density wave (S.D.W.) (or charge-density wave (C.D.W.)). Overhauser (1962) suggested that the optical anomaly in K, discovered by Mayer and El Naby (1963), could be explained by assuming the existence of a spin-density wave for the ground state in K (Cohen and Phillips 1964, Cohen 1964, Overhauser 1964). However, the experiments on cyclotron resonance by Grimes and Kip (1963) and on the de Haas–van Alphen effect by Shoenberg and Stiles (1964) failed to reveal the anisotropy of the Fermi surface of K which is required by Overhauser's hypothesis. It is possible to explain these results if one assumes that the wave vector of the S.D.W. (C.D.W.) orients itself parallel to an applied magnetic field. If this assumption is correct, the de Haas–van Alphen measurements, which only provide a measure of the extremal cross-sectional area of the Fermi surface normal to the applied magnetic field, would show no anisotropy, which is in agreement with experiment.

The suggestion that the electronic ground state of K is an S.D.W. (C.D.W.) which aligns itself in the direction of a sufficiently large external magnetic field can be shown to cause a distortion of the spherical Fermi surface to a lemon shape having conical points, with the major axes of the lemon along the direction of the magnetic field, see fig. 8. Measurement of the angular correlation of γ-rays produced by positron annihilation in K by Gustafson and Barnes (1967) were able to be interpreted on the S.D.W. (C.D.W.) model. Overhauser and Rodriguez (1966) and McGroddy *et al.* (1966) studied the theory of the propagation of helicon waves in K near the Kjeldaas edge; comparison of the calculated surface impedance of K with the experimental curves obtained by Taylor (1964, 1965) appeared to give better agreement with the S.D.W. (C.D.W.) model than with the free-electron model. However, the difference between these two models gives only a small effect and these experiments therefore do not provide conclusive evidence of an S.D.W. (C.D.W.) ground state in K. The difference between the nature of the Kjeldaas absorption edge for ultrasonic shear waves in a metal whose Fermi surface is spherical and a metal having an S.D.W. (C.D.W.) ground state in a magnetic field parallel to the direction of sound propagation has been discussed theoretically by Alig *et al.* (1965, 1966) with particular reference to K. They predicted that, at sufficiently high frequencies ($\sim$ 150 MHz) the position of the Kjeldaas edge in the S.D.W. (C.D.W.) model should be substantially lower than for the free-electron model. If the S.D.W. (C.D.W.) is oriented parallel to the d.c. field the value of $v_{\max}$ in K should be about 17% smaller for the S.D.W. (C.D.W.) state than v_F, its value for the free-electron model. Thus, by studying the position of the Kjeldaas edge, one should be able to determine the value of $v_{\max}$ and test Overhauser's conjecture. These experiments were performed by Greene *et al.* (1966) on single crystals of K using longitudinal acoustic waves at liquid helium temperatures for acoustic frequencies ranging from 20 to 140 MHz. Their results appeared to show good agreement with the free-electron model and to rule out the existence of an S.D.W. (C.D.W.) state if it is aligned parallel to the magnetic field. Other attempts by Thomas and Bohm (1966) and Blaney (1968) to observe these effects of an S.D.W. (C.D.W.) on the attenuation of ultrasonic shear waves in K, calculated by Alig *et al.* (1966), produced no conclusive evidence for the existence of an S.D.W. (C.D.W.) in K. It is possible that the magnetic fields which were employed were not sufficiently strong to reorient an S.D.W. (C.D.W.) from some preferred crystallographic direction such as the ⟨123⟩ direction suggested by Thomas and Bohm. However, it is difficult to reconcile such a preferred orientation of the S.D.W. (C.D.W.)

with the de Haas–van Alphen results. Further ultrasonic work by Thomas and Bohm (1966) indicated that S.D.W. (C.D.W.) effects may easily be masked in their experiments by small strains.

It has also been suggested that the 0·5 G splitting of the conduction–electron spin resonance in K at 4·2 kG observed by Walsh *et al.* (1966) can be explained by assuming that the conduction electrons are in an S.D.W. (C.D.W.) ground state (Overhauser and de Graaf 1968). The Fermi-surface distortion, see fig. 8, leads to an anisotropic conduction–electron g factor, depending on the angle between $\mathbf{H}$ and the wave vector $\mathbf{Q}$ of the S.D.W. (C.D.W.). The extremal values of g correspond to $\mathbf{Q}\perp\mathbf{H}$ and $\mathbf{Q}\parallel\mathbf{H}$ and the explanation of the experimental splitting requires the sample to have a macroscopic domain structure, caused by thermal stress and plastic flow when the sample is cooled. $\mathbf{Q}$ is then assumed to be parallel to $\mathbf{H}$ in the stress-free regions but to be approximately normal to $\mathbf{H}$ in the high-stress regions.

Koch and Wagner (1966) attempted to use the r.f. size effect (see § 3.8) to investigate the possible existence of an S.D.W. (C.D.W.) ground state in K. With the field tilted at an angle θ, signals are observed at values of H_n such that

$$d=\frac{n\hbar c}{eH_n}\left(\frac{dA}{dk_{\mathrm{H}}}\right)\sin\theta, \qquad (3.9.1)$$

where (dA/dk_{H}) is the rate of change of orbit area with k_{H}; such signals will be observed when (dA/dk_{H}) has an extremal value. For a spherical Fermi surface and for small tipping angles (θ up to 15° or so) this condition becomes, to a very good approximation:

$$H_n(\theta)=\frac{n2\pi\hbar c}{ed}k_{\mathrm{F}}\theta, \qquad (3.9.2)$$

so that the anomalies should therefore be periodic in H with period related to k_{F}. H_n, as a function of θ, will therefore move to higher fields in a linear fashion with slope $(n2\pi\hbar ck_{\mathrm{F}}/ed)$. However, the behaviour of (dA/dk_{H}) on Overhauser's S.D.W. (C.D.W.) theory is more complicated, see fig. 8, and in particular, while the H_n for $n=1, 2, 3, \ldots$ should move towards higher H as θ increases, the $n=0$ peak should move towards *lower* H as θ increases. Koch and Wagner (1966) observed no such peak that moved to lower H as θ increased. This was taken as showing the absence of the S.D.W. (C.D.W.) distortion of the Fermi surface of K (see also Peercy *et al.* (1968)). On the other hand, the S.D.W. (C.D.W.) theory does predict a linear behaviour for the magnetoresistance of K as a function of magnetic field (Reitz and Overhauser 1968) which is in agreement with the measurements of Penz and Bowers (1967); it also predicts that the Hall coefficient of K should decrease with increasing magnetic field, as has been observed experimentally by Penz (1968). However, it should be noted that other explanations of the linear magnetoresistance of K in terms of strain in the sample have also been suggested (Penz and Bowers 1968, Jones 1969).

Fig. 8

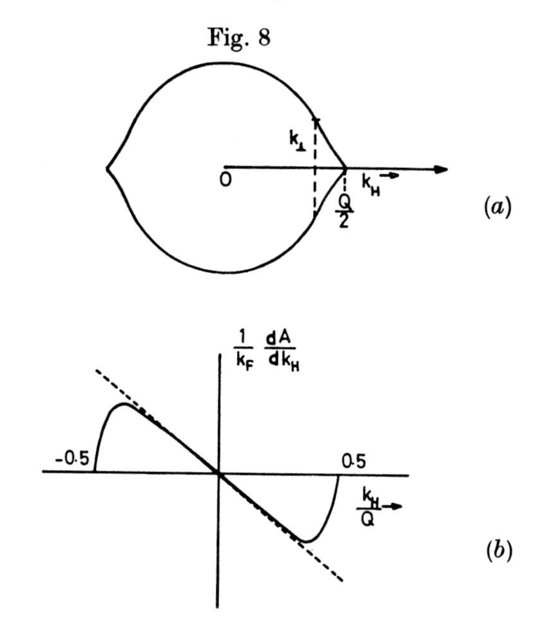

Fermi surface of K with the distortion predicted by Overhauser (1964): (*a*) cross section, (*b*) comparison of (dA/dk_H) for this model (continuous line) with the free-electron model (broken line) (Koch and Wagner 1966).

It would therefore seem to be a fair summary of the present state of the hypothesis of an S.D.W. (or C.D.W.) ground state in an alkali metal, to say that while it has been invoked to explain several phenomena the evidence in favour of the hypothesis is a little indirect and these S.D.W.s (or C.D.W.s) have proved rather too elusive to observe in a direct and convincing way.

3.10. *Conduction–electron Spin Resonance* (*C.E.S.R.*)

The theory of the departure of the value of the g factor observed in conduction–electron spin resonance (C.E.S.R.) in a metal from the free-electron g value has been considered by Dyson (1955). Extensive calculations of this g shift in alkali metals, using various different approximations were performed by Bienenstock and Brooks (1964); these authors were able to separate their expression for δg into two terms. The first of these can be written entirely in terms of radial wave functions and their derivatives evaluated at the surface of the Wigner–Seitz sphere; hence it can be evaluated by the quantum defect method. The second term is a volume integral that cannot be reduced to surface terms and therefore cannot be evaluated by the quantum defect method. The second term can be shown to be negligible for all the alkalis except Li. For Li the two terms are of the same order of magnitude (10^{-5}) but of opposite sign, with the result that the calculated value of δg is of order 10^{-6} with an uncertainty of the same order. Not even the sign of δg can be evaluated with certainty.

The various experimental values of the conduction electron g shift in the alkalis have been collected by Schultz and Shanabarger (1966) together with the results of the calculations of Bienenstock and Brooks. Subsequent to the compilation of data by Schultz and Shanabarger (1966) the conduction electron g shift in Li relative to that of the free electron has been accurately measured by VanderVen (1968), by including a correction for demagnetizing fields, as $\delta g = (-6{\cdot}1 \pm 0{\cdot}2) \times 10^{-5}$. The difference between this and the value $(-4 \pm 4) \times 10^{-6}$ obtained previously by Pressley and Berk (1965) is most likely due to their neglect of demagnetizing effects. The implication of the measurement of VanderVen is that the surface term in δg, which is negative in the calculation of Bienenstock and Brooks, must dominate the volume term which is positive. A re-calculation of δg for Li by Overhauser and de Graaf (1969) using a single O.P.W. conduction electron wave function led to $\delta g \doteqdot -5{\cdot}4 \times 10^{-5}$.

§ 4. s-block : Group IIA : the Alkaline Earth Metals

Be	Beryllium	2.2
Mg	Magnesium	2.8.2
Ca	Calcium	2.8.8.2
Sr	Strontium	2.8.18.8.2
Ba	Barium	2.8.18.18.8.2

In considering the alkali metals in the previous section we found it convenient to consider the various experimental techniques in succession rather than to consider all the results for each metal together. This was convenient because topologically the Fermi surfaces of the alkali metals are very simple and all consist of a sphere that is slightly distorted. However, for group IIA metals, since the important experimental methods have all been mentioned and because the Fermi surfaces are more complicated and the structures of the various metals are not all the same, it is more convenient to discuss the Fermi surfaces of the group IIA elements metal by metal.

4.1. *Structures*

Be and Mg have the h.c.p. structure, Ca and Sr are f.c.c., although they do undergo phase transitions to h.c.p. and b.c.c. structures, respectively, at high temperatures (Jayaraman *et al.* 1963 a), and Ba has the b.c.c. structure. The electrical resistivities of Ca, Sr and Ba exhibit various large changes at very high pressure (Drickamer 1965, Stager and Drickamer 1963 a); attempts have been made to explain this behaviour in terms of structural phase changes (Drickamer 1965) and in terms of changes in the electronic band structure (Altmann and Cracknell 1964, Vasvari *et al.* 1967, Vasvari and Heine 1967) without any change from the f.c.c. crystallographic structure.

4.2. *Beryllium*

Historically, Be was the first metal to which the O.P.W. method of calculating the electronic band structure was applied (Herring and Hill 1940). The l-dependent pseudopotential interpolation scheme which was used by Cornwell (1961) for the alkali metals, and which was mentioned in § 3.2. was also applied to Be. The interpolation was based on the use of the energy levels at the points of symmetry determined by Herring and Hill. Interpolated values of $E_{\mathbf{k}}$ and hence the density of states curve and Fermi energy were calculated. The first band was found to be nearly full, with 0·023 unoccupied states per atom and there was found to be an equal number of occupied states in the second band near the point K. This calculated Fermi surface differed appreciably from that for free electrons in an h.c.p. lattice.

Loucks and Cutler (1964) used a self-consistent O.P.W. method to calculate the band structure of Be at a large number of general points in the Brillouin zone and thence to find the shape of the Fermi surface. Their density of states was found to be in agreement with soft X-ray emission and absorption data and with the experimental low temperature specific heat coefficient (see below). The Fermi surface was found to consist of (i) a region of holes in the first band in the double zone which resembles a coronet and (ii) two identical pockets of electrons in the second band in the double zone similar in shape to a cigar with a triangular cross section. The de Haas–van Alphen frequencies in $1/H$ predicted from the computed Fermi surface were found to be in good agreement with those measured experimentally. The pieces of Fermi surface are illustrated in fig. 9 and

Fig. 9

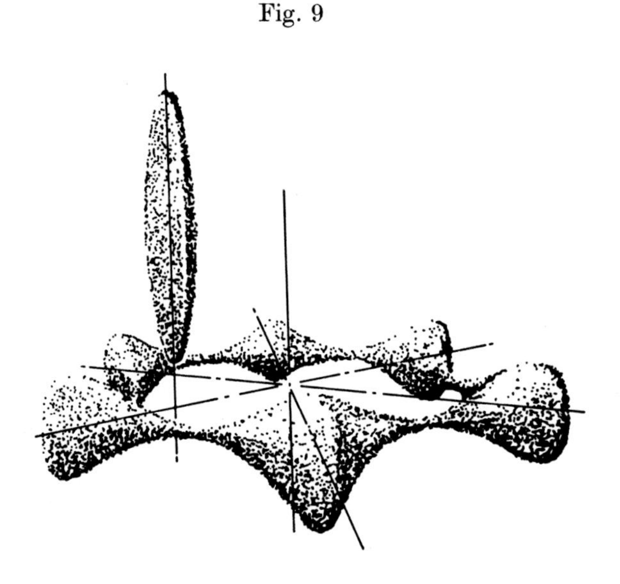

The Fermi surface of Be; cigar and coronet (Loucks and Cutler 1964).

Fig. 10

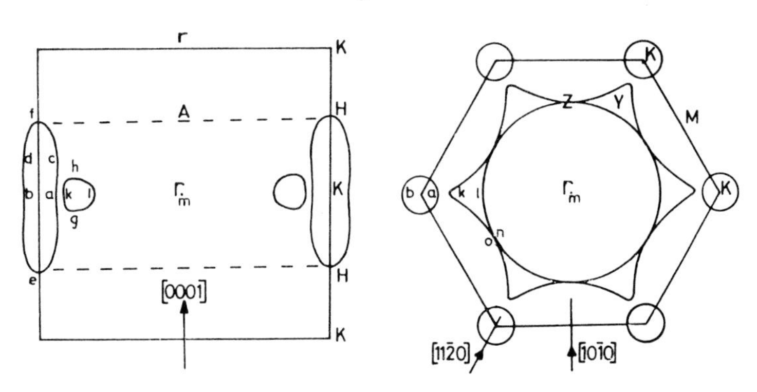

Two sections through the Fermi surface of Be (Watts 1963).

their cross sections are illustrated in fig. 10. The actual location of the cigars and coronet were deduced by Watts from re-mapping the free-electron Fermi surface in the extended zone scheme into the reduced zone scheme; the distortions of the free-electron Fermi surface which were then required to produce the observed cigars and coronet are then quite small (see fig. 9 of Watts (1964)). An A.P.W. calculation of the energy bands, Fermi energy, density of states and the electronic contribution to the specific heat of Be was performed by Terrell (1966). Agreement was found with the other existing band structure calculations of Herring and Hill (1940) and Loucks and Cutler (1964) and with the de Haas–van Alphen measurements of Watts (1964). A comparison is made by Terrell between the band structures of Be and Mg. Calculations of the shape of the cigars, etc. based on this A.P.W. band structure calculation in Be were given by Terrell (1964).

Emission and absorption x-ray spectra were measured for metallic Be by Lukirskii and Brytov (1964) and the width of the filled part of the valence band was measured and compared with previous results. The density of states curve for Be was determined by Phillips and Weiss (1968) from Compton line-shape measurements. Various experimental specific heat measurements have been made for Be and values of γ include:

$0{\cdot}23$ mJ mole^{-1} deg^{-2} Hill and Smith (1953)

$(0{\cdot}184 \pm 0{\cdot}02)$ mJ mole^{-1} deg^{-2} Gmelin (1964)

$(0{\cdot}1714 \pm 0{\cdot}0008)$ mJ mole^{-1} deg^{-2} Ahlers (1966).

The older band structure calculations of Herring and Hill (1940) gave a value of $0{\cdot}23$ mJ mole^{-1} deg^{-2} but the more recent calculations by Loucks (1964) using 80 000 points in the Brillouin zone yielded a value of γ of $0{\cdot}166$ mJ mole^{-1} deg^{-2} which is in much better agreement with the recent accurate value of Ahlers (1966). This agreement suggests that electron–phonon interactions do not make a large contribution to γ in Be. The phonon dispersion relations in Be have been calculated by Sahni *et al.*

(1966) using the pseudopotential method and the results were compared with the experimental data of Schmunk *et al.* (1962).

The galvanomagnetic properties of Be were investigated by Gruneisen and Adenstedt (1938), Erfling and Gruneisen (1942) and Borovik (1952); it was observed that the resistance increased in a magnetic field almost quadratically both for **H** parallel to [0001] and for **H** in the (0001) plane. These results implied that Be has a closed Fermi surface. These early experiments were carried out in weak magnetic fields and an investigation of the magnetoresistance of Be at higher fields was performed by Alekseevskiĭ and Egorov (1963 a, b, 1964). Up to about 35 kG the increase of the resistance in the magnetic field is close to quadratic for all directions of the magnetic field, so that up to 35 kG Be behaves like a metal with a closed Fermi surface. However, in fields of about 50 kG and above, the dependence of the resistance on the field in the [10$\bar{1}$0] direction exhibits a saturation tendency. This can be interpreted by assuming that in fields larger than 50 kG magnetic breakdown occurs and open orbits appear along the hexagonal axis of Be. At 78°K the change in the magnetoresistance due to magnetic breakdown occurs for a wider range of fields. As a result of thermal broadening of the Fermi level the motion of electrons along open orbits (or extended orbits) appears at somewhat lower fields. Evidence of magnetic breakdown in Be was also obtained in further magneto-resistance work (Alekseevskiĭ *et al.* 1967, Alekseevskiĭ and Egorov 1968 a) and in de Haas–van Alphen measurements (Alekseevskiĭ and Egorov 1968 b); a value of 130 kG was obtained for the breakdown field parameter.

The Knight shift in Be has long been known to be very small (Townes *et al.* 1950, Knight 1956). A small, but negative, result was obtained by Barnaal *et al.* (1967) and by Anderson, Ruhlig and Hewitt (1967). The conduction–electron contributions to the Knight shift can be considered to arise from four magnetic interactions (*a*) the Fermi contact interaction (direct contact interaction) with electrons in partly filled s states, (*b*) the contact interaction with electrons in filled s states (core electrons) due to core polarization, (*c*) the spin–dipolar interaction with electrons in partly filled non-s states, and (*d*) the orbital interaction with electrons in partly filled non-s states; in all cases only those conduction–electron states near the Fermi level are involved. The first two interactions are responsible for positive and negative contributions, respectively, to the Knight shift, and it is the possible mutual cancellation of these terms which presumably lies behind the explanation of the extremely small shift in Be. Detailed calculations of the direct and core polarization contributions ((*a*) and (*b*)) to the Knight shift in Be metal were made for a number of symmetry points near the Fermi surface by Shyu, Gaspari and Das (1966) using wave functions for the conduction electrons obtained by the O.P.W. method. The interactions (*c*) and (*d*) give rise to an anisotropy in the Knight shift in non-cubic metals which is typically of the order of 10% of the isotropic shift. The explanation of the negative value of the isotropic shift requires the inclusion of orbital effects (Jena *et al.* 1968).

Low frequency de Haas–van Alphen oscillations were observed in Be, by using the torque method at 2°K, for a few directions of the magnetic field by Verkin *et al.* (1950 a, b) and Verkin *et al.* (1955). The pulsed magnetic field method (Shoenberg 1957, 1962) has been used by Watts (1963, 1964) to measure the de Haas–van Alphen effect in Be and thence to deduce the shape of the Fermi surface which we have already described and illustrated (see fig. 10).

The agreement between the experimental de Haas–van Alphen results of Watts and the subsequent band structure calculations (Loucks and Cutler 1964, Terrell 1964, 1966) was quite good, see table 6. The accuracy of the

Table 6. Comparison of de Haas–van Alphen measurements with theoretical calculations for Be

Orbit	Direction of **H**	Theory (Terrell 1964)		Theory (Loucks and Cutler 1964)		Experiment (Watts 1964) Frequency (10^{-6} G^{-1})
		Area (A.U.)$^{-2}$	Frequency (10^{-6} G^{-1})	Area (A.U.)$^{-2}$	Frequency (10^{-6} G^{-1})	
Cigar	[11$\bar{2}$0]	0·152	56·7	0·141	52·9	53·7
	[10$\bar{1}$0]	0·140	52·3	0·149	55·7	53·7
	[0001]	0·026	9·5	0·0245	9·2	9·6
Coronet						
Inner circle	[0001]	1·07	400	1·04	389	381
Outer circle	[0001]	1·36	509	1·47	550	Not given
Belly	[11$\bar{2}$0]	0·031	11·6	0·038	14·2	12·5
Belly	[11$\bar{2}$0]	0·041	14·9	0·055	20·7	15·0
Neck	[11$\bar{2}$0]	0·0001	0·03	0·0006	0·23	0·11
Neck	[11$\bar{2}$0]	0·0002	0·07	0·0014	0·53	0·24
Neck	[10$\bar{1}$0]	0·0001	0·04	0·0008	0·30	0·13

Fermi surface calculations of Loucks and Cutler (1964) was subsequently improved and a slightly better agreement with the experimental results of Watts (1963, 1964) was obtained (Loucks 1964). The main surviving difference was that the theoretical cigar was shorter than the experimental one and had an almost triangular cross section rather than a circular one. Measurements of the effect of hydrostatic pressures up to 4 kbar on three prominent de Haas–van Alphen frequencies in Be have been made by O'Sullivan and Schirber (1967). The observed results, expressed in the form of pressure derivatives of the de Haas–van Alphen frequencies could not be explained even qualitatively on the nearly-free-electron model and these authors attempted a more sophisticated approach involving a Heine–Abarenkov (1964) type pseudopotential fit as a function of unit cell volume. More recent measurements made by Schirber and O'Sullivan

(1969) were in substantial agreement with calculations of the pressure dependence of the Fermi surface of Be using a non-local pseudopotential (Tripp *et al.* 1967, 1969).

Magnetothermal oscillations, which were first observed in 1962 in Bi (Kunzler *et al.* 1962), have been observed in Be by Lepage *et al.* (1964), Condon (1966), and Sullivan and Seidel (1967). These magnetothermal oscillations are the temperature oscillations of an adiabatically isolated sample as the magnetic field is changed and they arise as a result of quantum oscillations in the heat capacity of the metal.

Measurements of the angular correlation of γ-ray photons from positron annihilation in Be have been made by Stewart *et al.* (1962) and by Berko (1962). The observed distributions were interpreted as being in reasonable agreement with the only qualitative ideas which existed at that time concerning the Fermi surface of Be; this was before the de Haas–van Alphen measurements of Watts (1963, 1964) or the band structure calculations of Loucks and Cutler (1964) and of Terrell (1964, 1966) were made. It is emphasized by Berko (1962) that the angular correlation curves are sensitive to the details of the conduction electrons' wave functions and do not yield directly cross-sectional areas of the Fermi surface. On this argument positron annihilation measurements cannot be used to deduce the topology of the Fermi surface in a direct way, although, given the conduction electrons' wave functions, they could be used to check some given proposed Fermi surface, see, for example, the work of Shand (1969) on Be using a local pseudopotential.

4.3. *Magnesium*

A band structure calculation for Mg using the O.P.W. method was performed by Falicov (1962) in a similar way to that of Heine (1957 a, b, c) on Al. The Fermi energy was calculated and the shape of the Fermi surface deduced, see fig. 11. While considerable differences from the free-electron model exist the topological relationship of the Fermi surface to the free-electron Fermi surface is still recognizable. A calculation of the spin–orbit splitting of the energy levels at the corner of the Brillouin zone in Mg was performed by Falicov and Cohen (1963) using a six orthogonalized-plane-wave approximation and the parameters from the previous band structure calculation of Falicov (1962). The effects of spin–orbit coupling are particularly important on the face AHL of the Brillouin zone, where the double-degeneracy is lifted everywhere except along the line AL. The splittings of the levels at H in Mg by spin–orbit coupling calculated by Falicov and Cohen (1963) are:

$$H_1\ E = 0{\cdot}708\ \text{Ry}, \quad \Delta E = 4{\cdot}16 \times 10^{-4}\ \text{Ry}$$
$$H_2\ E = 0{\cdot}686\ \text{Ry}, \quad \Delta E = 5{\cdot}52 \times 10^{-7}\ \text{Ry}$$
$$H_3\ E = 0{\cdot}898\ \text{Ry}, \quad \Delta E = 4{\cdot}20 \times 10^{-4}\ \text{Ry}.$$

Fig. 11

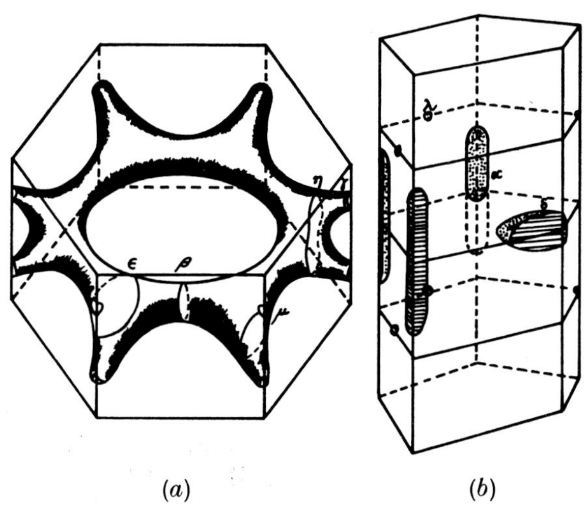

The Fermi surface of Mg calculated by Falicov (1962) in the double zone scheme: (*a*) multiply-connected hole surface in bands 1 and 2, and (*b*) electron surfaces in bands 3 and 4.

When spin–orbit coupling is neglected the band structure is labelled with the single-valued irreducible representations of the space group of the crystal. If time-reversal symmetry is included all the energy bands are four-fold degenerate (including spin degeneracy) all over the face AHL of the Brillouin zone. When spin-dependent terms are included in the Hamiltonian the band structure has to be described in terms of the double-valued irreducible representations of the space group of the crystal. In this case the four-fold degeneracy only survives along the line ARL; at the other points of symmetry as well as at a general point B on the face AHL the four-fold degeneracy is lifted to produce two two-fold degenerate levels, one corresponding to spin up and the other to spin down; the remaining double degeneracy everywhere on the face AHL is due to time-reversal symmetry and is only lifted in the presence of an external magnetic field or in a magnetic metal (Falicov and Ruvalds 1968, Cracknell 1969). The separation which therefore may appear between the spin-up and spin-down energy bands when spin–orbit coupling is included will clearly have important effects on the connectivities of the Fermi surface of a metal. All those properties of a metal which depend on the local or topological features of the Fermi surface are essentially changed. Since the bands no longer stick together all over the hexagonal top face of the Brillouin zone when spin–orbit coupling is included, the double-zone scheme that is commonly used is no longer valid. For example, the piece of the Fermi surface corresponding to the holes in the second band of Mg, which without spin–orbit coupling can only sustain open orbits perpendicular to the c axis, with spin–orbit coupling changes its connectivity and also permits open orbits parallel to the c axis (see fig. 3). However, because of the smallness of some of the spin–orbit splittings, magnetic breakdown effects must be expected to occur at relatively low magnetic fields, in which case the electron orbits will ignore the gap and thereby restore the previous topology which existed in the absence of spin–orbit coupling. This happens for Mg where relatively small fields, of the order of 200 G, are large enough to produce magnetic breakdown.

The de Haas–van Alphen effect in Mg was first observed by Verkin *et al.* (1950 a, b) using the torque method. Later work by Verkin *et al.* (1955) showed that several periods were present but no systematic study was undertaken. Shoenberg (1957) used the pulsed-field method and showed that much faster periods existed. The first systematic study was made by Gordon *et al.* (1960) using the torque method, however, since the fields then used were relatively low, their results cover only long periods. The de Haas–van Alphen effect in Mg has been studied by the pulsed-field method by Priestley (1963). The results were able to be explained on the theoretical model from the O.P.W. calculation of Falicov (1962) provided magnetic breakdown effects are taken into account, see fig. 11. This model is topologically equivalent to the nearly-free-electron model (see fig. 2 (c)) although the sizes of the various pieces of Fermi surface are rather different. From the results of Priestley (1963) the lens-shaped electron Fermi surface around Γ in the third band was found to be an ellipsoid of revolution about the c axis with an experimental radius in the basal plane of $(0{\cdot}217 \pm 0{\cdot}003)$ A.U. which is to be compared with $(0{\cdot}25 \pm 0{\cdot}05)$ A.U. on the model of Falicov (1962). The experimentally deduced length of the cigar (0·27 A.U.) was significantly less than that on the Falicov model (0·55 A.U.), while the experimental value for the area (0·0065 A.U.) of the horizontal cross section of the cigar is rather less than the value calculated by Falicov (0·010 A.U.). The remaining details of the relationship between the experimental de Haas–van Alphen results and the Fermi surface calculated by Falicov are described in the paper by Priestley (1963). The existence of a very short period, called G by Priestley, corresponding to an area about twice as large as that round the outside of the 'monster' and larger than the area of cross section of the Brillouin zone provided the first experimental evidence of magnetic breakdown (Priestley *et al.* 1963). The theory of the galvanomagnetic properties of Mg was given by Falicov *et al.* (1966) for magnetic fields parallel to the c axis and taking into account magnetic breakdown. The transverse magnetoresistance showed the expected transition from electron–hole compensation at low fields to a non-compensated state at high fields, i.e. a transition from quadratic behaviour to saturation. The Hall resistance showed the corresponding behaviour. In addition, all the transverse components of both conductivity and resistivity tensors show strong de Haas–van Alphen type oscillations. Although the de Haas–van Alphen data of Gordon *et al.* (1960) and Priestley (1963) and the anomalous skin effect data of Fawcett (1961) seemed to be compatible with Falicov's model, the limited data which were available for the galvanomagnetic properties of Mg (Alekseevskiĭ and Gaĭdukov 1960) suggested a Fermi surface topology which appeared to be incompatible with that model. The galvanomagnetic properties were, therefore, investigated again by Stark *et al.* (1964) (see also Stark *et al.* (1962)). Their results were able to be interpreted with Falicov's model for the Fermi surface of Mg if the model is suitably modified to include the effects of spin–orbit coupling (Falicov and Cohen 1963) and magnetic breakdown (Priestley *et al.* 1963). Quantum oscillations in the ultrasonic attenuation in Mg at magnetic fields between 5 kG and 20 kG were investigated by Fenton and Woods (1966) and the observed periods agreed with those observed in the de Haas–van Alphen effect in Mg by Gordon *et al.* (1960) and by Priestley (1963). The accuracy of these magnetoacoustic results was thought not to be quite as great as that of the de Haas–van Alphen results, but within the experimental accuracy the periods were found to be identical.

It therefore seemed that the Fermi surface calculated by Falicov (1962) for Mg, which was topologically the same as the free-electron Fermi surface, was also substantially correct in its predicted dimensions for the Fermi surface. However, a new and much more detailed study of the Fermi surface of Mg by using magnetoacoustic geometric resonances (Ketterson and Stark 1967) and the de Haas–van Alphen effect (Stark 1967) showed that the Fermi surface dimensions of Mg are much closer to those of the free-electron model than in Falicov's predictions. The free-electron model has sheets in the first four bands which are illustrated in fig. 2 (*c*) and previous experimental work had verified the existence of all these parts of the Fermi surface. All these pieces of Fermi surface are also present in the results of the band structure calculation of Falicov (1962). A total of twenty different geometric resonance branches were obtained by Ketterson and Stark and the dimensions of these various pieces of Fermi surface are compared in table 7 with the dimensions of the free-electron Fermi surface and the Fermi surface calculated by Falicov (1962). Some of the caliper dimensions determined by Ketterson and Stark were used by Kimball *et al.* (1967) to construct a non-local model pseudopotential. This produced a Fermi surface that was in very good agreement with the de Haas–van Alphen data of Stark (1967), see table 8. A Green's function formulation for the calculation of the amplitudes of the de Haas–van Alphen effect has been applied to the example of Mg, for magnetic fields parallel to [0001], by Falicov and Stachowiak (1966). The results, giving the magnetic field dependence of the amplitude of various important periods at $T = 1\,°K$, were found to be in agreement with preliminary experimental data.

To conclude this section we note several miscellaneous pieces of work related to the electronic properties of Mg. We have mentioned in § 4.2 the angular correlation experiments by Berko on the γ-rays emitted from positron annihilation in Be metal; similar experiments have also been performed on Mg metal (Berko 1962). The derivatives of the angular

Table 7. Fermi surface dimensions of Mg (atomic units)

Dimension		Free-electron model	Calculated (Falicov 1962)	Experimental (Ketterson and Stark 1967)
		Second band monster		
Inside	$k^{\Gamma K}$	0·341	0·405	0·370
	$k^{\Gamma M}$	0·341	0·415	0·370
Outside	$k^{\Gamma K}$	0·634	0·606	0·622
	$k^{\Gamma M}$	0·476	0·452	0·476
Waist	$C^{\Gamma M}$	0·135	0·037	0·100
	$k^{\Sigma' L}$	0·110	0·062	0·064
		Third band lens		
	$k^{\Gamma A}$	0·085	0·058	0·080
	$k^{\Gamma K}$	0·341	0·253	0·312
	$k^{\Gamma M}$	0·341	0·255	0·312
		Third band cigar		
	$C^{\Gamma KM}$	0·093	0·117	0·100
	$k^{K\Gamma}$	0·062	0·073	0·067
	k^{KM}	0·031	0·044	0·033
		Third band butterfly		
	k^{LH}	0·252	0·043	0·184
	$k^{L\Sigma'}$	0·252	—	0·206
		Fourth band pocket		
	k^{LM}	0·090	0·029	0·043

(Adapted from Ketterson and Stark (1967).)

distributions from Be show fine structure and marked anisotropy compared with the observed nearly isotropic behaviour of Mg. The surface conductance under anomalous skin effect conditions of polycrystalline samples of Mg was measured by Fawcett (1961) and the total area of the Fermi surface was calculated. The Hall effect in Mg was measured by Alty and Stringer (1969) and their results were able to be interpreted almost entirely by using the free-electron Fermi surface for Mg. A value of $g = (2{\cdot}009 \pm 0{\cdot}002)$ was obtained in the conduction–electron spin resonance (C.E.S.R.) measurements of Orchard-Webb and Cousins (1968). The electrical resistivity of Mg as a function of pressure was measured by Stager and Drickamer (1963 a) and it shows shallow maxima and minima along a given isotherm,

Table 8. Calculated and experimental cross-sectional areas of the Fermi surface of Mg

	θ coordinates of **H**†	Symbol‡	Areas (10^2 A.U.)	
			A_{exp} (Stark 1967)	$A_{\text{non-local}}$ (Kimball *et al.* 1967)
Second band monster	0 90 48·5 90	$\mu_1{}^1$ $\mu_1{}^5$ $\mu_2{}^5$ $\mu_2{}^7$	0·215 0·721 3·80 4·52	0·216 0·707 3·87 4·48
Third band lens	0 90 90	$\lambda_1{}^1$ $\lambda_1{}^1$ $\lambda_2{}^1$	30·8 7·26 7·27	30·9 7·32 7·34
Third band cigar	0 90	$\gamma_1{}^1$ $\gamma_2{}^1$	0·598 3·13	0·612 3·14
Third band butterfly	0 0 90	$L_1{}^1$ $C_1{}^1$ $C_1{}^1$	2·34 3·72 2·08	2·33 3·68 2·06

† The subscript number on each area symbol refers to the crystallographic plane containing **H** as defined by Stark (1967). The $(10\bar{1}0)$ plane is designated by 1, the $(11\bar{2}0)$ plane is designated by 2 and the (0001) plane is designated by 3.
‡ As defined by Stark (1967).

(Adapted from Kimball *et al.* (1967).)

but there are no striking features. In the previous section we noted that the Knight shift in Be metal was exceptionally small, presumably as a result of the cancellation of positive and negative contributions, (*a*) and (*b*). However, Mg exhibits a much larger Knight shift (+0·111%) so that it appears that the direct-contact contribution (*a*) to the Knight shift is substantial and dominant†. The Fermi surface of Mg differs from that of Be by having an extra lens-shaped piece of Fermi surface of electrons near Γ in band 3 and this, which includes a $\Gamma_4{}^-$ level, is thought to contribute a relatively large positive direct-contact contribution to the Knight shift (Shyu, Gaspari and Das 1966, Barnaal *et al.* 1967). Optical interband and intraband transitions in Mg have been studied experimentally by Jones and Lettington (1967) and by Graves and Lenham (1968). The temperature coefficient of the electronic specific heat of Mg has been measured by Friedberg *et al.* (1952) and found to be 1·36 mJ mole^{-1} deg^{-2} (see also Estermann *et al.* 1952) and by Martin (1961 a) to be (1·23 ± 0·01) mJ mole^{-1} deg^{-2}.

† See note added in proof.

The phonon dispersion relations in Mg have been investigated experimentally by Collins (1962), Maliszewski *et al.* (1963), Iyengar *et al.* (1965a, b), and Squires (1966). The results of Squires were examined for evidence of the Kohn effect, that is, of the influence of the Fermi surface on the phonon dispersion relations (Kohn 1959). The only region where there is any appearance of discontinuity in the dispersion curves is for the highest frequency branch in the ΓK range at $aq = 0{\cdot}53 \pm 0{\cdot}02$. The only part of the Fermi surface of Mg calculated by Falicov (1962) that could give rise to this value of aq is a pocket of electrons in the third band; the calculated diameter in the ΓK direction for this part of the Fermi surface corresponds to $aq = 0{\cdot}49 \pm 0{\cdot}10$. These two values are not inconsistent. On the other hand, the Kohn effect is expected to be small in Mg and there appears to be no other evidence for the effect in the measurements. The lattice dynamics of Mg have also been discussed by Roy and Venkataraman (1967) using a model pseudopotential approach. The model pseudopotential thus derived should be useful for calculating other properties of the metal, see § 8.1.

4.4. *Calcium, Strontium and Barium*

The Fermi surfaces of the divalent f.c.c. metals Ca and Sr in the free-electron approximation are illustrated in fig. 2(*a*). Each of them consists of a multiply-connected hole surface in band 1 and a set of lens-shaped pockets of electrons at L on the surface of the Brillouin zone in band 2.

The de Haas–van Alphen effect in Ca was studied by Condon and Marcus (1964) by the torsion method in magnetic fields up to 33 kG. The technical problems involved in obtaining and handling pure single crystals of such an active metal as Ca are not insignificant. Three distinct orbits on the Fermi surface were observed with maximum periods in $(1/H)$ of $3{\cdot}05 \times 10^{-7}\,\text{G}^{-1}$, $0{\cdot}77 \times 10^{-7}\,\text{G}^{-1}$ and $0{\cdot}57 \times 10^{-7}\,\text{G}^{-1}$. The longest period exhibited an angular dependence characteristic of a hyperboloidal piece of Fermi surface. The longest and shortest periods appeared to be correlated, while there was a negative correlation between the longest period and the $0{\cdot}77 \times 10^{-7}\,\text{G}^{-1}$ period. The cyclotron masses were found to be $0{\cdot}35 \pm 0{\cdot}04$, $0{\cdot}62 \pm 0{\cdot}09$ and $0{\cdot}65 \pm 0{\cdot}10$, respectively. The results were interpreted in terms of the Fermi surface suggested by Harrison (1963c) on a nearly-free-electron model, see fig. 12, but there appear to be more orbits predicted on Harrison's model than the three actually observed by Condon and Marcus.

The first band structure calculation for Ca was performed by Manning and Krutter (1937). A cellular calculation of the band structure of Ca was performed by Altmann and Cracknell (1964) and the density of states and shape of the Fermi surface were constructed. A much less free-electron-like Fermi surface was produced than that calculated by Harrison

Fig. 12

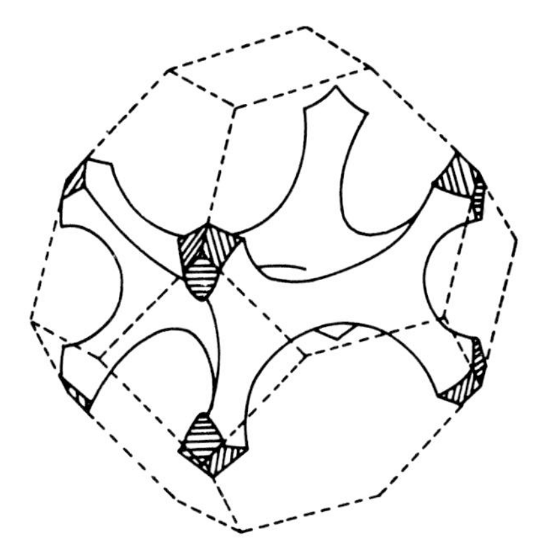

The multiply-connected hole Fermi surface for Ca in band 1 proposed by Harrison (1963 c).

(1960 b, 1963 c). The first band was found to be full nearly all over the Brillouin zone except for very small kidney-shaped pockets of holes at K and U, see fig. 13; the second band is nearly empty and only contains saucer-shaped pockets of electrons at L. The metal is therefore compensated, having equal numbers of electrons and holes. Altmann and

Fig. 13

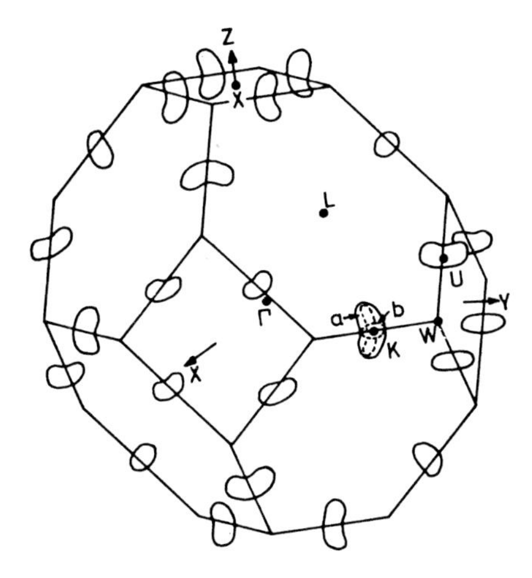

The hole Fermi surface of Ca in band 1 calculated by Altmann and Cracknell (1964).

Cracknell (1964) showed that this model was capable of explaining the de Haas–van Alphen measurements of Condon and Marcus (1964) more satisfactorily than did the Fermi surface postulated by Harrison. More recent calculations by Vasvari *et al.* (1967) using a model pseudopotential have been performed for Ca, Sr and Ba (Abarenkov and Heine 1965, Animalu and Heine 1965); for Ca these authors suggested a similar Fermi surface with pockets of electrons at L in band 2 and small pockets of holes in band 1. However, there is one important difference, in that the pockets of holes in band 1 would be at W rather than at K and U (Vasvari 1968), see fig. 14. Pockets at W would of course have a larger

Fig. 14

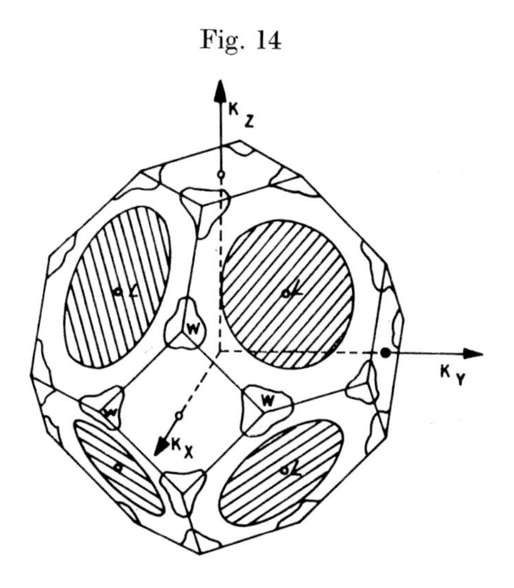

The Fermi surface of Ca calculated by Vasvari (1968); unshaded pockets represent holes in band 1 and shaded pockets represent electrons in band 2.

volume than pockets at K and U. Further experiments are necessary to determine the shape of pockets of holes in band 1 and, hopefully, to determine their positions. Further band structure calculations on Ca suggested that the holes in band 1 are in fact at W (Altmann 1970, Johansen 1970, private communications)†.

A cellular calculation of the band structure of Sr was performed by Cracknell (1967) who suggested that in Sr, like Ca, the first band is nearly full and contains a few small pockets of holes, while the second band is nearly empty and contains a few small pockets of electrons, although the pockets of holes in band 1 could form a multiply-connected Fermi surface. It was suggested that the pockets of electrons in band 2 are centred at the point X and of radius approximately $\frac{1}{3}$XW. The pockets of holes in band

† See note added in proof.

1 were predicted to be centred at W again with a radius approximately $\frac{1}{3}$WX. However, the first band at K was sufficiently high that these pockets might be multiply-connected to form a tubular Fermi surface all round the edges of the Brillouin zone through W, K and U. The band structure of Sr was re-computed by Vasvari *et al.* (1967) and the Fermi surface suggested by their band structure is very similar to their results for Ca, namely pockets of electrons around L in band 2, in disagreement with Cracknell (1967), and in band 1, as suggested by Cracknell, either isolated pockets of holes around W or else a multiply-connected ' monster ' through W, K and U. No direct experimental measurements of Fermi surface dimensions in Sr appear to have been made so far. A later relativistic A.P.W. calculation for Sr by Johansen (private communication) produced a band structure that was essentially in agreement with that of Vasvari *et al.* (1967). The band structure for the hypothetical f.c.c. phase of Ba determined by Vasvari *et al.* was very similar to those obtained by them for Ca and Sr. One might therefore expect that, in the real b.c.c. phase, Ba would exhibit similar semi-metallic properties to those of Ca and Sr so that its Fermi surface would consist of very small isolated pockets of holes in band 1 and of electrons in band 2 at appropriate points in the b.c.c. Brillouin zone. However, the behaviour of the electrical resistance of Ba at very high pressures is different from that of Ca and Sr (see below), which would seem to suggest a qualitatively or topologically different Fermi surface for Ba from those of Ca and Sr. Maybe the pockets of holes in band 1 in Ba are not isolated but become multiply-connected†. Optical reflection spectra observed by Müller (1966) in Ba exhibited structure in the spectral dependence of $\sigma(\omega)$ at energies slightly higher than 1 ev and around 2·5 ev ; these were assigned to optical interband transitions.

The behaviour of the electrical resistance of Ca, Sr and Ba at high pressures is interesting (Drickamer 1965, Stager and Drickamer 1963 a). The electrical resistance of Ca rises steeply at a pressure of about 150 kbar followed by a fall at about 300 kbar, see fig. 15 (*a*); this behaviour was explained by Stager and Drickamer in terms of a phase transition to a semi-conducting or semi-metallic state at 150 kbar, followed by a second phase transition near 300 kbar to a metallic state. The high value of the resistance was accompanied by a negative temperature coefficient, $\partial R/\partial T$, of the resistance. The behaviour of the electrical resistance of Sr as a function of pressure is similar to that of Ca although the sharp rise in electrical resistance as well as the existence of a negative $\partial R/\partial T$ occur at a considerably lower pressure in Sr ($\sim$ 35 kbar) than in Ca, see fig. 15 (*b*) (see also Jayaraman *et al.* (1963 a, b) and McWhan and Jayaraman (1963)). Although the resistance of Ba as a function of pressure does exhibit some sharp rises, these are by no means so disastrous as those exhibited by Ca and Sr, see fig. 15 (*c*). Although this high pressure behaviour of the electrical resistance of the alkaline earth metals was previously explained

† See note added in proof.

Fig. 15

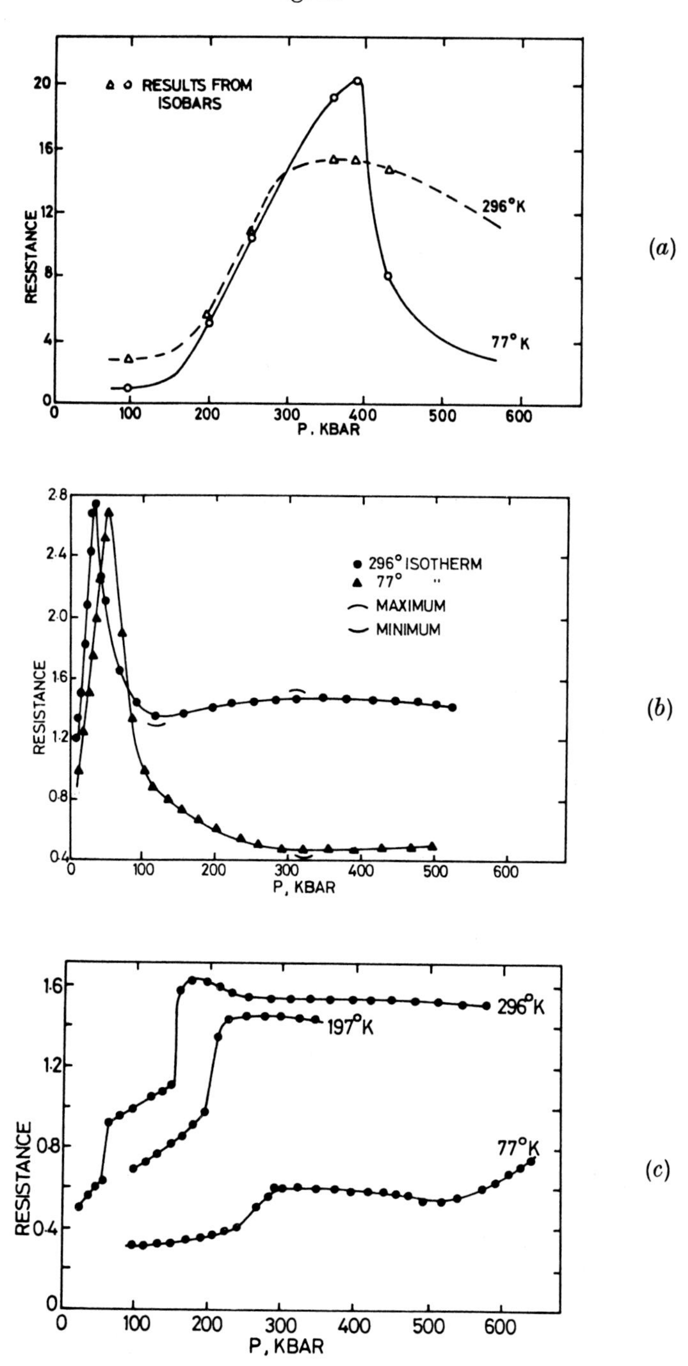

The resistance versus pressure of (*a*) Ca, (*b*) Sr, and (*c*) Ba (Stager and Drickamer 1963 a).

in terms of assumed structural phase changes, there was no experimental x-ray evidence for such changes. By calculating the band structure of Ca at various pressures, i.e. by altering the lattice constant, Altmann and Cracknell (1964) showed that as the pressure is increased so the size of the pockets of holes in band 1 and of electrons in band 2 diminished and eventually disappeared. Were it not for a degeneracy at one point along the line LW in the Brillouin zone (Cracknell 1964, Vasvari *et al.* 1967, Vasvari and Heine 1967) there would be a complete separation of bands 1 and 2 and a gap would appear thereby making Ca into an intrinsic semiconductor or even an insulator. However, because of the existence of the isolated degeneracy along the line LW there is never actually a gap between bands 1 and 2 as the pressure is raised but there always remains this isolated point of contact and Ca is a semi-metal rather than a semiconductor. In such a situation band 1 is completely full while band 2 is completely empty and the Fermi level passes through this point of contact so that, strictly speaking, there is no Fermi surface. In practice there will be a small overlap of order 0·003 Ry somewhere along this line LW. Vasvari and Heine (1967) suggested that this could also explain the negative temperature coefficient $\partial R/\partial T$ of Ca at room temperature at pressures of about 300–400 kbar, although it is not known for sure that Ca remains f.c.c. at this high pressure. The model pseudopotential calculations of the electronic band structure of Ca and Sr by Vasvari *et al.* (1967) were performed as a function of atomic volume up to 60% compression. The appearance of the sharp rise in the electrical resistance at a lower pressure in Sr suggests that the pockets of holes and electrons constituting the Fermi surface of Sr are smaller than those in Ca. The rise in the electrical resistance of Ca and Sr was analysed by Animalu (1967) in terms of a structural phase change from the f.c.c. to the b.c.c. structure rather than in terms of the properties of the conduction electrons and the Fermi surface. Animalu analysed the transitions induced by temperature and pressure between the f.c.c. and b.c.c. phases by computing the differences in the Gibbs free energy between the two phases in the nearly-free-electron and harmonic approximations. At the absolute zero of temperature and pressure, the observed f.c.c. structure in Sr and b.c.c. structure in Ba were found to have the lower internal energy; in Ca, however, identical analysis led to lower energy in b.c.c. rather than f.c.c., which is in contradiction with experience. An f.c.c. to b.c.c. transition in Sr was predicted by Animalu (1967) by this method at a pressure of about 10 kbar at 0°K or at a temperature of about 150°K at zero pressure. Both these results agree only qualitatively with the observed values of ~35 kbar and ~830°K, respectively. The existence of a pressure-induced transformation to a structure with lower packing density, i.e. f.c.c.→b.c.c., is at first sight puzzling. However, it should be remembered that the atoms are not infinitely rigid and it can be shown that, for a model of compressible spheres, it is possible, at sufficiently high pressures, for a polymorph with lower packing density to possess higher absolute density (Weddeling 1965).

§ 5. p-BLOCK : GROUP IIIB : BORON GROUP

[B	Boron	2.(2, 1)]
Al	Aluminium	2.8.(2, 1)
Ga	Gallium	2.8.18.(2, 1)
In	Indium	2.8.18.18.(2, 1)
Tl	Thallium	2.8.18.32.18.(2, 1)

5.1. *Structures*

The following information is given by Smithells (1967). Al has the face-centred cubic structure ($a=4{\cdot}04$ Å) and In has a tetragonal structure (F4/mmm, D_{4h}^{17}) which is simply related to the face-centred cubic structure by a distortion along the c axis ($a=4{\cdot}58$ Å and $c=4{\cdot}94$ Å). Ga has a complicated orthorhombic structure (Abma, D_{2h}^{18}) which possesses eight Ga atoms in the conventional unit cell at

$$[000\,;\ 0\tfrac{1}{2}\tfrac{1}{2}] \pm [x0z,\ \tfrac{1}{2}+x,\ \tfrac{1}{2},\ \bar{z}]$$

where $x=0{\cdot}079$ and $z=0{\cdot}153$ ($a=4{\cdot}52$ Å, $b=4{\cdot}51$ Å, $c=7{\cdot}65$ Å). A discussion of the reasons for the existence of this complicated structure in terms of the behaviour of the electrons is given by Heine (1968). Tl has the (slightly distorted) h.c.p. structure below 230°C ($a=3{\cdot}45$ Å, $c=5{\cdot}51$ Å) and the b.c.c. structure above 230°C ($a=3{\cdot}87$ Å); these two phases are sometimes referred to as α-Tl and β-Tl respectively. Although Tl has been observed to have an anomaly in the pressure coefficient of the superconducting transition temperature, $\partial T_c/\partial P$, at pressures near 1·2 kbar (see § 5.5) Barrett (1958) found no evidence of any instability of the h.c.p. phase at zero pressure down to 5°K.

5.2. *Aluminium*

The Fermi surface of Al is now well established and the history of its determination over a period of about ten years is a particularly good example of the use of both experiment and theory, in balanced proportions, by a number of workers in a concerted attack on the problem.

Al contains three conduction electrons and therefore, in the free-electron approximation, band 1 is full everywhere, band 2 is nearly all full but the Fermi surface does not touch the Brillouin zone boundary; in band 3 the Fermi surface contains sets of arms that intersect the Brillouin zone boundary and form a connected 'monster' and band 4 contains a few very small isolated pockets of electrons (see fig. 2 (*a*)).

The initial investigations of the Fermi surface of Al were made by means of the de Haas–van Alphen effect and the early work was reviewed by Shoenberg (1952). Gunnerson (1957) observed the de Haas–van Alphen effect in Al in fields up to 15·4 kG and found three high frequency periods for each field direction in addition to some low frequency periods. This work was followed by an O.P.W. band structure calculation performed by Heine (1957 a, b, c) in which many different contributions to the potential were included. The energy eigenvalues obtained were quite close to the free-electron values. A numerical error was discovered by Behringer (1958) in the construction of one of the contributions to the potential used in this band structure calculation, but it was estimated that the effect of this correction was to raise all Heine's energy levels by about 0·1 Ry, but to shift them relative to one another by only about 0·03 Ry, which, roughly, doubles the error given by Heine but does not change the general appearance of the energy bands. The eigenvalues determined by Heine were later used by Harrison as the starting point of a pseudopotential interpolation scheme to deduce the shape of the Fermi surface of Al. For the Fermi surface computed in this way the band 4 pockets of electrons of the free-electron model disappeared and the dimensions of the band 2 Fermi surface were altered slightly although that surface was not altered topologically. The Fermi surface in band 3 was also slightly distorted from the free-electron model. The surface proposed by Harrison (1959) showed similar connectivities to those of the free-electron Fermi surface, see fig. 16, but, in fact, there remained some doubt as to whether all the arms were connected together or not. Magnetoresistance measurements by Alekseevskiĭ and Gaĭdukov (1959 a) and Lüthi (1960) suggested that there are no open orbits possible on the Fermi surface of Al.

Fig. 16

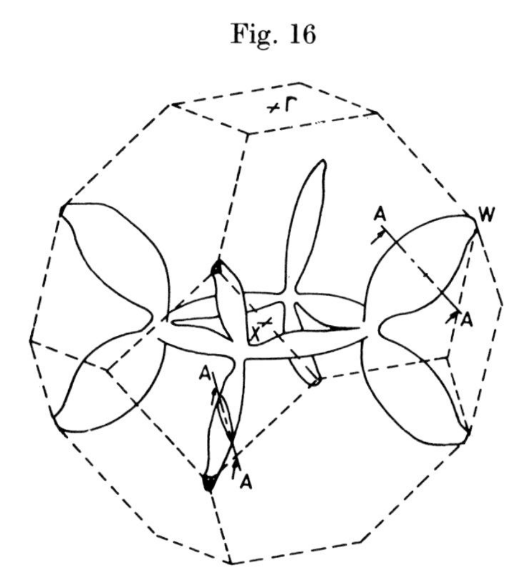

The Fermi surface of electrons of Al in band 3 proposed by Harrison (1959); note that X, and not Γ, is at the centre of the zone.

A further calculation of the band structure of Al was performed by Segall (1961) by the Green's function method and various cross sections of the Fermi surface in band 2 were calculated (Segall 1963). These calculated cross sections in band 2 agree well with the experimental magnetoacoustic measurements of Kamm and Bohm (1962, 1963). Detailed comparisons are given in table I of Kamm and Bohm (1963); a sample of the experimental results of Kamm and Bohm is shown in fig. 17. However, the accuracy of Segall's calculation in band 3 was inadequate in the crucial regions near the corners of the Brillouin zone at W to fix the connectivities of the Fermi surface in band 3. More recent *ab initio* band structure calculations for Al by Snow (1967) were also insufficiently accurate in the region near W. Neither were the results of Kamm and Bohm (1963) able to lead

Fig. 17

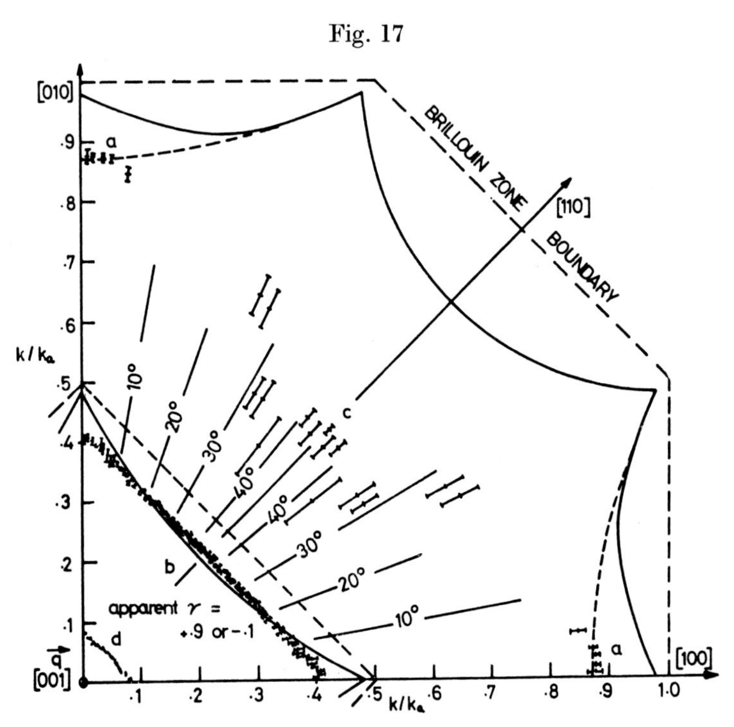

One of the Fermi surface cross sections of Al determined by magnetoacoustic measurements (Kamm and Bohm 1963). The free-electron Fermi surface is shown in outline (continuous line) and its central section (broken line).

to a definitive statement about the connectivities in band 3. There have also been several other studies of magnetoacoustic effects in Al and most of these confirmed the general appearance of the Fermi surface in band 2. Although the oscillations associated with the Fermi surface in band 2 have been observed, it is difficult to measure the period, and hence to follow its anisotropy, with sufficient accuracy to draw definite conclusions about the Fermi surface in band 3 (Roberts 1960, Bezuglyĭ, Galkin and Pushkin 1962, 1963, Bezuglyĭ, Galkin, Pushkin and Khomchenko 1962, Balcombe *et al.* 1964, Fossheim and Olsen 1964, Jones 1964, Beattie and Uehling 1966, Miller 1966, Aubauer 1967).

Further de Haas–van Alphen measurements on Al were made by Priestley (1962) and by Shepherd *et al.* (1964). These results, together with the earlier results of Gunnerson (1957), were carefully studied by Ashcroft (1963 a, b). The available experimental de Haas–van Alphen data, which were listed by Ashcroft (1963 b), were compared with calculated periods arising from extremal cross sections of a Fermi surface defined by a fourth-order secular equation. Several different models for the Fermi surface were found to be possible, depending on the values of the two Fourier coefficients V_{111} and V_{200} of a weak pseudopotential. By adjusting V_{111} and V_{200} until the best fit with the experimental de Haas–van Alphen data was obtained, the Fermi surface which was finally calculated differed only slightly from the nearly-free-electron Fermi surface proposed by Harrison (1959, 1960 a). The only essential difference between the conclusions of Ashcroft and the model proposed by Harrison (see fig. 16) is that, according to Ashcroft, band 3 at W is very slightly above the Fermi level. This alters the connectivities and the 'monster' in band 3 becomes 'dismembered' by becoming disconnected at W, see fig. 18. The determination by means of the de Haas–van Alphen effect of the connectivity at

Fig. 18

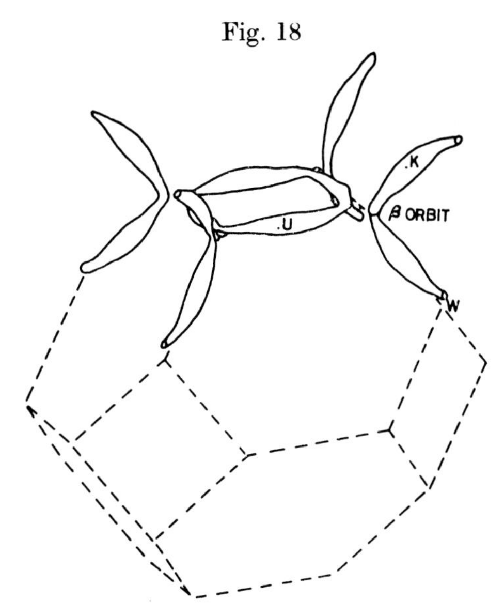

The Fermi surface of electrons in band 3 of Al proposed by Ashcroft (Larson and Gordon 1967).

W is complicated by the possibility of magnetic breakdown. If the connectivity at W is broken, the dismembered 'monster' in band 3 becomes reduced to a set of square rings in various orientations where each ring consists of four sausage-like arms joined together. The subsequent low-field de Haas–van Alphen measurements of Larson and Gordon (1967) contained several orbits that provided confirmation of the dismembering of the 'monster' in band 3. The measurements by Vol'skiĭ (1964) of the periods of the quantum oscillations of the surface resistance of Al in a magnetic field also demonstrated fairly convincingly the dismembering of the 'monster' (see figs. 5–8 of Vol'skiĭ (1964)). These quantum oscillations of the surface resistance have the same periodic behaviour as the de Haas–van Alphen oscillations and yield the same information. Cyclotron masses were also determined for the Fermi surface in band 3 by Larson and Gordon (1967) and by comparison with the band structure masses the enhancement of the band structure effective mass due to electron–phonon interactions was found to be between 30 and 50% of the band structure masses. A new band structure calculation for Al by Faulkner (1969) using the Korringa–Kohn–Rostoker method was able to reproduce the connectivities at W determined by Ashcroft.

Although it was mostly the experimental work on the de Haas–van Alphen and magnetoacoustic effects, together with the associated theoretical work already described, which was chiefly responsible for establishing the Fermi surface of Al, we nevertheless mention briefly the substantial body of other related experimental work that gives general supporting evidence for this topology.

The effect of pressures up to 7 kbar on portions of the Fermi surface in band 3 of Al has been investigated by Melz (1966) by using the field-modulation method for observing the de Haas–van Alphen oscillations. The rate of change of cross-sectional area was substantially larger than that predicted by the nearly-free-electron model. Semi-quantitative agreement was obtained with a model based on using the pressure dependence of the pseudopotential coefficients. The de Haas–van Alphen effect has also been studied in dilute alloys of Zn, Si, Ge, Mg, and Ag in Al by Shepherd and Gordon (1968). The variation of the average value of k_F with pressure up to 100 kbar was determined by Burton and Jura (1968) from experiments performed on the angular correlation of γ-rays produced by the annihilation of positrons. Other positron annihilation measurements have also been performed on Al (Bell and Jørgensen 1960, Berko and Plaskett 1958, Colombino *et al.* 1964). The use of pseudopotentials in the analysis of positron annihilation experiments has been introduced for Al by Stroud and Ehrenreich (1968).

Measurements have been made by Wooten *et al.* (1965) of the energy distribution of electrons emitted from Al as a function of photon energy in the range 5 to 11·5 ev. A direct experimental determination of the density of states for Al was performed by Phillips and Weiss (1968) using Compton scattering line shapes. Experimental values of γ, the temperature coefficient of the electronic specific heat, include

1·46 mJ mole^{-1} deg^{-2}	Giauque and Meads (1941)
(1·35 ± 0·01) mJ mole^{-1} deg^{-2}	Phillips (1959)
(1·360 ± 0·001) mJ mole^{-1} deg^{-2}	Dixon *et al.* (1965)
(1·362 ± 0·003) mJ mole^{-1} deg^{-2}	Dicke and Green (1967).

Determinations of cyclotron masses have been performed by several workers (Moore and Spong 1962, Galkin *et al.* 1963, Grimes *et al.* 1963, Naberezhnÿkh and Tolstoluzhskiĭ 1964, Spong and Kip 1965, and Mina *et al.* 1966). Doppler-shifted cyclotron resonance of helicon waves has been observed by Stanford and Stern (1966) and the Gaussian radius of curvature of the Fermi surface, normal to a few important directions, was determined. The early experimental data on the magnetoresistance of Al as a function of magnetic field had exhibited saturation at high fields (Borovik 1952, Yntema 1953, Borovik and Volotskaya 1960) and this was taken to mean that the Fermi surface of Al was closed, thereby having no open orbits. This is, of course, in agreement with the conclusions of the de Haas–van Alphen and magnetoacoustic experiments mentioned already. However, in more recent work on the magnetoresistance for various orientations and on the Hall constant, also as a function of orientation, for samples of Al of very high purity evidence was found for the existence of open orbits for nearly every orientation (Lüthi and Olsen 1956, Balcombe 1963, Volotskaya 1963a, Borovik *et al.* 1963, Borovik and Volotskaya 1965). These results should be interpreted in terms of Ashcroft's model together with magnetic breakdown in the Fermi surface in band 3 rather than in terms of Harrison's 'monster'. Low frequency helicon resonances in Al were observed by Amundsen and Seeberg (1969) (see also Alstadheim and Risnes (1968) and Goodman (1968)). It is interesting to note that these helicon measurements give a magnetoresistance which shows signs of saturation in high fields, in agreement with the normal behaviour for closed Fermi surfaces. The strain dependence of the magnetoresistance of Al was measured by Stevenson (1967). Calculations of the Hall constant of Al by Feder and Lothe (1965), assuming a free-electron Fermi surface, were in good agreement with the experimental measurements; but their calculations of the magnetoresistance, while qualitatively in agreement with experiments at low magnetic fields, were not in agreement with the non-saturation results just mentioned for very pure Al at high magnetic fields. Calculations of the infra-red absorptivity of Al, also based on a free-electron Fermi surface, gave a value 30% lower than the measured value (Biondi and Guobadia 1968).

The phonon dispersion relations in Al have been determined experimentally by Stedman and Nilsson (1965, 1966) and examined for evidence of Kohn anomalies. These anomalies in Al are too small to be observed by a casual inspection of a set of phonon dispersion curves and are only manifested as weak anomalies in the slope of the dispersion curves so that a fairly accurate set of experimental results will be needed to observe the Kohn effect. Stedman and Nilsson were able to identify eight points on the Fermi surface by a rather careful analysis of their data and they found good agreement with the Fermi surface determined by other methods (see Björkman *et al.* (1967) for a discussion of the electron–phonon interactions in Al in relation to the results of Stedman and Nilsson). The electron–phonon interactions in a metal can be investigated by studying the ultrasonic attenuation without external magnetic fields present. For a metal that is a superconductor it is possible to determine the electronic part of the attenuation by exploiting the fact that α_s/α_n, the ratio of the attenuation in the superconducting and normal states, is equal to zero at 0°K. It is then possible to deduce the magnitude of the electron–phonon interaction and the electron mean free path from the frequency dependence of α_n. If the metal is not superconducting, or if the transition temperature is inconveniently low, these two parameters can be determined from the difference between the attenuation in an infinite magnetic field and in zero field. This method yields the magnitude of the electron–phonon interaction in an infinite field in addition to the interaction in zero field and the electronic mean free path. Berre and Olsen (1965) used this method to determine these three quantities for Al. The results for the electronic mean free path were consistently smaller than the values obtained by Førsvoll and Holwech (1962, 1963) from the size effect measurements of the electrical resistance. The experimental results yielded values of the electron–phonon interaction constant A_q which differed from theoretical estimates by factors of between 1·5 and 2.

The eddy current method suggested by Cotti (1963, 1964) has been used to determine directly the ratio $\sigma/\bar{\lambda}$ (=electrical conductivity/average mean free path at Fermi surface) for Al (Brändli *et al.* 1964, Cotti *et al.* 1964). This can be used to determine the area, S, of the Fermi surface since

$$\frac{\sigma}{\bar{\lambda}} = \frac{e^2}{12\pi^3\hbar} S, \qquad (5.2.1)$$

and, for Al, S was found to be only 59% of the area of the free-electron Fermi surface. These authors suggest that this low value is not entirely due to the distortions of the Fermi surface (which seem unlikely to decrease the area quite so much) but is also due to large variations in λ over the Fermi surface. The temperature dependence of the mean free path was determined from the ultrasonic attenuation measurements (without magnetic field) of Wang and McCarthy (1969). The d.c. size effect of a thin film in a magnetic field perpendicular to the current and in the plane of the film (sometimes called the MacDonald effect) consists of the observation of a sharp kink in the magnetoresistance at the value of the magnetic field for which the extremal orbit diameter is equal to the sample thickness; this has been observed by Holwech and Risnes (1968) for Al and the theory has been verified for the Fermi surface in band 2. Similar effects have also been observed with the magnetic field normal to the film (Førsvoll and Holwech 1964 a, b, c, Druyvesteyn 1968). A calculation of both the direct and core-polarization contributions to the Knight shift in Al using O.P.W. wave functions by Shyu, Das and Gaspari (1966) yielded a calculated shift of 0·163% as compared with the value of 0·168% obtained by Sagalyn and Hofmann (1962). An attempt was made by Jones and

Williams (1964) to observe quantum oscillations as a function of magnetic field in the Knight shift of Al, but only a small number of oscillations was observed.

5.3. *Gallium*

As mentioned in § 5.1 the conventional orthorhombic unit cell of Ga contains eight Ga atoms. The fundamental unit cell therefore contains four Ga atoms so that, with a valence of three, an equivalent of six energy bands must be occupied throughout the Brillouin zone. Because Ga does not have one of the three common metallic structures its Fermi surface in the free-electron approximation will not be found among the illustrations in fig. 2. We therefore begin our consideration of this metal by discussing the form of its free-electron Fermi surface which has been investigated by Slater *et al.* (1962) and, with spin–orbit coupling, by Koster (1962). We may take the unit vectors of the crystal lattice of Ga to be:

$$\left.\begin{aligned} \mathbf{t}_1 &= a\mathbf{i}, \\ \mathbf{t}_2 &= \tfrac{1}{2}b\mathbf{j} + \tfrac{1}{2}c\mathbf{k}, \\ \mathbf{t}_3 &= -\tfrac{1}{2}b\mathbf{j} + \tfrac{1}{2}c\mathbf{k}, \end{aligned}\right\} \quad \ldots\ldots \quad (5.3.1)$$

and in the construction of the Brillouin zone the exact positions of the Ga atoms, given this lattice, are not important. The reciprocal lattice vectors defined by $\mathbf{t}_i \cdot \mathbf{g}_j = 2\pi\delta_{ij}$ are then:

$$\left.\begin{aligned} \mathbf{g}_1 &= \frac{2\pi}{a}\,\mathbf{i}, \\ \mathbf{g}_2 &= \frac{2\pi}{b}\,\mathbf{j} + \frac{2\pi}{c}\,\mathbf{k}, \\ \mathbf{g}_3 &= -\frac{2\pi}{b}\,\mathbf{j} + \frac{2\pi}{c}\,\mathbf{k}. \end{aligned}\right\} \quad \ldots\ldots \quad (5.3.2)$$

The Brillouin zone is illustrated in fig. 19 where the points and lines of symmetry are labelled in the notation of Slater *et al.* (1962). The character tables for the group of the wave vector $\mathbf{k}$, together with symmetrized sets of plane waves, are given by Slater *et al.*, who also consider the possibility of extra degeneracies in the band structure due to the existence of time-reversal symmetry. The energy bands in the free-electron approximation have also been constructed and in the neighbourhood of the Fermi level they are exceedingly complicated (see fig. 5 of Slater *et al.* (1962)) with the result that there are pieces of Fermi surface in several bands. For example, at Γ nine bands are occupied but proceeding along ΓT this number drops in stages until at T there are only three occupied bands. At X there are only two occupied bands. Not only is this free-electron Fermi surface exceedingly complicated, involving bands 3, 4, 5, 6, 7, 8 and 9, but also

Fig. 19

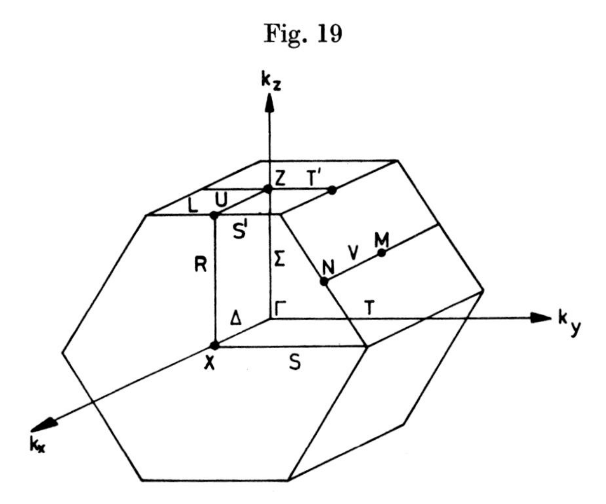

The Brillouin zone for Ga (Slater *et al.* 1962).

it is clear that any quite small changes to the free-electron bands, as the result of using a non-zero potential, will make drastic changes to the topology of the Fermi surface. Clearly, also, the presence of pieces of Fermi surface in so many bands is bound to make it very difficult to interpret any experimental data on the de Haas–van Alphen, magnetoacoustic, and other 'effects'. The various pieces of Fermi surface are shown in fig. 20 and sections through all these pieces for various planes in the Brillouin zone normal to $\mathbf{g}_1$ are given in fig. 6 of Slater *et al.* (1962). The Fermi surface in band 6 shown in fig. 20 is obviously multiply-connected in a rather complicated fashion. It seems likely therefore that even when a more realistic potential is used this multiple connectivity will survive and therefore band 6 can be expected to lead to open orbits. Some evidence of open orbits in Ga was obtained in the magnetoresistance measurements of Alekseevskiĭ and Gaĭdukov (1959 a, b). In the experiments of Reed and Marcus (1962) it was found that with the current parallel to *c* the magnetoresistance was quadratic in the magnetic field for all directions of the magnetic field in the *ab* plane, but the magnetoresistance saturates when both the current and field are in the *ab* plane. This suggests a Fermi surface for Ga with the topology of a cylinder with its axis parallel to the *c* axis, which is not inconsistent with the Fermi surface in band 6 shown in fig. 20. From fig. 20 we see that the Fermi surface in band 5 in the free-electron approximation also leads to the possibility of open orbits. The pieces of Fermi surface in the remaining bands (3, 4, 7, 8 and 9) in the free-electron approximation are not multiply-connected. It will be noticed that the free-electron Fermi surface in fig. 20 appears to exhibit hexagonal symmetry. This hexagonal symmetry is accidental and not exact and arises as a result of the special values of *b* and *c*; the actual values of the angles on the hexagonal face of the Brillouin zone are not 60° but 59° and 61° (Reed and Marcus 1962).

A band structure calculation for Ga has been performed by Wood (1966) using the A.P.W. method. In this calculation energies were determined at 115 points in one-eighth of the Brillouin zone and the calculated bands could be realistically regarded as perturbed free-electron bands. The Fermi level was determined, by counting states, to be 0·807 Ry referred to Γ_1^+ as zero and two Fermi surfaces were constructed taking E_F to be 0·802 Ry and 0·812 Ry. Both Fermi surfaces showed considerable departures from the free-electron Fermi surface. Figure 21 shows the Fermi surface constructed by Goldstein and Foner (1966) from the sections given by Wood (1966). There are many differences from the free-electron Fermi surface of fig. 20. The small pieces of Fermi surface in bands 3, 4 and 9 have disappeared and several pieces of Fermi surface in bands 7 and 8 have also disappeared. Of general interest perhaps is the fact that the pseudohexagonal symmetry which was such a marked feature of the free-electron model is not present in the results of the A.P.W. calculation. The legs of the band 5 Fermi surface in the free-electron model, which rather resembles the new Liverpool Cathedral, have also vanished and the appearance of the band 6 Fermi surface has changed drastically and is described by Goldstein and Foner (1966) as a 'six-legged-two-headed-camel', see fig. 21. In addition to the removal of the pseudohexagonal symmetry this Fermi surface in band 6 is only multiply-connected in the k_a direction and not in the other directions in the $k_b k_c$ plane shown in fig. 20. This would appear to be in conflict with the high-field galvanomagnetic measurements of Reed and Marcus (1962) which indicated that the Fermi surface was multiply-connected in a direction in the $k_b k_c$ plane rather than in the k_a direction. In the absence of spin–orbit coupling the energy bands in Ga are four-fold degenerate all over the hexagonal face of the Brillouin zone and along the line MN (see fig. 19) and are two-fold degenerate everywhere else (Slater *et al.* 1962). If spin–orbit coupling is included the four-fold degeneracy only survives at M and along the line XL and the energy bands are two-fold degenerate everywhere else (Koster 1962). All these degeneracies include the spin degeneracies.

We have described in considerable detail the Fermi surface of Ga obtained as a result of the A.P.W. calculations of Wood (1966), because it provides a useful basis for the discussion of the experimental work on Ga (see below) and is almost certainly more realistic than the free-electron model. However, as will emerge shortly, because of the large number of pieces of Fermi surface in various bands, the interpretation of the results of the experimental work on Ga is not easy and it is often difficult to establish firm conclusions. On the theoretical side it is admitted by Wood that the construction of the potential used in his calculations leaves something to be desired and there are uncertainties in the value of the Fermi energy. Small changes in either of these quantities can lead to quite large changes in the calculated Fermi surface. The density of states curve for Ga was also calculated by Wood (1966) from which the value of γ, the temperature coefficient of the electronic specific heat, of 0·9 mJ mole^{-1} deg^{-2} was

Fig. 20

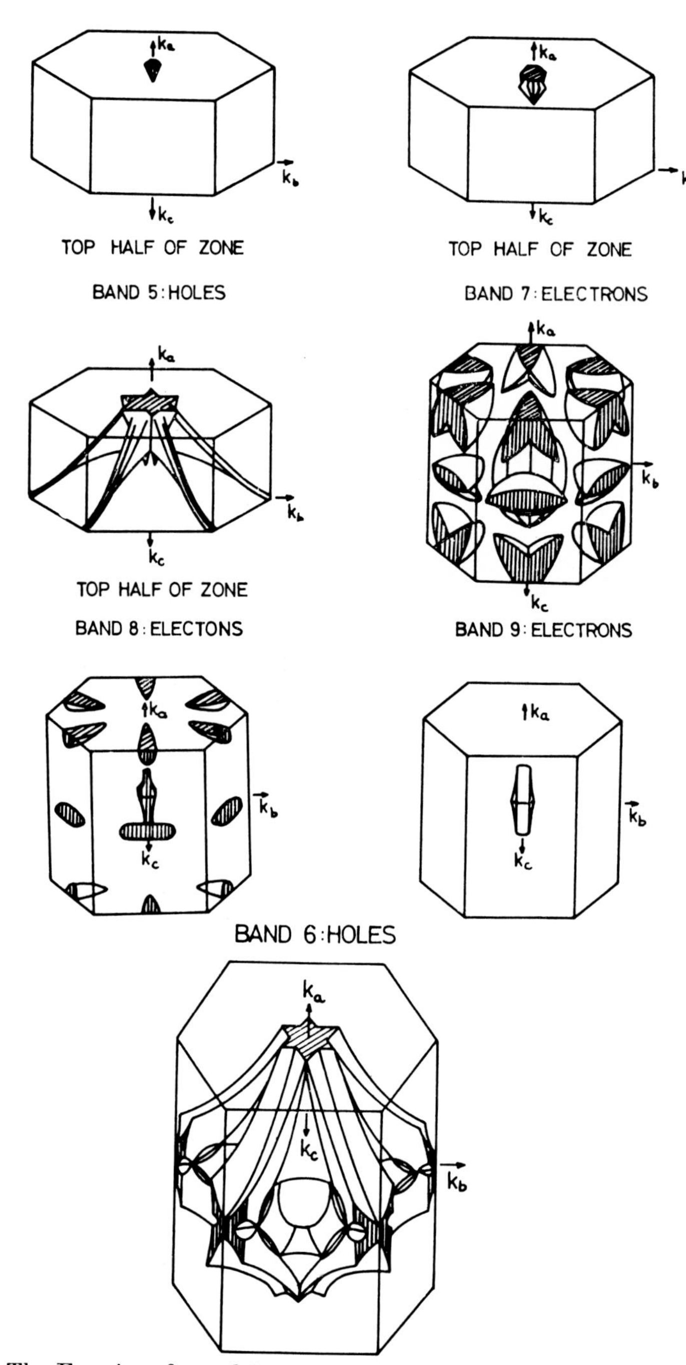

The Fermi surface of Ga in the free-electron approximation (Reed and Marcus 1962).

Fig. 21

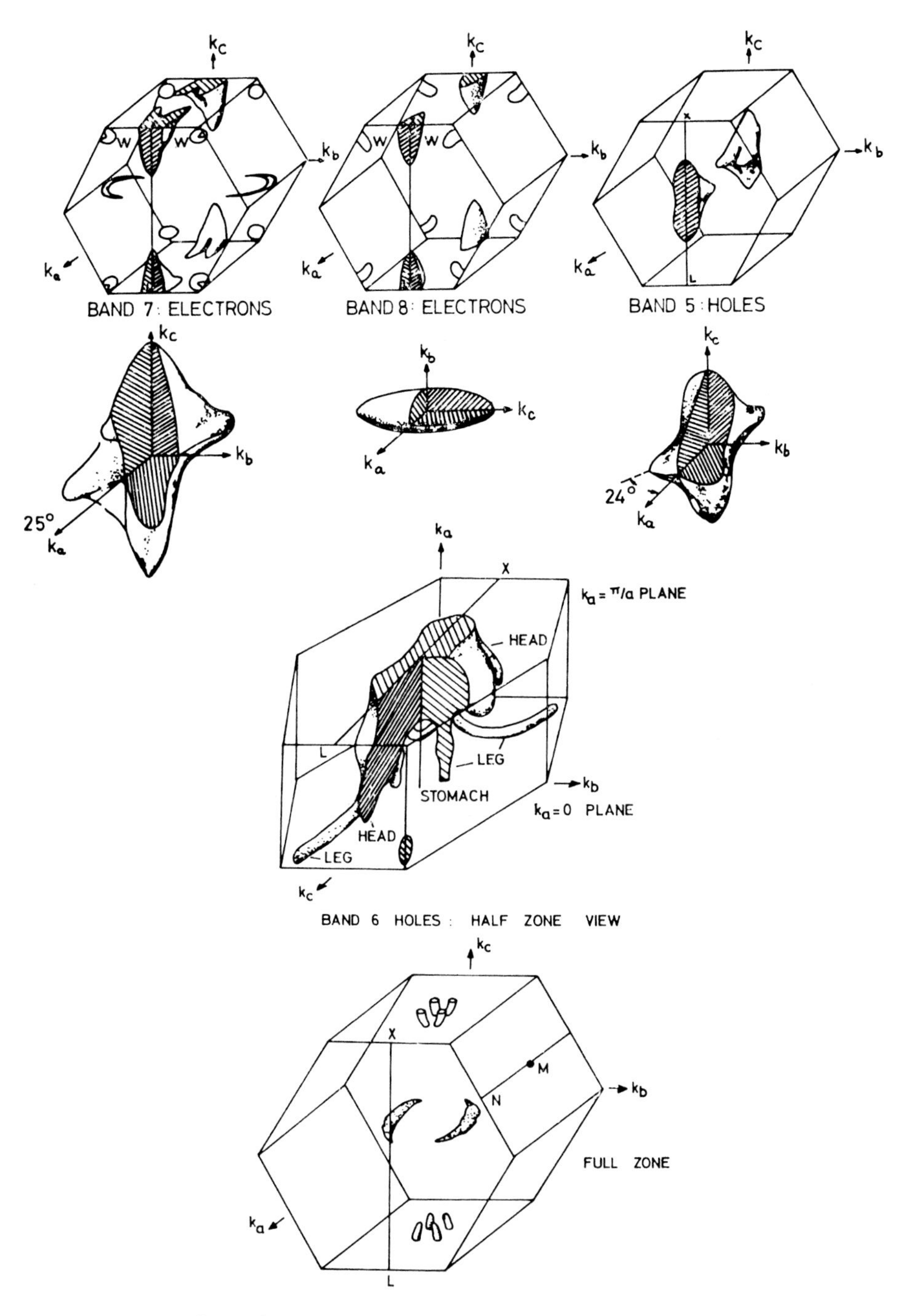

The Fermi surface of Ga constructed by Goldstein and Foner (1966) from the results of the A.P.W. calculation of Wood (1966).

calculated. The corresponding experimental values of γ are:

$0{\cdot}75$ mJ mole^{-1} deg^{-2} Wolcott (1956),
$0{\cdot}601$ mJ mole^{-1} deg^{-2} Seidel and Keesom (1958),
$(0{\cdot}596 \pm 0{\cdot}005)$ mJ mole^{-1} deg^{-2} Phillips (1964).

The initial work on the de Haas–van Alphen effect in Ga has been described by Shoenberg (1952) who attempted to describe the various periods observed in terms of a Fermi surface composed of three ellipsoids, with their axes along crystallographic axes. If this model represented the complete Fermi surface the longitudinal magnetoresistance for all three axes would have to be zero, this was later observed not to be the case in practice (Reed and Marcus 1962). More recent de Haas–van Alphen measurements have been made by Condon (1964) and by Goldstein and Foner (1966). A large number of periods have been found, see table 9, and some tentative assignments and qualitative comparisons with the A.P.W. Fermi surface were made, see table 10. Some additional periods were found in the magnetothermal oscillations by Goy *et al.* (1967). The general feature that there are a large number of pieces of Fermi surface in different bands appears to be confirmed by the de Haas–van Alphen results. Although the A.P.W. Fermi surface of fig. 21 agrees more closely with the de Haas–van Alphen results than does the free-electron model, quantitative agreement was not obtained nor does the A.P.W. Fermi surface possess the correct connectivity; some modifications to the Fermi surface in fig. 21 are therefore clearly necessary.

Table 9. The de Haas–van Alphen frequencies for Ga (megagauss)

Axis	Goldstein and Foner (1966)			Shoenberg (1952)	Shapira (1964 a, b)
a	0·135	—	—	0·24	—
	0·495	0·505	0·500	0·50†	—
	0·855	0·865	0·87	0·91	—
	23·5	23·2	23·0	—	—
	56·7	—	—	—	—
b	0·345	0·335	0·342	0·33†	0·336
	0·725	0·735	—	—	0·71
	19·2	18·7	19·0	—	~18
	22·5	22·0	23·0	—	—
	30·0	30·0	30·7	—	~31
	63·5†	—	—	—	63
c	0·20	0·20	0·209	0·20†	—
	0·220	0·225	0·232	0·22†	0·21
	0·765	0·76	0·76	—	0·83
	8·3	8·5	8·5	—	8·3
	13·0	12·8	12·8	—	—
	20·5	20·5	20·6	—	—

† Extrapolated to axis.
(Adapted from Goldstein and Foner (1966).)

Table 10. Extremal areas of cross section of Fermi surface of Ga

(i) A.P.W. calculation (Wood 1966)

Normal to area	Electron bands 7	Electron bands 8	Hole bands 5	Hole bands 6
	Area (Å^{-2})			
a	0·11	0·11	0·11	0·11, ~0·24†
b	0·33	0·15	0·19	0·60
c	0·22	0·028	0·12	~0·14†

(ii) de Haas–van Alphen measurements

Direction of **H**	High frequency		Intermediate frequency	
	Freq. (MG)	Area (Å^{-2})	Freq. (MG)	Area (Å^{-2})
a	23·5	0·224	0·855	0·00816
b	30·0	0·286	~6·9†	~0·066
c	12·8	0·122	0·76	0·0073

† Estimated.
(Goldstein and Foner (1966).)

Fig. 22

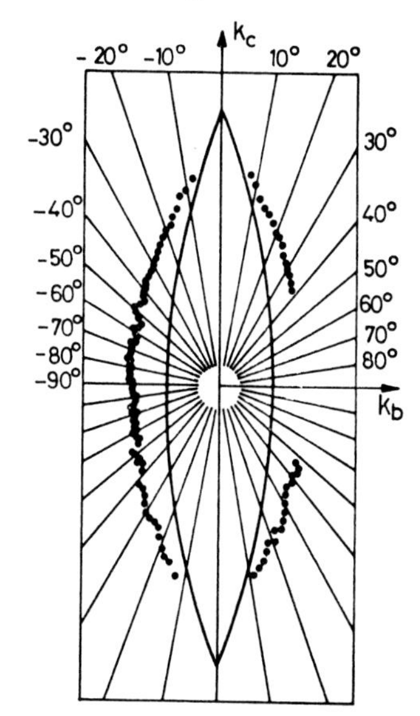

The intersection of the Fermi surface of Ga in band 8 with the plane $k_a = \pi/a$, from the magnetoacoustic data of Bezuglyĭ *et al.* (1964) compared with the free-electron model (continuous line).

A magnetoacoustic investigation of the Fermi surface of Ga by Bezuglyĭ *et al.* (1964, 1965) was interpreted by these authors in terms of the free-electron model which was the only theoretical Fermi surface available at that time. Although it was possible to identify many of their results with pieces of the free-electron Fermi surface it is not really surprising that considerable discrepancies in detail were found. For example, the intersection of the band 8 electron Fermi surface with the plane $k_a = \pi/a$ is shown in fig. 22 together with the experimental results of Bezuglyĭ *et al.* (1964), in comparison with the free-electron model the experimental cross section is shorter and fatter which is in agreement with the trend shown in the A.P.W. Fermi surface in band 8 (Wood 1966). By re-assigning to the band 6 hole Fermi surface (the camel), some of the experimental points that Bezuglyĭ *et al.* (1964) assigned to the band 7 electron Fermi surface, Wood showed that good agreement (10%) with the shoulder of this animal could be obtained. Good agreement was also found between the dimensions of the A.P.W. Fermi surface in bands 7 and 8 (butterflies and cigars) and the experimental measurements of Bezuglyĭ *et al.* (1964) except that the head of the butterfly is located about 40% farther from the Brillouin zone boundary in the A.P.W. Fermi surface than in the experimental results. In particular the pseudohexagonal symmetry vanishes, whereas in bands 7 and 8 in the free-electron model one would expect to see six sections in each band on the big hexagonal face of the Brillouin zone (see fig. 20) only two were observed experimentally in each band which is topologically the same as in the A.P.W. model (see fig. 21). The cross sections normal to k_c of the Fermi surface in bands 7 and 8 were determined from ultrasonic geometric resonances together with a part of the cross section of a third piece of Fermi surface that was tentatively assigned to band 5 (Lewiner 1967, Lewiner and Biquard 1967).

Preliminary radio-frequency size effect (Gantmakher effect) measurements on Ga were made by several workers (Sparlin and Schreiber 1964, Cochran and Shiffman 1965). More exhaustive radio-frequency size effect results have been obtained by Fukumoto and Strandberg (1966, 1967). These authors compared their results with the A.P.W. predictions of Wood (1966) with $E_F = 0{\cdot}807$ Ry and close agreement was found for several bands. The A.P.W. Fermi surface in bands 5 and 6 was found to agree with these experiments to within 5% in the plane normal to b; the band 5 Fermi surface was also measured in the plane normal to c and the A.P.W. prediction in the k_b direction was found to be 25% narrower than the measured value. The results of measurements on the band 6 Fermi surface (the A.P.W. two-headed-six-legged camel) in the plane normal to a as well as that already mentioned suggested that some modifications to the shape of this animal are required, but only semi-quantitative modifications were suggested (see page 692 of Fukumoto and Strandberg (1967)). The A.P.W. band 7 Fermi surface was verified in the three planes normal to a, b and c and agreement within 10% was obtained nearly everywhere; only at the butterfly's head was the error as high as 30%. This is in substantial agreement with Wood's analysis of the magnetoacoustic measurements of Bezuglyĭ *et al.* which we have already mentioned. The A.P.W. band 8 Fermi surface (cigar) was measured in the plane normal to *c* and agreement was obtained to within 10%.

Therefore we see that, so far, de Haas–van Alphen, magnetoacoustic and radio-frequency size effect measurements have shown substantial agreement with the A.P.W. Fermi surface of Ga in bands 5, 7 and 8 whereas some modification (not too drastic) of the band 6 camel appears to be necessary. The modification of the band 6 camel would also seem to be necessary in connection with the existence of open orbits along the *c* axis shown by the magnetoresistance results of Reed and Marcus (1962) mentioned above. While it appears that some very small pieces of Fermi surface do exist in Ga (Shoenberg 1952, Condon 1964) it is not, as yet, clearly established whether they are the remains of the pieces of free-electron Fermi surface in bands 3, 4 and 9 or whether, as seems much more likely, they are some of the small pieces in bands 6, 7 and 8 of the A.P.W. Fermi surface shown in fig. 21.

We conclude this section by mentioning briefly several other experiments related to the study of the Fermi surface that have been performed on Ga. The fine structure of the extended K absorption edge of a Ga single crystal was measured by Alexander *et al.* (1965). The Knight shift has been measured for Ga metal by Valič *et al.* (1968) and a value of $(0{\cdot}115 \pm 0{\cdot}005)\%$ at $4{\cdot}2°$K was obtained. Magnetic breakdown of cyclotron resonance orbits in Ga has been reported and the principal effective masses determined (Moore 1967, 1968); cyclotron resonance line shapes have been studied in Ga by Surma (1968). Ultrasonic cyclotron resonance, which had formerly only been observed in Bi (Reneker 1959) was observed in Ga by Roberts (1961) and was later studied by Munarin (1968). Roberts also observed geometric (magnetoacoustic) resonances with orbit cross sections consistent with regions of electrons in band 7 of the nearly-free-electron Fermi surface. Giant quantum oscillations were observed in the ultrasonic attenuation in Ga by Shapira and Lax (1964) and the spin-splitting of these oscillations was also observed (Shapira 1964 b). The line shapes of these oscillations were investigated by Shapira and Neuringer (1967) who were able to use them to provide a rather interesting direct verification of the Fermi–Dirac distribution of the electrons in the metal. Various galvano-magnetic experiments have been performed on Ga (Yahia and Marcus 1959, Cochran and Yaqub 1965, Munarin and Marcus 1965, Yaqub and Cochran 1965, Munarin *et al.* 1968). The temperature and frequency dependence of the amplitude of the radio-frequency size effect in Ga has been investigated by Haberland and Shiffman (1967) and the magnetic field dependence of the line width by Haberland *et al.* (1969). Boundary scattering effects in the electrical resistivity (Neighbor and Shiffman 1967, Newbower and Neighbor 1967) and thermal resistivity (Boughton and Yaqub 1968) have been observed in small specimens of Ga. An interesting experiment on the direct measurement of the Fermi velocity in Ga has been performed by von Gutfeld and Nethercot (1967). This was done by simply timing the traverse of a heat pulse across a specimen of Ga metal. In most normal metals the transport of heat is mainly by the electrons. Heat-pulse measurements on metals can, therefore, be expected to give information on the electron-scattering processes and for very pure metals at very low temperatures, where the electron mean free path is very long, it is then possible to determine the Fermi velocity by measuring the transit time of a heat pulse. The experimental arrangement described by von Gutfeld and Nethercot involved the generation of heat pulses by the absorption of optical radiation from a chloroaluminium phthalocyanine laser side-pumped with a giant-pulse ruby laser. The heat pulses were detected by an In–Sb alloy thin-film superconducting bolometer. It is assumed that the initially hot electrons rapidly attain the Fermi velocity v_F and then traverse the crystal of Ga with no further collisions. In the temperature region used ($\sim 1{\cdot}8°\text{K}$–$4{\cdot}0°\text{K}$) electron–phonon collisions are unlikely in the distance concerned and von Gutfeld and Nethercot were able to check that impurity collisions were also unlikely by using samples of two or three different lengths. Values of v_F of between $5{\cdot}5 \times 10^7$ cm sec^{-1} and $6{\cdot}0 \times 10^7$ cm sec^{-1} for Ga were obtained. Values of the mean free path were also obtained from the shapes of the heat pulses for different thicknesses of sample.

5.4. *Indium*

The structure of In is that of a face-centred cubic metal with a slight distortion in the c direction which reduces its symmetry to that of the tetragonal space group given in § 5.1. In the absence of this distortion the free-electron Fermi surface would be that shown in fig. 2 (a). The distortion from the f.c.c. structure consists of a stretching by about 8% in the

Fig. 23

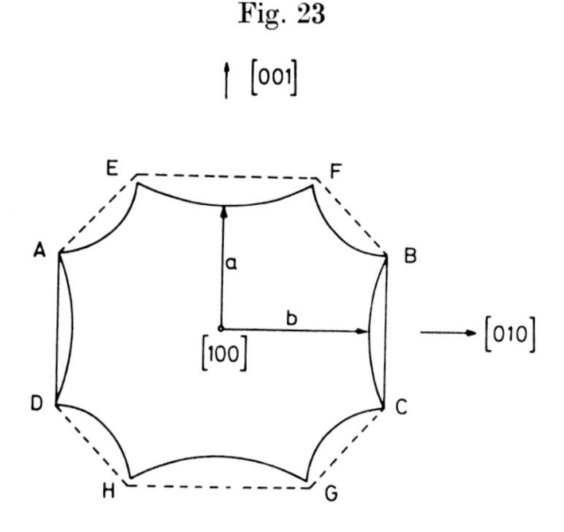

Cross section of the Fermi surface of In in band 2 in the free-electron model (Rayne 1963 a).

c axis direction, and consequently in reciprocal space the Brillouin zone is slightly contracted in the k_z direction. This has an important effect on the connectivity of the Fermi surface in band 2 in the free-electron model which now just touches the Brillouin zone boundary at those W points in the planes normal to k_x and k_y, see fig. 23, and, therefore, becomes multiply-connected in the k_z direction. In the free-electron model the Fermi surface in band 3 in In has the same connectivity as in an undistorted face-centred cubic metal. However, various galvanomagnetic experiments have failed to reveal any evidence of open orbits in In (see, for example, Alekseevskiĭ and Gaĭdukov (1959 a), Lüthi (1960)) and it therefore seems likely that the Fermi surface in band 2 does not touch the Brillouin zone boundary after all and that the Fermi surface in band 3 becomes disconnected in some way (Olsen 1958, Borovik and Volotskaya 1960, Volotškaya 1963 b, Gaĭdukov 1965, Lück 1966 a).

Preliminary cyclotron resonance measurements in In were made by Bezuglyĭ and Galkin (1959) at 9·3 GHz and they found m^* to be 0·8–0·9. Further cyclotron resonance measurements at 9·1 GHz in In by Castle *et al.* (1961) revealed several values of the effective mass. Preliminary de Haas–van Alphen data on In have been reported by Verkin *et al.* (1950 b) and Shoenberg (1952). Later experimental work on the de Haas–van Alphen, magnetoacoustic and radio-frequency size effects showed that the free-electron model is a fairly good first approximation to the Fermi surface of In although the changes are sufficiently large to alter the connectivities and to remove all possibilities of open orbits (in the absence of magnetic break-down). In band 2 the dimensions of the closed Fermi surface have been determined by the de Haas–van Alphen effect (Brandt and Rayne 1963, 1964, O'Sullivan *et al.* 1968), by the magnetoacoustic effect (Rayne and Chandrasekhar 1962, Rayne 1962 b, 1963 a), by the radio-frequency size effect (Gantmakher and Krylov 1964, 1965, Krylov and Gantmakher 1966), and by the use of cyclotron resonance (Mina and Khaĭkin 1966) ; for the actual dimensions see table I of Rayne (1963 a) and table 3 of Mina and Khaĭkin (1966). The linear dimensions of the band 2 Fermi surface are about 10% smaller than in the free-electron model. Because of the tetragonal distortion from the f.c.c. structure the band 3 Fermi surface of In in the free-electron model consists of two different kinds of arms, commonly called α arms, in the $\langle 101 \rangle$ directions, and β arms, in the $\langle 110 \rangle$ directions. On the basis of their magnetoacoustic data Rayne and Chandrasekhar claimed to have found some evidence for the existence of both the α and β arms in the band 3 Fermi surface of In (Rayne 1962 b, Rayne and Chandrasekhar 1962). Then on the basis of their de Haas–van Alphen results Brandt and Rayne (1964) claimed to have proved the definite existence of both α and β arms ; they found the β arms to be consistently smaller than on the free-electron model and were also very close to being cylindrical along a considerable part of their length, while the α arms were consistently larger than on the free-electron model, see table 11. The connectivities of the α and β arms were not established by this work.

Quantum oscillations in the ultrasonic attenuation in In single crystals were observed by Balcombe *et al.* (1964). For 30 MHz longitudinal sound waves propagated in approximately the [001] direction and with **B** normal to the propagation direction, two distinct periods were visible. The short period, which was attributed to the β arms, agrees fairly closely with the

Table 11. Band 3 de Haas–van Alphen results for In

Field direction	Period (10^{-6} G^{-1})	Type of arm	Extremal area (experimental) (Å^{-2})	Extremal area (free-electron model) (Å^{-2})
		[001] suspension		
[100]	4·69	α	0·046	0·035
	6·16	β	0·058	0·094
[110]	6·03	α	0·057	0·040
	4·55	β	0·044	0·079
		[100] suspension		
[010]	4·60	α	0·044	0·035
	6·20	β	0·059	0·094
[001]	5·40	α	0·052	0·033
[011]	4·80	α	0·046	0·029
	8·30	β	0·079	0·132

(Brandt and Rayne (1964).)

de Haas–van Alphen periods of Brandt and Rayne (1964) while there are small systematic deviations for the other periods. The effective masses of the current carriers in In were measured by cyclotron resonance by Mina and Khaĭkin (1965a) and their anisotropy in the (010), (001) and (111) planes was studied. The band 3 Fermi surface tubes in the [110] and [1$\bar{1}$0] directions (the β arms) were found to be connected into square rings but the tubes parallel to the [101] directions (the α arms) were found to be disconnected. In fact Mina and Khaĭkin (1966) found no evidence for the existence of the α arms. In the results of the radio-frequency size effect experiments of Gantmakher and Krylov (1965) the dimensions of the β arms were clearly established and further evidence was also provided to show that they are connected in square rings as suggested by the cyclotron resonance data of Mina and Khaĭkin (1966). Gantmakher and Krylov (1965) found no evidence of the α arms observed by Brandt and Rayne (1963, 1964); they also attempted to reproduce the results of Brandt and Rayne by using the quantum oscillations of the surface impedance but again were without success. Because of the apparent inconsistency that the α arms had only been seen in the de Haas–van Alphen work of Brandt and

Rayne and were calculated to exist in the band structure calculation of Gaspari and Das (1968) but had eluded observation by means of the cyclotron resonance experiments and the radio-frequency size effect experiments further de Haas–van Alphen experiments were performed to try to observe the α arms (Hughes and Lettington 1968, Hughes and Shepherd 1969). Hughes and Lettington used five different samples of In and were able to reproduce the β arm frequencies of Brandt and Rayne (1964) and Gantmakher and Krylov (1965). However, no α arm oscillations were observed in any of the large number of orientations investigated. It was therefore concluded that the α arms are either very small or, more probably, are non-existent. Hughes and Shepherd used four of their de Haas–van Alphen orbit sizes to construct a Fermi surface model based on an empirical pseudopotential, see table 12. Ashcroft and Lawrence (1968) showed that

Table 12. Extremal dimensions of the Fermi surface of In

Plane	Direction	Mina and Khaĭkin (1966)	Gantmakher and Krylov (1965)	Hughes and Shepherd (1969)	Band
{001}	7° from [100]	0·915 ± 0·015	0·93 ± 0·05	0·895	2
$\{1\bar{1}0\}$	[001]	0·780 ± 0·010	0·80 ± 0·05	0·815	2
$\{1\bar{1}0\}$	[110]	—	0·22 ± 0·01	0·209	3
$\{1\bar{1}0\}$	[001]	—	0·17 ± 0·01	0·157	3
$\{1\bar{1}0\}$	24° from [110]	—	0·22 ± 0·01	0·206	3
$\{1\bar{1}0\}$	56° from [110]	—	0·18 ± 0·01	0·171	3

(Adapted from Hughes and Shepherd (1969).)

their pseudopotential form factor, which was consistent with the transport properties and which accurately reproduced the experimental dimensions of the β arms, was sufficiently strong to remove all but minute disconnected remnants of the α arms near the points K. It can be seen from table 11 that the areas assigned by Brandt and Rayne to α arms are very close to those of the β arms and it has been suggested that all the experimental areas in table 11 actually belong to β arms, some of which were in misoriented parts of the specimen. The present balance of evidence then seems to suggest that the band 3 Fermi surface of In contains no α arms but only the β arms and that these β arms are connected into square rings. The actual dimensions of the band 2 and band 3 Fermi surfaces in In are given in the paper of Gantmakher and Krylov (1965).

We have mentioned already in connection with Al the relationship between S, the area of the Fermi surface, and the ratio $\sigma/\bar{\lambda}$. Various experimental measurements of $\sigma/\bar{\lambda}$ have been made for In and most of them confirm that the area of the Fermi surface is substantially less than in the free-electron approximation (see table 13).

Table 13. $\sigma/\bar{\lambda}$ for In

Method	$\sigma/\bar{\lambda}$	Reference
Free-electron model	18·9	—
Anomalous skin effect	$18{\cdot}0 \pm 1{\cdot}1$	Dheer (1961)
d.c. size effect	7·41	Aleksandrov (1962)
Eddy-current size effect	7·88	Cotti (1964)
Anomalous skin effect†	$9{\cdot}0 \pm 1{\cdot}6$	Lyall and Cochran (1967)

† See note added in proof.

(Adapted from Lyall and Cochran (1967).)

The effect of impurities on the behaviour of T_c, the superconducting transition temperature, under pressure can be used to give some information about the topology of the Fermi surface of a metal. This was first applied to Tl by Lazarev *et al.* (1965) (see § 5.5 below). The non-linear dependence of $\partial T_c/\partial P$ on the impurity concentration is connected with a change in the number of depressions on the Fermi surface of the metal (Makarov and Bar'yakhtar 1965). This effect was investigated in In by

Fig. 24

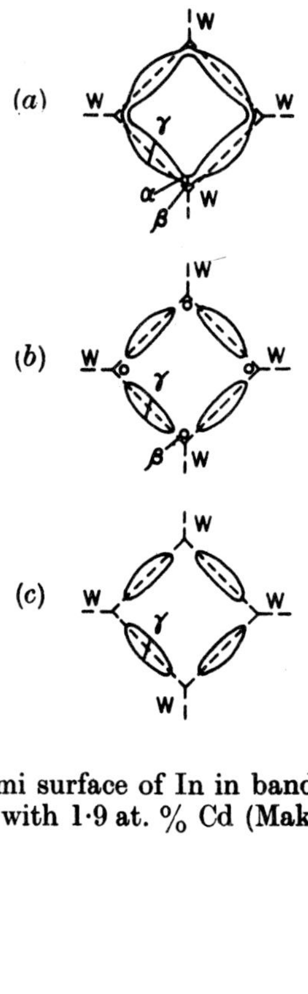

Cross sections of the Fermi surface of In in band 3, β arms: (*a*) pure, (*b*) with 1·2 at. % Cd, and (*c*) with 1·9 at. % Cd (Makarov and Volynskii 1966).

Makarov and Volynskii (1966) and non-linearity in $\partial T_c/\partial P$ with impurity concentration was observed. The topological changes were assigned to changes in the Fermi surface in band 3 and it was assumed that the change involved was the breaking up of the toroids, which exist in pure In, into separated ellipsoids, see fig. 24.

We conclude this section by noting a few miscellaneous results for In. The temperature coefficient of the electronic contribution, γ, to the specific heat of In has been measured by several workers:

1·81 mJ mole^{-1} deg^{-2}	Clement and Quinnell (1953)
(1·60 ± 0·01) mJ mole^{-1} deg^{-2}	Bryant and Keesom (1961)
1·69 mJ mole^{-1} deg^{-2}	O'Neal and Phillips (1965).

A 60% enhancement of the band structure value of γ, due to electron–phonon interactions, was calculated by Ashcroft and Lawrence who then obtained a value of γ very close to the experimental value of Clement and Quinnell. The Knight shift in In was measured by Torgeson and Barnes (1962) who obtained an isotropic shift of (0·82 ± 0·04)% and an anisotropic contribution of –(0·14 ± 0·04)%; a value of 0·81% for the Knight shift was obtained from the band structure calculation of Gaspari and Das (1968). The effect of pressure on the Fermi surface of In has been studied by O'Sullivan *et al.* (1967) using the de Haas–van Alphen effect. Measurements of the variation of the cross-sectional area of the arms in band 3 of the Fermi surface of In as a function of pressure were expected to provide a sensitive test of the ability of a model pseudopotential to predict the volume dependence, and hence the pressure dependence, of the lattice potential. Ultrasonic attenuation measurements, with no magnetic field, by Bliss and Rayne (1966, 1967) and Fossheim and Leibowitz (1966) were generally consistent with existing ideas of the Fermi surface of In. Doppler-shifted cyclotron resonance has been observed in In by Mina and Khaĭkin (1965 b). The d.c. size effect, which was mentioned in § 5.2 in connection with Al, has also been observed in In by Blatt *et al.* (1967) and a value for the average value of k_F was deduced.

5.5. *Thallium*

In the free-electron model the form of the Fermi surface of Tl, an h.c.p. metal (below 230°c) with valence 3, can be seen from fig. 2 (*c*), where the double-zone scheme is used. In this double-zone scheme band 1 and band 2 are completely full, there is a complicated multiply-connected Fermi surface in band 3 and band 4 and the Fermi surface in band 5 and band 6 consists of isolated pockets of electrons.

The band structure and Fermi surface of Tl were calculated by Soven (1965 a, b) and the results were compared with the experimental data on the Fermi surface of Tl which then existed. Soven used a relativistic modification of the O.P.W. method. The O.P.W. method was chosen because experiments had suggested that the free-electron model gave a good approximation to the true Fermi surface of Tl. Estimates of the gaps produced at the hexagonal face, AHL, of the Brillouin zone showed that these are of the same order of magnitude as the crystal field splittings found in metals similar to Tl. This is unlike the case of Mg which we have discussed previously see §§ 2.2 and 4.3, where these gaps were about two or three orders of magnitude smaller than the crystal field splittings. The perturbation calculations of the spin–orbit coupling effects which were used in Mg (Falicov and Cohen 1963) are therefore inappropriate for Tl. It was for this reason that Soven (1965 a) decided that for Tl it was necessary to use a proper relativistic extension of the O.P.W. method to the solution of the Dirac equation for an electron in a metal. The details of the relativistic modification of the O.P.W. method do not concern us here as we are only interested in the results of the calculation (the interested

Table 14. Spin-splitting at K and H in Tl

Symmetry without spin	Symmetry with spin	Energy (Ry)	Spin-splitting (Ry)
K_5	K_9	0·477	0·006
—	K_8	0·483	
K_1	K_7	0·603	—
H_2	H_4+H_6	0·548	< 0·001
—	H_8	0·548	
H_3	H_5+H_7	0·655	0·018
—	H_9	0·673	
H_1	H_9	0·713	0·022
—	H_8	0·735	

(Soven (1965 a).)

reader should consult the paper by Soven (1965 a)). The bands have to be labelled according to the double-valued representations of the space group P6_3/mmc ($D_{6h}{}^4$) of the h.c.p. structure. These labels are used in table 14 where the magnitudes of the spin-splittings are given. The inclusion of spin effects and the consequent use of the double group may allow, in principle, the possibility of substantial modifications to the connectivity of the Fermi surface, but there always exists the possibility that, if the splittings are small enough, magnetic breakdown may restore the original connectivity when a magnetic field is present. The most important consequence of the introduction of spin effects is the lifting of the two-fold degeneracy on the hexagonal face AHL of the Brillouin zone, except along the line AL. The spin-splitting of bands 3 and 4 and of bands 5 and 6 is largest at H and vanishes completely at the points A and L. As noted in § 4.3 in connection with Mg, the lifting of this degeneracy means that the double-zone scheme is no longer appropriate. The Fermi energy, and hence the calculated Fermi surface, was determined by Soven (1965 a) by comparing the dimensions of various calculated constant-energy surfaces with some of the Fermi surface dimensions that had been determined experimentally. Soven, therefore, did not calculate the density of states curve as an intermediate step in determining the Fermi energy and, consequently, did not predict a value of γ for comparison with the experimental data that are available:

$1{\cdot}53$ mJ mole^{-1} deg^{-2}	Maxwell and Lutes (1954),
$(2.56 \pm 0{\cdot}35)$ mJ mole^{-1} deg^{-2}	Snider and Nicol (1957),
$(1{\cdot}47 \pm 0{\cdot}02)$ mJ mole^{-1} deg^{-2}	Van der Hoeven and Keesom (1964).

The Fermi surface of Tl calculated by Soven consisted of six non-equivalent sheets distributed among band 3, band 4, band 5 and band 6 in the single-zone scheme. It is convenient to describe this Fermi surface in terms of a Brillouin zone centred on A instead of Γ and it consists of:

(i) in band 3, a large closed surface, centred at A, bounding unoccupied states; this is described by Soven as a hexagonal 'cookie' see fig. 25. This surface has the symmetry of the point group 6/mmm(D_{6h}) and dimensions:

r_{0001}	$r_{10\bar{1}0}$	$r_{11\bar{2}0}$
0·13 AΓ	0·38 AL	0·48 AH

where the indices refer to directions in the reciprocal lattice;

(ii) in band 3 a smaller closed sheet, bounding unoccupied states, centred at M with the symmetry of the point group mmm(D_{2h}) and dimensions:

r_{0001}	$r_{10\bar{1}0}$	$r_{11\bar{2}0}$
0·01 ML	0·03 MΓ	0·16 MK;

(iii) in band 4, a honeycomb-like network, bounding occupied states, spanning the rectangular faces of the Brillouin zone, see fig. 26. The dimensions from A in the middle AHL section are:

$r_{10\bar{1}0}$	$r_{11\bar{2}0}$
0·38 AL	0·505 AH.

There are 12 posts protruding from the network (iii) which are roughly parallel to [0001], but some doubt remained as to whether or not these posts were long enough to join up with each other to form a Fermi surface that is multiply-connected in the [0001] direction as well as in the horizontal plane. The resolution of this difficulty experimentally is, obviously, complicated by the fact that, if the posts are not joined together, the size

Fig. 25

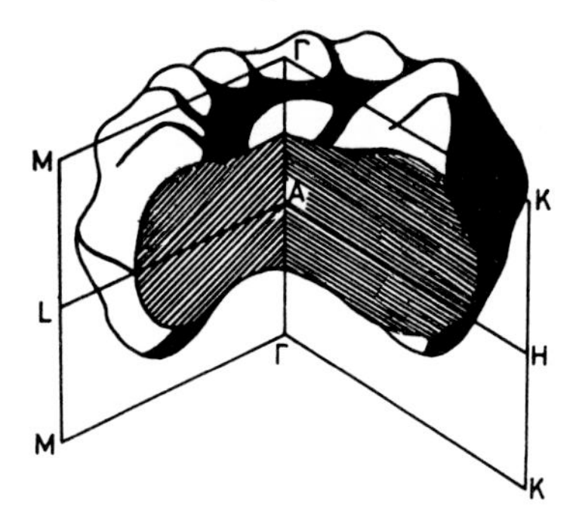

The Fermi surface of Tl in band 3 (Soven 1965 a).

Fig. 26

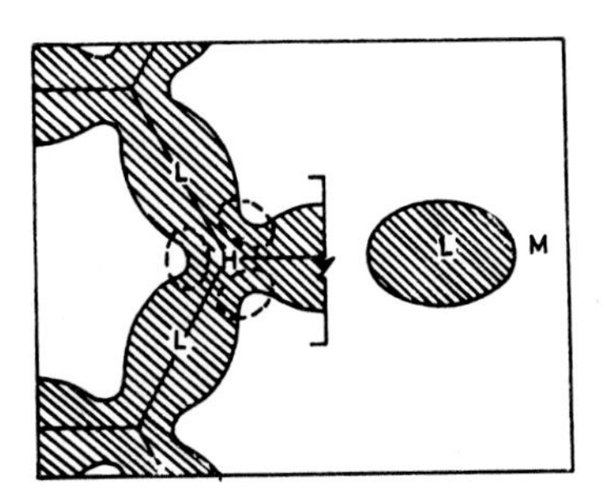

Cross section AHL of the Fermi surface of Tl in band 4 (Soven 1965 a).

of the gap would be sufficiently small that there would be a finite probability of magnetic breakdown (Soven 1965 b) in experiments in which a magnetic field is applied;

(iv) in band 4, if the posts in (iii) are not multiply-connected in the [0001] direction then small occupied pockets remain on the ΓMK planes between adjacent posts;

(v) in band 5, small pockets of electrons at H, with approximate dimensions,

r_{0001}	$r_{12\bar{3}0}$	$r_{11\bar{2}0}$
0·15 HK	0·06 HL	0·08 HA

and

(vi) in band 6, smaller pockets of electrons at H, with approximate dimensions:

r_{0001}	$r_{12\bar{3}0}$	$r_{11\bar{2}0}$
0·07 HK	0·05 HL	0·07 HA.

This calculated Fermi surface was then compared by Soven (1965 b) with the experimental results which existed at that time; these results consisted of galvanomagnetic, de Haas–van Alphen and magnetoacoustic work. The transverse magnetoresistance of Tl had been measured by Alekseevskiĭ and Gaĭdukov (1962) and Mackintosh *et al.* (1963) and their results showed that open orbits exist for several orientations of the magnetic field. The interpretation of the results is complicated by the existence of magnetic breakdown. According to Alekseevskiĭ and Gaĭdukov (1959 a, b, 1962) open orbits in Tl are located only in the (0001) plane. For orientations of the magnetic field within the (0001) plane open orbits may be possible on the multiply-connected Fermi surface in band 4. Open orbits are possible in the [10$\bar{1}$0] direction involving the apparently closed Fermi surface in band 3; this occurs because electrons can pass backwards and forwards between band 3 and band 4 as a result of the essential degeneracy along AL; these rather special open orbits have actually been observed in the magnetoacoustic work of Coon *et al.* (1967). Alekseevskiĭ and Gaĭdukov (1962) and Mackintosh *et al.* (1963) described the open orbits in Tl according to the free-electron Fermi surface but Soven (1965 b) was able to show that their experimental results could equally well be explained in terms of the Fermi surface calculated by the relativistic O.P.W. method. This is a general feature of magnetoresistance measurements that while they give a good indication of whether or not open orbits exist, the measurements are often unable to distinguish between several

Table 15. Comparison of predicted and experimental extremal areas of the Fermi surface of Tl

Cross section	Orbit	Period	Areas (A.U.)		
			Free electron	Soven (1965 a)	Priestley (1966)
Third band crown					
Central (11$\bar{2}$0)	γ	P_3	0·219	0·25†	0·249
Non-central (11$\bar{2}$0)	β	P_2 ?	0·292	0·275	0·307 ?
Extended (11$\bar{2}$0)	λ	P_2 ?	None	0·25–0·35	0·307 ?
Central (10$\bar{1}$0)	δ	P_4	0·338	0·273	0·264
Central (0001)	α_2	P_{1b}	0·599	0·548	0·558
Fourth band network					
Central (11$\bar{2}$0)	ζ	P_5'	0·083	0·11†	—
Central (10$\bar{1}$0)	ζ	P_5'	0·094	0·12†	0·101
Non-central (11$\bar{2}$0)	ϵ	P_5	0·043	0·069	0·073
P_6 extremal value 20° from [10$\bar{1}$0]	η	P_6	None	0·12†	0·107
Central (0001)	α_1	P_{1a}	0·599	0·582	0·583

† Estimated by Priestley (1966) from the surfaces given by Soven (1965 a). (Priestley (1966).)

different proposed Fermi surfaces. The de Haas–van Alphen effect in Tl was first observed by Shoenberg (1952) and was studied in detail in the (0001) and $(10\bar{1}0)$ planes by Priestley (1966) using the pulsed-field method. The dimensions predicted by Soven (1965 a) for the 'cookie' in band 3 were substantially confirmed, see table 15. Fairly good agreement was also obtained for several orbits on the Fermi surface in band 4, also see table 15. In connection with the posts in band 4 the results of Priestley were not completely decisive but the absence of any large observed area corresponding to extended orbits parallel to [0001], together with the form of the angular dependence of one of the periods which was assigned to the posts, seemed to indicate that the posts were not multiply-connected in the [0001] direction. Some further magnetoresistance measurements with magnetic fields up to 100 kG also failed to detect open orbits parallel to [0001], thereby adding further weight to the view that the posts in band 4 are not multiply-connected in the [0001] direction in the absence of a magnetic field (Milliken and Young 1966). In connection with the small pockets of electrons in band 5 and band 6 several de Haas–van Alphen periods of the correct order of magnitude were observed by Priestley (1966) but they were of very low amplitude and a detailed study of them was not made; similar de Haas–van Alphen periods were observed by Ishizawa and Datars (1969) who could not obtain quantitative agreement with any of the small pieces of Fermi surface in Soven's model. The first magneto-acoustic work on Tl was performed by Rayne (1962 a, 1963 b) and interpreted in terms of the free electron model, since that was the only theoretical Fermi surface available at that time. Rayne's results were re-considered by Soven (1965 b) in the light of the relativistic O.P.W. Fermi surface of Tl; while general agreement was obtained with the Soven model there were some difficulties in connection with open-orbit resonances. In particular Rayne had reported an open orbit parallel to [0001] which disagreed with Soven's model in requiring the posts in band 4 to be multiply-connected while some predicted open orbits in the basal plane were not observed. Eckstein *et al.* (1966) and Coon *et al.* (1967) performed some further magnetoacoustic work on Tl to investigate these discrepancies and both geometric and open-orbit resonances were observed at frequencies up to 460 MHz. The shape and dimensions of the Fermi surface in band 3 were confirmed in detail (see table 16) while the experimental data associated with the Fermi surface in band 4 was concluded to be 'at least consistent' with the model of Soven (1965 a). Eckstein *et al.* (1966) and Coon *et al.* (1967) also failed to observe any open orbits parallel to [0001], thereby strengthening the view that the posts in band 4 are not multiply-connected in this direction. We therefore conclude that the balance of evidence of magnetoresistance, de Haas–van Alphen, and magnetoacoustic work is that the posts in band 4 are not multiply-connected, although in sufficiently large fields magnetic breakdown can be expected to occur between the posts. Tl is in fact a good metal in which to study magnetic breakdown *per se* because the energy gap across which breakdown occurs can be varied from zero to about 0·01 Ry by varying the orientation of the magnetic field in the basal plane (Priestley 1966). Breakdown fields were determined for several orientations in the magnetoresistance work of Young (1967) up to 220 kG. Alternatively one can investigate the angular variation of breakdown field for a constant energy gap by tilting the field out of the basal plane. In connection with the pockets of electrons in band 5 and band 6 Rayne (1962 a, 1963 b) observed a roughly isotropic long period corresponding to a diameter of 0·074 in the basal plane which is in order of magnitude agreement with one or other of these pockets in the relativistic O.P.W. model. Cyclotron masses determined by Dahlquist and Goodrich (1967) were assigned to pieces of Fermi surface in bands 3, 4 and 5 of Soven's model. In connection with In in the previous section we mentioned that the existence of a discontinuity in $\partial T_c/\partial P$ (the pressure dependence of the superconducting transition temperature) as a function of impurity concentration indicates a change in the number of depressions on the Fermi surface as the average number of conduction electrons per atom is changed by varying the amounts of impurity present. This effect was first observed in Tl (Lazarev *et al.* 1965, Ignat'eva *et al.* 1968).

Table 16. The Fermi surface of Tl in band 3

Extremal calipers (A.U.)				
Direction	Measured		Calculated	
	Eckstein *et al.* (1966)	Rayne (1962 a, 1963 b)	Soven (1965 a)	Free-electron model
$[10\bar{1}0]$AL	0·38	0·39	0·38	0·41
$[11\bar{2}0]$AM	0·48	0·44	0·48	0·51
[0001]AΓ	0·17	0·16	0·13	0·14
	0·14	—	—	—
Extremal areas (A.U.)				
	de Haas–van Alphen (Priestley 1966)	Eckstein *et al.* (1966)	Soven (1965 a)	Free-electron model
(0001)	0·56	0·54	0·548	0·599
$(11\bar{2}0)$	0·245	0·21	~0·25	0·219
$(10\bar{1}0)$	0·262	0·29	0·273	0·338

(Eckstein *et al.* (1966).)

We, therefore, make the following conclusions about the relativistic O.P.W. Fermi surface calculated by Soven (1965 a, b). The band 3 'cookie' is well established experimentally, see tables 15 and 16. The existence of some small pockets of Fermi surface is confirmed by several experiments, but it is not clear whether these are (ii) the band 3 pockets of holes at M, or (v) the band 5 pockets of electrons at H, or (vi) the band 6 pockets of electrons at H; it is not established which, or how many, of all these actually exist. The general features of the predictions of Soven about the Fermi surface in band 4 have been moderately well confirmed (see table 15) and it seems fairly certain that the posts are not multiply-connected in the [0001] direction.

§ 6. p-block : Group IVB : Carbon Group

C	Carbon	2.(2, 2)
[Si	Silicon	2.8.(2, 2)]
Ge	Germanium	2.8.18.(2, 2)]
Sn	Tin	2.8.18.18.(2, 2)
Pb	Lead	2.8.18.32.18.(2, 2)

6.1. *Structures*

C in the form of diamond is an insulator while Si, Ge and grey Sn, which have the same structure, are semiconductors; we shall not be concerned with any of these materials here. In the form of graphite C is a semi-metal and has the well-known hexagonal structure described by the space group $P6_3mc$ ($C_{6v}{}^4$) with atoms at (0, 0, 0), (0, 0, $\frac{1}{2}$), ($\frac{1}{3}$, $\frac{2}{3}$, z), and ($\frac{2}{3}$, $\frac{1}{3}$, $z+\frac{1}{2}$) where $z \approx 0$, see fig. 27. If z is exactly equal to zero the space-group symmetry is increased to that of $P6_3/mmc$ ($D_{6h}{}^4$) with $a = 2{\cdot}46$ Å and $c = 6{\cdot}71$ Å (Smithells 1967). Metallic Sn has a tetragonal structure and belongs to the

Fig. 27

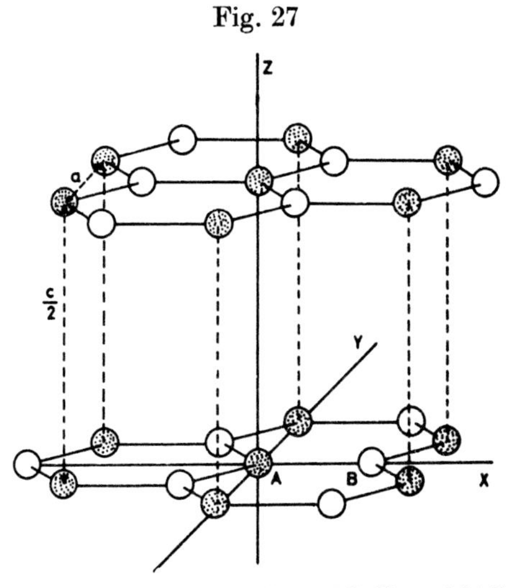

The structure of graphite (McClure 1964).

space group I4$_1$/amd (D_{4h}^{19}) with $a = 5{\cdot}83$ Å and $c = 3{\cdot}18$ Å and with atoms situated at (0, 0, 0), ($\frac{1}{2}, \frac{1}{2}, \frac{1}{2}$), ($\frac{1}{2}$, 0, $\frac{1}{4}$) and (0, $\frac{1}{2}$, $\frac{3}{4}$). The transition temperature between the low-temperature form (grey Sn) and the high-temperature form (metallic Sn) is 13·2°C but it is a matter of everyday experience that it is not uncommon for metallic Sn to exist in a metastable state below this temperature. In this article wherever we refer to Sn we mean metallic (white) Sn. Pb has the face-centred cubic (cubic close-packed) structure belonging to the space group Fm3m (O_h^5) with $a = 4{\cdot}95$ Å (Smithells 1967).

6.2. *Carbon* (*Graphite*)

The electrical resistivity of graphite is considerably higher than that of most metals and graphite can be regarded as a semi-metal, along with As, Sb and Bi (see §§ 7.1–7.3). The distinction between a semi-metal and a semiconductor is that in a semiconductor there is a finite gap between the valence band, which is full at absolute zero, and the conduction band, which is empty at absolute zero, whereas in a semi-metal there is a small but finite overlap of the valence and conduction bands. Of course the semi-metal must have an even number of conduction electrons per unit cell. The distinction between metal and semi-metal is not clear-cut but depends on an arbitrary decision as to how small is a 'small overlap'; thus graphite, As, Sb and Bi are regarded as semi-metals and there is some case for regarding Ca, Sr and (?) Ba as semi-metals too (see § 4.4). The Fermi surface of a semi-metal can then be expected to consist of sets of small isolated pockets of holes in the valence band together with sets of small pockets of electrons in the conduction band; assuming there are no impurities present the total number of holes will be equal to the total number of electrons and the material is said to be *compensated*.

The electronic structure and Fermi surface of graphite have been studied rather extensively by all the usual methods and a comprehensive review article was published by Haering and Mrozowski in 1960, by which time the general features of the shape of the Fermi surface of graphite were well established. We shall concentrate on summarizing that review and mentioning the additional work which has been done since then. We shall not repeat all the extensive list of references given by Haering and Mrozowski (1960).

The separation of adjacent horizontal sheets of C atoms in the graphite structure is nearly 2½ times as large as the separation between neighbouring C atoms in a given horizontal sheet. Consequently it has been very common to consider the behaviour of the electrons in graphite as a two-dimensional problem involving the motion of electrons in a single layer; interactions between the layers can then be introduced afterwards as a perturbation Of the four valence electrons on each C atom, three form covalent (σ) bonds with the three nearest-neighbour C atoms in the same horizontal sheet. The remaining electron is less tightly bound and goes into the valence band. Since there are two C atoms per unit cell there are just enough electrons to fill the valence band if there is no overlap between the valence band and the conduction band. The two-dimensional approach initially met with some considerable success since its introduction by Wallace (1947) and Coulson (1947).

The two-dimensional problem suggested that there was a point of contact between the conduction band and the valence band at the corners of the Brillouin zone. When the proper three-dimensional structure is considered the Brillouin zone has the same general appearance as that of the h.c.p. structure although the axial ratio c/a is much larger than that of the h.c.p. structure; consequently the Brillouin zone for graphite is very much flatter (i.e. less tall) than the h.c.p. Brillouin zone. Slonczewski and Weiss (1958) performed a perturbation calculation to determine the band structure of three-dimensional graphite by starting with the four tight-binding wave functions with wave vector **k** corresponding to one of the corners of the Brillouin zone for the two-dimensional problem of a single layer. The region of particular interest is the vertical line HKH which joins two corners of the Brillouin zone. The expressions obtained by Slonczewski and Weiss for the energy bands in the region of HKH represented a set of four hyperboloids of revolution with $E_{\mathbf{k}}$ actually given by

$$\left.\begin{aligned} E_{\mathbf{k}} &= \tfrac{1}{2}(E_1+E_3) \pm [\tfrac{1}{4}(E_1-E_3)^2+(\gamma_0\sigma)^2]^{1/2}, \\ E_{\mathbf{k}} &= \tfrac{1}{2}(E_2+E_3) \pm [\tfrac{1}{4}(E_2-E_3)^2+(\gamma_0\sigma)^2]^{1/2}, \end{aligned}\right\} \quad . \quad . \quad (6.2.1)$$

where

$$\begin{aligned} E_1 &= \Delta + 2\gamma_1 \cos(\tfrac{1}{2}\xi), \\ E_2 &= \Delta - 2\gamma_1 \cos(\tfrac{1}{2}\xi), \\ E_3 &= 2\gamma_2 \cos^2(\tfrac{1}{2}\xi), \\ \sigma &= \tfrac{1}{2}\sqrt{3}a\kappa, \\ \xi &= k_z c, \end{aligned}$$

and $\boldsymbol{\kappa}$ is the wave vector measured from the line HKH in a horizontal plane. Δ, γ_0, γ_1 and γ_2 are adjustable parameters. The bands given by Slonczewski and Weiss (1958) along HKH are shown for one choice of these parameters in fig. 28. On this set of energy bands there will be expected to be a thin vertical cigar-shaped pocket of holes in the valence band centred at K together with a pocket of electrons at each end of the cigar; these regions of electrons and holes will just touch because of the crossing of the bands in fig. 28.

The problem of the determination of the best set of values for the parameters Δ, γ_0, γ_1 and γ_2 (as well as γ_3 and γ_4 which were omitted from eqn. (6.2.1)) to use in the Slonczewski–Weiss model to give the best fit to all the various pieces of experimental data is discussed at length by Haering and Mrozowski (1960). The experimental data involved included principally the de Haas–van Alphen effect, the Hall effect and cyclotron resonance (for references see Haering and Mrozowski (1960)). Using the de Haas–van Alphen data of Shoenberg (1952) and Berlincourt and Steele (1955) values of γ_1, γ_2 and Δ were calculated by McClure (1957) for a range of values of γ_0.

Fig. 28

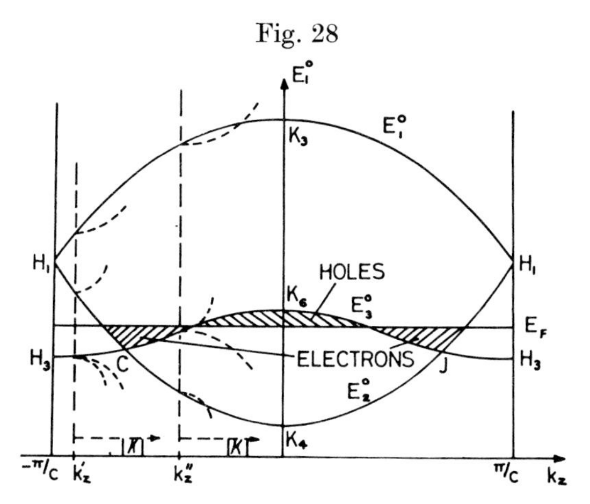

The band structure of graphite calculated by Slonczewski and Weiss (1958).

The Fermi surface deduced by McClure is shown in fig. 29, where it is to be noted that the pockets of electrons do not quite reach the point H itself†. At the stage of the writing of the article of Haering and Mrozowski (1960) the agreement between the measured value of the temperature coefficient of the electronic specific heat γ ($=25{\cdot}2\,\mu$J mole^{-1} deg^{-2} (Keesom and Pearlman 1955, De Sorbo and Nichols 1958)) with the calculated values (from about $8{\cdot}4\,\mu$J mole^{-1} deg^{-2} to about $13{\cdot}6\,\mu$J mole^{-1} deg^{-2}) was not very good. This has subsequently been improved with a new experimental value of $\gamma = 13{\cdot}8\,\mu$J mole^{-1} deg^{-2} obtained by Van der Hoeven and Keesom (1963) and a new theoretical value of $13\,\mu$J mole^{-1} deg^{-2} (McClure 1964). An extensive discussion by McClure (1958) of the existing experimental measurements on the Hall effect and the magnetoresistance enabled the carrier densities and mobilities (at 4·2°K, 77°K and 300°K) to be determined (see either McClure (1958) or table 2 of Haering and Mrozowski (1960) for the details of the results).

We now summarize the additional relevant results which have been obtained since the review by Haering and Mrozowski (1960) was written. A claim was made by Soule (1964) to have observed a new very small ellipsoidal piece of Fermi surface which was tentatively ascribed to holes; it was suggested that these new small ellipsoids were arranged in threes around the major hole surface. These very small ellipsoids had not been observed in the previous de Haas–van Alphen work of Spry and Scherer (1960). Other de Haas–van Alphen experiments have also been performed on graphite, either to compare the results for pyrolytic graphite with the results for natural single crystal graphite (Williamson, Foner and Dresselhaus 1964, 1965, 1966) or else to investigate the effect of pressure on the shape of the Fermi surface (Anderson, O'Sullivan, Schirber and Soule 1967). A theoretical treatment of the energy bands of graphite for various

† See note added in proof.

Fig. 29

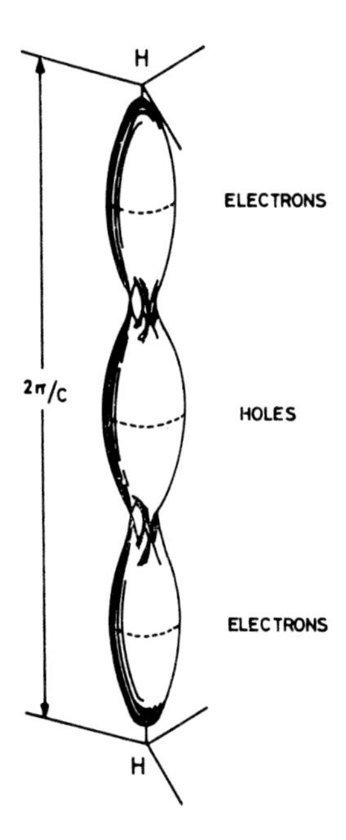

The Fermi surface of graphite proposed by McClure (1957).

changes in temperature and pressure, using the relaxation time approximation and the Slonczewski–Weiss model, was given by Arkhipov *et al.* (1963); their results predicted a 23% *increase* in the total number of carriers in graphite at a pressure of 10 kbar. Good agreement with this theory was obtained in the experimental measurements of the resistance, Hall effect and magnetoresistance made by Likhter and Kechin (1963). This is the reverse of the case of Ca and Sr (and Bi, see § 7.2) where increasing the pressure reduces the number of carriers. A diagram illustrating the deformed Fermi surface of graphite at a pressure of 30 kbar is given by Likhter and Kechin (1963). Measurements of the Shubnikov–de Haas oscillations in graphite at pressures up to 8 kbar by Itskevich and Fisher (1967 a) gave results in agreement with the theory of Arkhipov *et al.* (1963), namely an experimental value for the ratio of $\gamma_1\gamma_2$ at 8 kbar to $\gamma_1\gamma_2$ at atmospheric pressure of 1·32 compared with a calculated value of 1·47.

The Fermi surface for electrons in graphite shown in fig. 29 does not quite reach to the corners H of the Brillouin zone. More recent determinations of the parameters in the Slonczewski–Weiss model involving the magnetoreflection measurements of Dresselhaus and Mavroides (1964 b, c, 1966) led to a band structure in which the two-fold degenerate level at H is below the Fermi energy and the pockets of electrons now reach to H so that there is a small pocket of electrons in band 3 at H. In the extended zone scheme this is represented by simply extending the pocket of electrons just beyond H, see fig. 30. A new set of measurements of the Shubnikov–de Haas effect in graphite by Soule *et al.* (1964) and on cyclotron resonance by Williamson, Surma, Praddaude, Patten and Furdyna (1966) provided direct confirmation of the principal features of the Fermi surface of graphite

Fig. 30

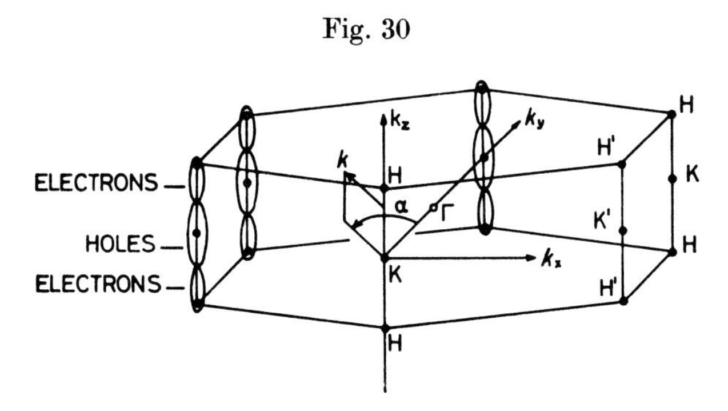

The Brillouin zone for graphite, showing the positions of the Fermi surface (McClure 1964). The Fermi surfaces are magnified by a factor of about four in the horizontal linear dimensions.

Fig. 31

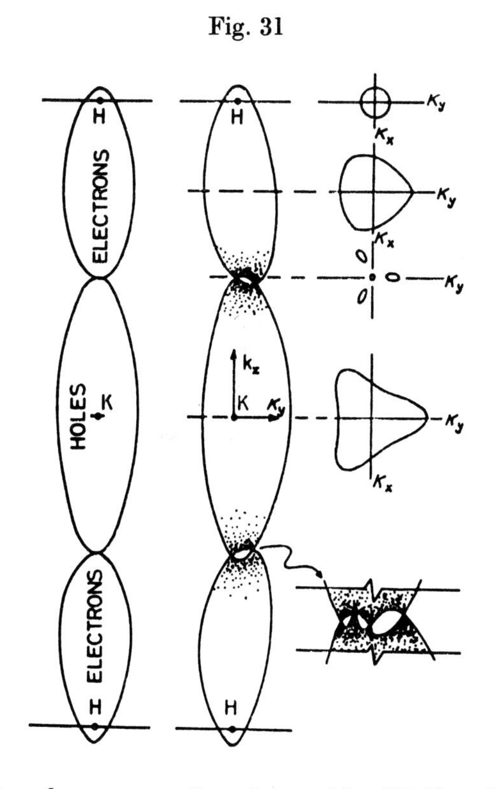

Fermi surface cross sections in graphite (McClure 1964).

as described and illustrated above, although the details of the region where the electron and hole Fermi surfaces meet (see figs. 29 and 31) are difficult to verify directly†.

The fact that the pockets of electrons and holes in graphite are so small means that for magnetic fields stronger than about 60 kG there is only one Landau level below the Fermi level in each band, that is, the electrons and the holes occupy only their lowest Landau levels; this is called the 'quantum limit régime' by Adams and Holstein (1959). In this régime it can be shown theoretically that the transverse magnetoresistance should be directly proportional to the field strength; this was verified experimentally over the range 60 kG to 160 kG by McClure and Spry (1968). Other magnetoresistance measurements have been made on graphite by Pospelov and Kechin (1963), Mills *et al.* (1965), Sugihara and Ono (1966), and at pressures up to 9 kbar, by Pospelov and Kechin (1963). A negative magnetoresistance, which became positive at large magnetic fields, was observed in *some* of the samples of pyrolytic graphite investigated by Takeya and Yazawa (1964) and a possible theoretical explanation of this in terms of aggregates of inhomogeneous crystallites of pyrolytic graphite was advanced by Saha (1965); this appears not to be a feature of pure single crystal graphite. A non-linearity in the resistance of graphite in a given magnetic field was observed by Goldsmid and Corsan (1964) and Mizushima and Endo (1968) similar to the non-linearity originally observed in Bi by Esaki (1962), see § 7.2.

Some further discussion of the band structure of two-dimensional graphite' has been given by Fukuda (1965), Bassani and Parravicini (1967), and Linderberg and Mäkilä (1967) and of three-dimensional graphite by Barriol (1960) and McClure (1960). The calculation of the Landau levels and their eigenfunctions and hence the conduction electron diamagnetism was carried out with the Slonczewski–Weiss model by McClure (1960) and Inoue (1962) and values of the band parameters γ_0, γ_1 and Δ were obtained from cyclotron resonance and de Haas–van Alphen data. According to Sato (1969) band-to-band transitions which are not included in the conventional Landau–Peierls treatment should be included for graphite.

Finally, we mention a miscellaneous collection of results related to the electronic band structure of graphite; these include experimental investigations of Alfvén wave propagation (Surma *et al.* 1964), the K emission band (Sagawa 1966), the spin susceptibility (Singer and Wagoner 1962, Wayne and Cotts 1963) and infra-red transmission and reflection spectra (Yasinsky and Ergun 1965, Sato 1968) and a theoretical calculation of the absorption of ultrasound using the known form of the phonon dispersion curves and the electronic band structure (Kaganov and Semenenko 1967).

---

† See note added in proof.

6.3. *Tin*

We shall not consider the semiconducting form (grey Sn) of this element. The Fermi surface of metallic (white) Sn, which we shall discuss, has been studied extensively by a variety of methods. Because Sn does not exhibit one of the common metallic structures its Fermi surface in the free-electron approximation is not given in the work of Harrison (1960 b) which was reproduced in fig. 2. The free-electron Fermi surface has to be constructed using the appropriate tetragonal Brillouin zone with the correct value of the axial ratio c/a and with four conduction electrons per atom. This is shown in fig. 32(a) (Gold and Priestley 1960) where the sharp corners on the free-electron Fermi surface have been rounded off. Since there are two Sn atoms per unit cell there are altogether eight conduction electrons per unit cell and therefore the equivalent of four bands must be filled. In the free-electron model band 1 is full throughout the Brillouin zone and band 2 and band 3 are very nearly full, except for the small pockets of holes at W in band 2 and the multiply-connected Fermi surface of holes in band 3 (see fig. 32 (a)). In band 4 there are two sheets of the Fermi surface, there is a closed surface around a region occupied by electrons centred at Γ, which is surrounded by a multiply-connected region of holes, and there is a multiply-connected Fermi surface marking the outside of this region of holes (see fig. 32 also). In band 5 there is an isolated piece of Fermi surface at Γ enclosing electrons and in band 6 there are two isolated pieces of Fermi surface which also enclose regions occupied by electrons. Finally in band 5 there is a multiply-connected piece of Fermi surface enclosing electrons and consisting of 'pears' centred at H and 'connecting pieces' centred at V. A substantial amount of the experimental work that has been done on Sn was done when the only theoretical Fermi surface available was this free-electron Fermi surface; consequently the various authors naturally attempted to explain their experimental results in terms of the free-electron Fermi surface. These experiments included the Shubnikov–de Haas effect, via helicon propagation (Hays and McLean 1965), the de Haas–van Alphen effect (Shoenberg 1953, Gold and Priestley 1960), galvanomagnetic properties (Alekseevskiĭ and Gaĭdukov 1959 a, b, 1961 b, Alekseevskiĭ *et al.* 1960, Lüthi 1960, Young 1965, 1966), magnetoacoustic attenuation (Galkin *et al.* 1960, Olsen 1960, 1963 a, b, Kearney *et al.* 1965, Miller 1966), anomalous skin effect (Fawcett 1955, Chambers 1956), and various cyclotron resonance and size effect experiments (Fawcett 1956, Bezuglyĭ and Galkin 1957, Kip *et al.* 1957, Gantmakher and Sharvin 1960, Khaĭkin 1960, 1961, 1962 a, b, Khaĭkin and Mina 1962, Koch and Kip 1962, Gantmakher 1962 b, 1963, 1964, Gantmakher and Kaner 1965).

The tight-binding method was used for Sn by Miąsek (1958) and a perturbation calculation of the band structure of Sn was performed using the O.P.W. method (Miąsek 1963); the energy eigenvalues were determined for several points of symmetry in the Brillouin zone for different choices of potential. The band structure and Fermi surface of Sn were calculated by Weisz (1966) using a local-pseudopotential approximation. Because

Fig. 32

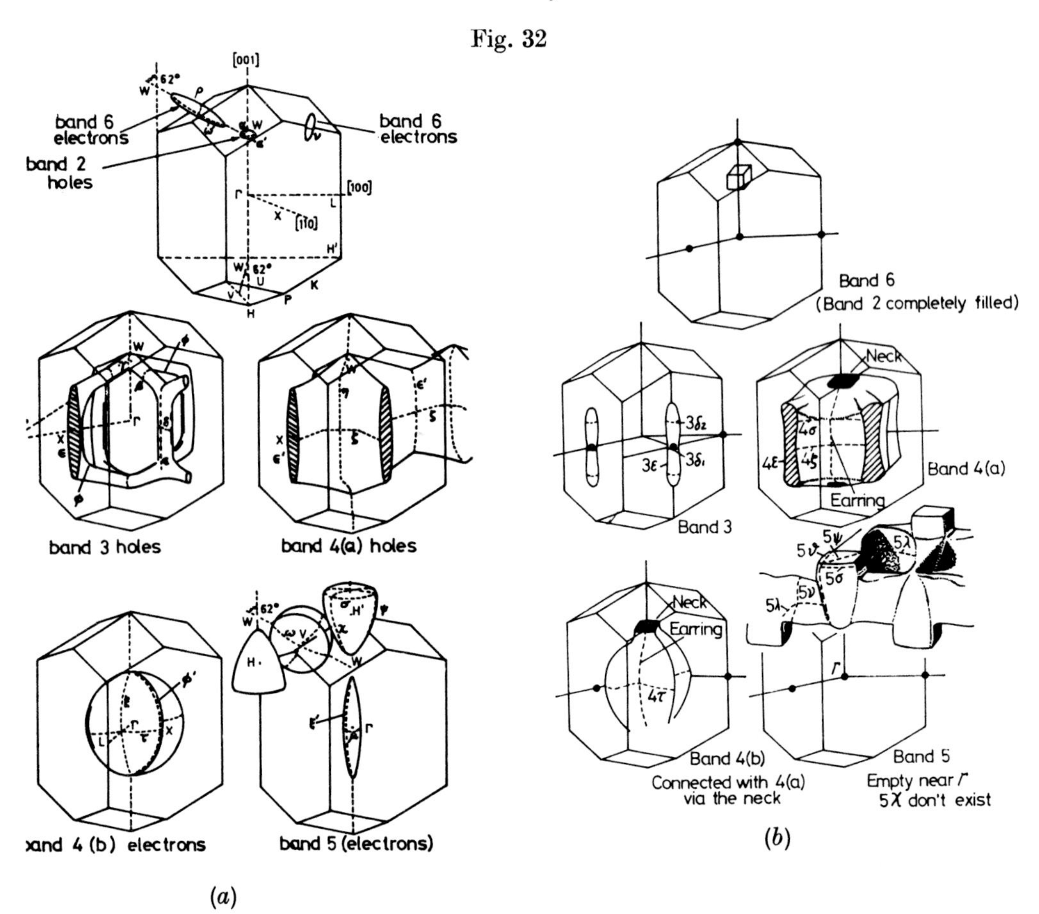

The Fermi surface of Sn: (*a*) on the free-electron model (with the sharp corners rounded off) (Gold and Priestley 1960), and (*b*) according to Stafleu and de Vroomen (1967 b).

the atomic number of Sn is relatively high (50) one would expect spin–orbit coupling effects to be important. Although the actual spin–orbit interaction in Sn is weaker than in the very heavy elements, such as Pb, its importance is enhanced by the fact that the crystal structure has relatively low symmetry. Several authors have studied the group-theoretical analysis of the electronic energy bands in Sn and the prediction of their essential degeneracies (e.g. Mase 1959 b, Miąsek and Suffczyński 1961 a, b, Suffczyński 1961). There are substantial reductions in the degeneracies of the energy bands when spin–orbit coupling is introduced. Because of the importance of spin–orbit coupling in Sn Weisz found a way of including the spin–orbit interaction into a pseudopotential secular equation in preference to introducing the spin–orbit interaction as a perturbation at the end of the calculation. This procedure is shown to be justified for Sn by studying the details of the bands obtained from the calculation. For the details of the method and results of the band structure calculation the reader should consult the paper by Weisz (1966). From this band structure Weisz obtained a Fermi surface that, although distorted, was still recognizably related to the free-electron Fermi surface of Sn.

Fig. 33

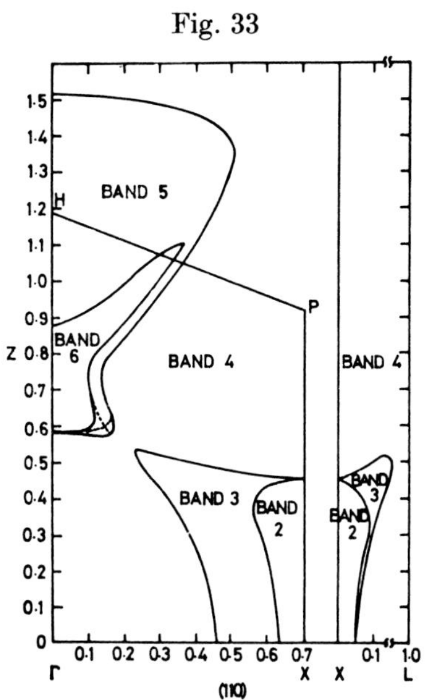

Cross sections of the Fermi surface of Sn calculated by Weisz (1966).

In the results of the calculation by Weisz the following modifications are made to the free-electron Fermi surface of Sn which we described above and which is illustrated in fig. 32 (*a*). The pocket of holes at W in band 2 vanishes and in band 3 the regions of holes around W disappear leaving isolated cylinders along the lines XP, see fig. 33. In band 4 some electrons are introduced along ΓH near W so that there is a neck connecting the previously isolated electron pocket at Γ in band 4 ((*b*) in fig. 32) to the other, already multiply-connected, region occupied by electrons in band 4 ((*a*) in fig. 32). In band 5 the cigar of electrons around Γ disappears but the network involving the pears is not substantially changed. In band 6 the small pockets of electrons near K (labelled ν in fig. 32 (*a*)) disappear while the cigars near V (labelled ρ in fig. 32 (*a*)) become amalgamated into a vertical shape with a square cross section with slight prongs in the directions of the original cigars; a cross section of this piece of Fermi surface is given in fig. 33. Some other cross sections of various pieces of the calculated Fermi surface will be found in figs. 7–10 of Weisz (1966).

Weisz was able to compare this calculated Fermi surface for Sn with the rather detailed radio-frequency size effect (Gantmakher effect) results which were then available (Gantmakher 1962 b, 1963, 1964). The agreement between the various caliper dimensions in the calculated Fermi surface of Weisz and the experimental results of Gantmakher is impressive, see fig. 34 and table 17; for a detailed account the reader is referred to the paper by Weisz (1966). Of course, it must be remembered that the pseudopotential method used by Weisz (1966) is not entirely an *ab initio* calculation but does depend on a few experimental parameters. Since these parameters were obtained by fitting certain of the caliper dimensions from Gantmakher's size-effect results it is not too surprising that good agreement with the remainder of the size-effect calipers was then obtained. This was not an unreasonable way to proceed with the calculation because the size-effect data were, at that time, the most complete and accurate

Fig. 34

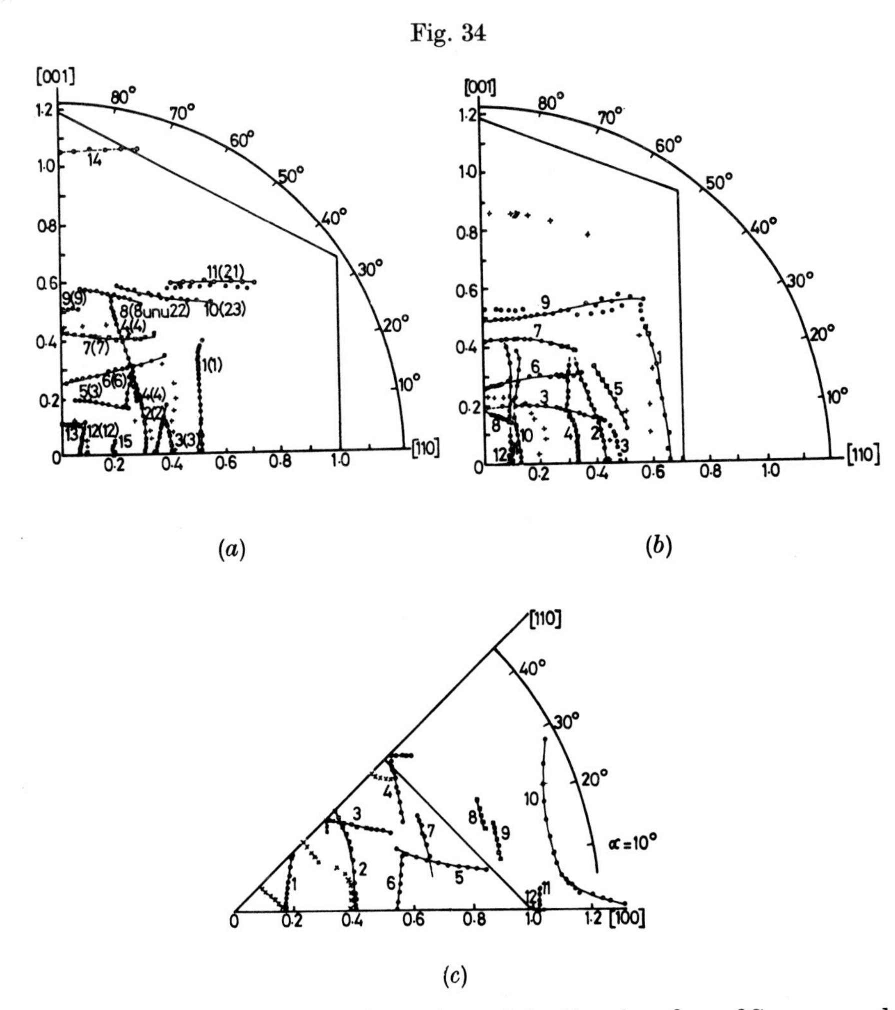

The experimental semi-caliper dimensions of the Fermi surface of Sn measured by the radio-frequency size effect (Gantmakher 1963, 1964). Crosses indicate the magnetoacoustic results of Olsen (1963 b). (The axis labelled [110] in (*a*) by Gantmakher (1963) should be labelled [100].)

Table 17. Experimental and theoretical caliper dimensions of the Fermi surface of Sn for **B** in the [001] direction

Figure of Gantmakher†	Curve No.	Expt.†	Calc.‡	Band label	Description
1	1	0·515	0·51	4	$z=0$
1	2	0·35	0·35	5	$z=0.92$
1	3	0·4	0·42	5	Pear crown
1	4	0·305	0·34	4	$z=0$
1	15	0·19	0·17	4	Neck
2	2	0·43	0·46	4	$z=0$
2	1	0·66	0·66	4	$z=0$
1	12	0·1	0·177	6	Smallest orbit
2	10	0·125	0·145	3	Greatest orbit
2	11	0·093	0·087	3	Greatest orbit
2	12	0·083	0·07	3	Largest caliper of smallest orbit

† Gantmakher (1963, 1964), for 1 see fig. 34 (*a*) and for 2 see fig. 34 (*b*).
‡ Weisz (1966).

(Weisz (1966).)

experimental data available ; since that time the available de Haas–van Alphen data have been considerably improved (see below (Stafleu and de Vroomen 1966, 1967 a, b, Craven and Stark 1968)). There is, of course, always the possibility of error in the assignment of a particular piece of experimental data to some given piece of theoretical Fermi surface. It was also pointed out by Weisz that the pseudopotential technique has the advantage that the form of the secular equation is a strong condition which does not allow, for reasonable pseudopotentials, as much arbitrariness as the number of fitting parameters might suggest. There are some discrepancies between the size-effect results and the calipers determined from the magnetoacoustic data of Olsen (1963 b) (indicated by crosses in fig. 34); however, the size-effect results are of greater accuracy. Of course, some of the calipers calculated by Weisz are very close to the free-electron values but some are significantly different. In the latter cases the only significant discrepancies between the calculations of Weisz and the experiments of Gantmakher arise in band 6 and a small part of band 5. The height of the band 6 Fermi surface calculated by Weisz was found by comparison with both the radio-frequency size effect and de Haas–van Alphen effect data, to be over-estimated by about 28%. Because of the very large number of pieces of the Fermi surface of Sn it is difficult to use the cyclotron resonance experiments that we mentioned above to distinguish between the free-electron Fermi surface and the Fermi surface calculated by Weisz. However, Weisz (1966) considered that the results of the cyclotron resonance experiments of Khaĭkin (1960, 1962 a, b) and Koch and Kip (1962) were not inconsistent with his own calculated Fermi surface. A large number of de Haas–van Alphen periods for various orientations of **B** had been observed by Gold and Priestley (1960). The use of the de Haas–van Alphen effect has the distinct disadvantage that it can only determine (extremal) areas of cross section of the Fermi surface normal to a given direction, whereas the radio-frequency size effect can actually determine a (linear) caliper dimension of the Fermi surface in some known direction. Several of the de Haas–van Alphen results of Gold and Priestley, although generally less accurate than Gantmakher's size-effect results, showed satisfactory agreement with the extremal areas of cross section of the theoretical Fermi surface calculated by Weisz. However, some other observed de Haas–van Alphen periods did not agree with the theoretical model of Weisz and also the theoretical model predicted more de Haas–van Alphen periods than had actually been observed experimentally. The existence of a multiply-connected Fermi surface, both in the free-electron model and in the calculations of Weisz, is in agreement with the magnetoresistance measurements mentioned above (Alekseevskiĭ and Gaĭdukov 1962) although all the precise details of the shape of that surface have not been determined in this way. Some of the dimensions of the band 4 hole Fermi surface were measured quite accurately by Khaĭkin (1962 b) by using size effects in cyclotron resonance in Sn.

After the calculation by Weisz (1966) of the Fermi surface of Sn an extensive series of further de Haas–van Alphen measurements was made by Stafleu and de Vroomen (1966, 1967 a, b) and by Craven and Stark (1968). Both sets of authors obtained results which were in good qualitative agreement with the Fermi surface calculated by Weisz rather than with the free-electron Fermi surface although quantitative agreement with the Weisz model was not obtained. Some orbits involving magnetic breakdown were observed. The areas of cross section of the Fermi surface determined by Craven and Stark differed fairly systematically by about 30% from those calculated by Weisz. Stafleu and de Vroomen (1967 b) used their results to determine a pseudopotential for Sn. The Fermi surface constructed by Stafleu and de Vroomen is reproduced in fig. 32 (*b*) which shows how the 'pears' and 'double pancakes' in band 5 are much more heavily connected than in the free-electron model. In the model of Stafleu and de Vroomen the Fermi surface in band 6 comprises a pocket around W which does not, unlike the Weisz model, extend to V. A detailed description and quantitative linear dimensions of this Fermi surface are given in tables 1 and 4 of Stafleu and de Vroomen (1967 b). An interpolation scheme, based on the relativistic A.P.W. method, has been applied to Sn using the radio-frequency size-effect data of Matthey and gave extremal areas of cross section that fitted the de Haas–van Alphen results of Craven and Stark to about 1·5%; only a preliminary account is so far available (Devillers and de Vroomen 1969).

We now turn our attention to miscellaneous results which give indirect information about the electronic structure of Sn. Interband transitions in Sn have been studied in the optical work of MacRae *et al.* (1967). Further cyclotron resonance experiments on Sn have been performed by Van Nieuwstadt and de Vroomen (1967) and Khaĭkin and Cheremisin (1968). The effect of magnetic breakdown, at various fields up to 150 kG, on the observed magnetoresistance of Sn (Young 1966, 1968 b, Anderson and Young 1968) in Shubnikov–de Haas oscillations (Woollam 1968, Young 1968 a) and in quantum oscillations in the thermoelectric e.m.f. (Woollam and Schroeder 1968) could be explained satisfactorily by assuming a Fermi surface with the general topological features proposed by Weisz. The period of the quantum oscillations in the magnetoresistance of Sn observed by Young (1966) was in reasonable agreement with one of the de Haas–van Alphen frequencies observed by Gold and Priestley (1960). γ was measured by Bryant and Keesom (1961) and a value of 1·80 mJ mole^{-1} deg^{-2} was obtained. The Knight shift in Sn has been measured by several workers, see table 18. Quantum oscillations in the Knight

Table 18. Values of the Knight shift parameters of Sn

Reference	Temperature (°K)	Field (G)	$K \times 10^4$	$K^1 \times 10^4$
McGarvey and Gutowsky (1953)	300	6050	70·5 ± 0·7	—
Bloembergen and Rowland (1953)	77	6100	75·7	3·3
Karimov and Shchegolev (1961)	4·2	900–5800	—	6·6 ± 0·6
Jones and Williams (1964)	1·15	10100	72·0 ± 0·1	5·4 ± 0·1
	4·2	10100	71·7 ± 0·1	5·4 ± 0·1
	1·15	6300	70·8 ± 0·2	5·4 ± 0·2

(Jones and Williams (1964).)

shift in Sn have been observed by Jones and Williams (1964) and Reynolds *et al.* (1966). Some kinks in the experimental phonon dispersion curves for Sn obtained by Rowe *et al.* (1965) were suggested to be Kohn anomalies, but a unique identification with various pieces of the Fermi surface could not be obtained. The temperature dependence of the isotropic and anisotropic Knight shift in Sn has been measured by Borsa and Barnes (1966)† and the pressure dependence by Matzkanin and Scott (1966) at room temperatures for hydrostatic pressures up to 12 kbar. The effect of tension and pressure on the band 3 hole extremal areas of cross section of the Fermi surface of Sn have been deduced from de Haas–van Alphen measurements

† See note added in proof.

(Hum and Perz 1969, Perz *et al.* 1969); the results were used to estimate the contribution of this piece of Fermi surface to the ultrasonic attenuation (in the absence of any applied magnetic field). The determination of $\sigma/\bar{\lambda}$ gives a measure of the total area of the Fermi surface, S, and several measurements have been made for Sn, see table II of Lyall and Cochran (1967). The temperature dependence of $\bar{\lambda}$ was determined by Gantmakher and Sharvin (1964) from the radio-frequency size-effect amplitudes. The anisotropy in the electrical conductivity, σ, and in the phonon-drag component of the thermoelectric power, S_g, has been studied by Klemens *et al.* (1964); a satisfactory qualitative explanation was obtained using the free-electron Fermi surface since these quantities are relatively insensitive to the finer details of the Fermi surface. The temperature dependence of the electronic absorption of ultrasound in Sn, in the absence of any magnetic field, has been measured by Shepelev and Filimonov (1966) and by Perz and Dobbs (1967); while the results could be explained in terms of the existing models of the Fermi surface it was again not possible to obtain any new information on the topology of the Fermi surface. Positron annihilation measurements on Sn have been made by Badoux *et al.* (1967). The phonon dispersion relations for Sn have been calculated by Brovman and Kagan (1966) within the framework of the Born–von Kármán model.

6.4. *Lead*

As in the case of Sn, the free-electron model gives a good first approximation to the Fermi surface of Pb. Pb has the face-centred cubic structure Fm3m (O_h^5), and its Fermi surface in the free-electron approximation is

Fig. 35

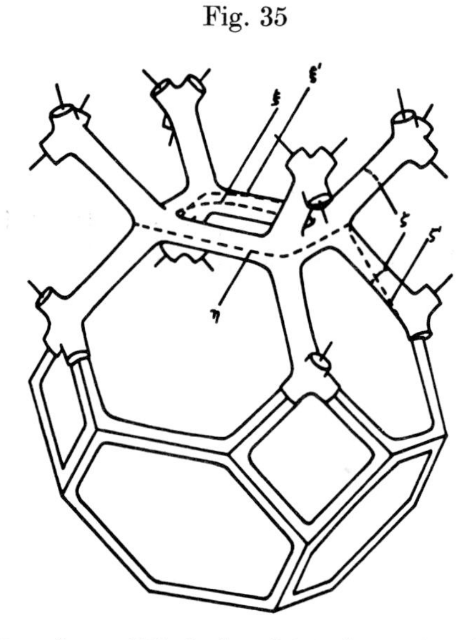

The Fermi surface of Pb in band 3 (schematic) (Gold 1958).

shown in fig. 2 (*a*). It is now generally agreed that metallic Pb has four conduction electrons per atom although it used to be thought to have only two (see, for example, Hume-Rothery (1959)).

The first observation of the de Haas–van Alphen effect in Pb was made by Shoenberg (1953) and the first extensive work on the Fermi surface of Pb was performed by Gold (1958) using the de Haas–van Alphen effect with the pulsed-field technique in fields up to 80 kG. The Fermi surface determined by Gold is very clearly related to the f.c.c. free-electron Fermi surface with four conduction electrons; band 1 is full all over the Brillouin zone and band 5 is empty all over the Brillouin zone, the Fermi surface in band 2 consists of a sphere which encloses a region of holes and is centred at Γ, the Fermi surface in band 3 is a multiply-connected set of tubes topologically similar to the free-electron Fermi surface in band 3 (see fig. 35) and the Fermi surface in band 4 contains six isolated pockets of electrons at the corners of the Brillouin zone. The actual dimensions of the Fermi surface in bands 2, 3 and 4 are given in table 1 of Gold (1958). Since there were, at that time, no reliable band structure calculations for Pb these de Haas–van Alphen measurements were used to deduce a schematic curve for the density of states in Pb (Gold 1960). The value for the total area of the Fermi surface deduced by Gold was $25{\cdot}5 \times 10^{16}\,\mathrm{cm}^{-2}$ which is substantially higher than the value of $14{\cdot}5 \times 10^{16}\,\mathrm{cm}^{-2}$ (46% of the free-electron value) obtained previously from the anomalous skin effect work of Chambers (1952). This discrepancy would be reduced if the band 4 Fermi surface disappeared (see below).

Cyclotron resonance measurements in Pb have been made by various workers (Bezuglyĭ and Galkin 1958, Khaĭkin 1959, Aubrey 1960, Khaĭkin and Mina 1962, Young 1962, Mina and Khaĭkin 1963; see also Agababyan *et al.* 1968). Aubrey obtained a value of S, the total area of the Fermi surface, of $(17{\cdot}1 \pm 1{\cdot}7) \times 10^{16}\,\mathrm{cm}^{-2}$. The results of Aubrey and of Khaĭkin and Mina suggested that band 4 is empty, which is contrary to the interpretation of the de Haas–van Alphen data by Gold (1958) It was found that the Fermi surface in band 2 is not a perfect sphere but resembles quite closely the band 2 Fermi surface in the free-electron model (Mina and Khaĭkin 1963). Important dimensions of the Fermi surface in band 3, which were found to agree well with Gold's results, were also determined in considerable detail (Khaĭkin and Mina 1962, Mina and Khaĭkin 1963). Magnetoresistance experiments gave results substantially in agreement with these conclusions about the Fermi surface in band 2 and band 3(Alekseevskiĭ and Gaĭdukov 1959 a, b, 1961 a, Lüthi 1960, Schirber 1963, Lück 1966 b, Caroline 1969); a quantitative comparison between the magnetoresistance data and Gold's de Haas–van Alphen results was made by Alekseevskiĭ and Gaĭdukov (1961 a) and some of the dimensions of the arms in the band 3 network were determined by Schirber (1963) who obtained results in close agreement with those of Gold (1958). Various magnetoacoustic effect experiments have been performed on Pb (Mackinnon *et al.* 1959, Mackinnon and Taylor 1960, Mackinnon *et al.* 1962, Mackintosh 1963, Rayne 1962 b,

1963a). The results of Rayne, which were interpreted assuming that band 4 is empty, gave detailed dimensions for the Fermi surface in bands 2 and 3 which were in quite good agreement with the results of the O.P.W. band structure calculation and de Haas–van Alphen experiments of Anderson and Gold (1965). These new de Haas–van Alphen measurements of Anderson and Gold were made by the pulsed-field method in fields up to 200 kG; the results were much more accurate than the older results and, in addition, several new periods were observed. Once again no evidence was found for the existence of any Fermi surface in band 4. The results of Anderson and Gold provide the most accurate detailed data on the dimensions of the Fermi surface of Pb in bands 2 and 3 and several of their dimensions are also in good agreement with the magnetoacoustic results of Rayne (1963a) (see table III of Anderson and Gold (1965)). Anderson and Gold also performed an O.P.W. band structure calculation for Pb and obtained values for the Fourier coefficients of the pseudopotential by fitting the results of the band structure calculation to certain of the areas of cross section of the Fermi surface determined by their de Haas–van Alphen results. This calculation further supported the view that there is no Fermi surface in band 4 in Pb. Also in this calculation the effects of spin–orbit coupling were found to be quite large at certain points in the Brillouin zone; for instance, one of the levels at W which is two-fold degenerate in the absence of spin–orbit coupling becomes split by about 0·124 Ry when spin–orbit coupling is included. While this band structure calculation by Anderson and Gold (1965) is clearly semi-empirical it gave a set of energy bands in quite good agreement with those obtained by Loucks (1965) in an *ab initio* relativistic band structure calculation for Pb.

There are several other pieces of experimental work related to the problem of the determination of the Fermi surface of Pb. γ has been measured by various authors

3·00 mJ mole^{-1} deg^{-2}	Meads *et al.* (1941), Horowitz *et al.* (1952)
3·06(2·99)† mJ mole^{-1} deg^{-2}	Decker *et al.* (1958)
2·98 mJ mole^{-1} deg^{-2}	Keesom and Van der Hoeven (1963)
3·00(3·02)† mJ mole^{-1} deg^{-2}	Phillips *et al.* (1964)
(3·00 ± 0·04) mJ mole^{-1} deg^{-2}	Van der Hoeven and Keesom (1965).

The contribution to γ from electron–phonon interactions was found by Ashcroft and Wilkins (1965) to be quite large. The temperature dependence of the isotropic and anisotropic Knight shift in Pb has been investigated in the range 4·2°K–450°K by Borsa and Barnes (1966); some

† The values in parentheses are revised values quoted by Neighbor *et al.* (1967) based on the same original data.

values are given in table 1 of Borsa and Barnes (1966). The pressure dependence of the Knight shift in Pb has been measured by Matzkanin and Scott (1966) at room temperature for hydrostatic pressures up to 12 kbar. The change in the Knight shift in Pb as a result of alloying with In was measured by Snodgrass and Bennett (1963). The effect of hydrostatic pressure on the de Haas–van Alphen oscillations in Pb was measured by Anderson, O'Sullivan and Schirber (1967) who found that the frequencies increased by about 0·3% per kbar which is more than twice the rate expected just from scaling. Quantitative estimates of the changes in the Fourier coefficients of the pseudopotential with pressure were made from these results. The optical constants of Pb in the range 0·7–12 μ were measured by Golovashkin and Motulevich (1963) and the frequency of electron–electron collisions determined. Experiments on ultrasonic attenuation in Pb, in the absence of a magnetic field, were performed by Fate (1968) to investigate the mean free path lengths and the electron–phonon interaction. Quantum oscillations in the contact potential which were predicted by Kaganov *et al.* (1957) have been observed in Pb by Caplin and Shoenberg (1965). Pb was used in the positron annihilation experiments of Briscoe *et al.* (1966) in which an attempt was made to see if the transition to the superconducting state caused any change in the topology of the Fermi surface of a metal; since many other experimental methods of determining Fermi surfaces cannot be used for metals in the superconducting state this is an important feature of positron annihilation techniques. The phonon dispersion relations of Pb have been investigated experimentally by neutron scattering by several authors and a search made for Kohn anomalies (for example, Brockhouse *et al.* 1961, 1962, Gilat 1965). Several Kohn anomalies were found by these authors in positions to be expected from the Fermi surface described already. About 20 points on the Fermi surface of Pb were identified in the comprehensive measurements of Kohn anomalies by Stedman *et al.* (1967); their positions were in good general agreement with the dimensions of the Fermi surface given by Anderson and Gold (1965). Of special interest, perhaps, is the observation of Kohn anomalies in Pb in the work of Paskin and Weiss (1962) who, unlike the other workers, used x-rays rather than neutrons in studying the phonon dispersion relations.

In conclusion, therefore, we may say with reasonable confidence that the Fermi surface of Pb is described to a good first approximation by the f.c.c. free-electron Fermi surface with four electrons per atom; more detailed experiments revealed that there is no Fermi surface in band 4 and that in bands 2 and 3 there are quantitative, but not qualitative, departures from the free-electron Fermi surface. Assuming band 4 to be empty the volumes of the (closed) Fermi surface of holes in band 2 and the (multiply-connected) Fermi surface of electrons in band 3 must be equal (Alekseevskiĭ and Gaĭdukov 1961 a). The detailed values of the dimensions of the Fermi surface of Pb in bands 2 and 3 are best obtained from the figures and tables of Anderson and Gold (1965).

§ 7. p-BLOCK : GROUP VB : NITROGEN GROUP

[N	Nitrogen	2.(2, 3)]
[P	Phosphorus	2.8.(2, 3)]
As	Arsenic	2.8.18.(2, 3)
Sb	Antimony	2.8.18.18.(2, 3)
Bi	Bismuth	2.8.18.32.18.(2, 3)

7.1 *Structures*

The three elements As, Sb and Bi all have the same crystal structure but with different values of the lattice constants. The crystal structure of each of them belongs to the trigonal space group R$\bar{3}$m ($D_{3d}{}^5$) with two atoms per unit cell at (x, x, x) and at $(-x, -x, -x)$; alternative descriptions of

Table 19. Lattice parameters for As, Sb and Bi

Coordinates : Rhombohedral (I) 2As(C_{3v}) : $\pm(xxx)$
Rhombohedral (II) 8As(C_{3v}) : [000 ; $\frac{1}{2}\frac{1}{2}0$; $\frac{1}{2}0\frac{1}{2}$; $0\frac{1}{2}\frac{1}{2}$] $\pm(xxx)$
Hexagonal (III) 6As(C_{3v}) : [000 ; $\frac{2}{3}\frac{1}{3}\frac{1}{3}$; $\frac{1}{3}\frac{2}{3}\frac{2}{3}$] $\pm(00x)$

	Rhomb. (I) $A=2$		Rhomb. (II) $A=8$		Hexagonal (III) $A=6$			c/a
	a	α	a	α	a	c	x	
As	4·12	54° 10′	5·57	84° 38′	3·75	10·50	0·226	2·80
Sb	4·50	57° 06′	6·20	87° 24′	4·30	11·24	0·233	2·62
Bi	4·74	57° 14′	6·57	87° 32′	4·54	11·84	0·237	2·61
Simple cubic	—	60°	—	90°	—	—	—	2·45

(Smithells (1967).)

Fig. 36

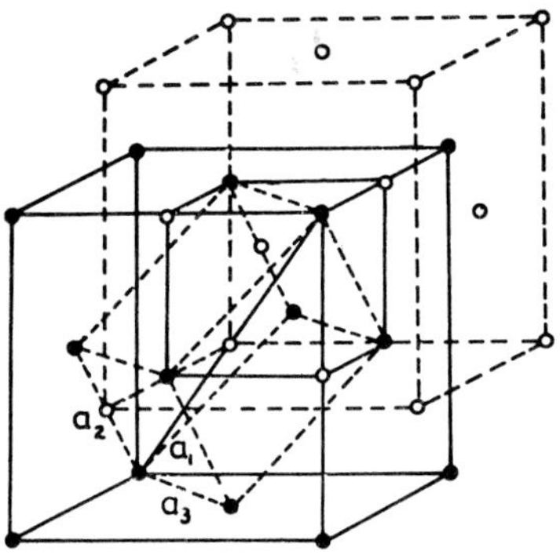

The relationship of the structures of As, Sb and Bi to the simple cubic lattice (Abrikosov and Fal'kovskiĭ 1962).

the unit cell of the crystal are also possible and the parameters are given in table 19 (from Smithells 1967). These lattices can be obtained from simple cubic lattices by slight displacements of the atoms. Suppose that we separate a simple cubic lattice into two face-centred sublattices, where the two sublattices are related by a translation along the body-diagonal of the original simple cube, see fig. 36. The structure of As, Sb or Bi can then be obtained by performing a slight trigonal distortion; the extent of this distortion can be seen to be quite small from the values of the angle α in table 19 which are quite close to the value of $60°$ for the undistorted simple-cubic lattice.

Since the Bravais lattice of the crystal structure of As, Sb or Bi is very closely related (by only a small distortion) to one of the face-centred cubic sublattices shown in fig. 36, the Brillouin zone of each of these structures can therefore be obtained by a slight distortion of the face-centred cubic Brillouin zone which is shown, for example, in fig. 2 (*a*). Of course, the principal axis of this distorted face-centred cubic Brillouin zone is a three-fold axis normal to one of the pairs of large hexagonal faces, see fig. 37. Because of the differences in the lattice constants for the three crystal structures of As, Sb and Bi the actual dimensions of the Brillouin zone will also be slightly different for the three different elements.

Fig. 37

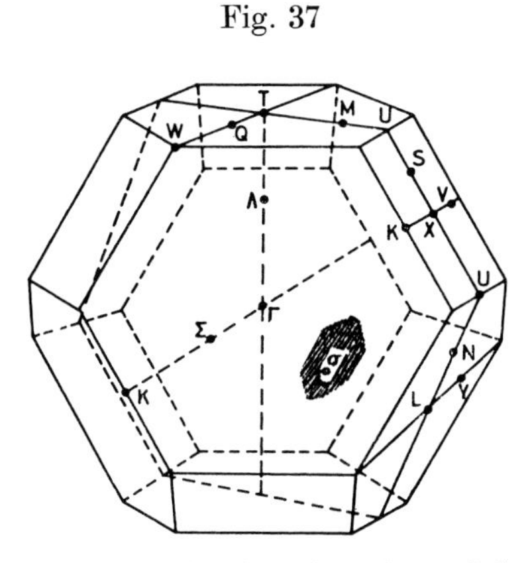

The Brillouin zone for As, Sb and Bi (Cohen 1961).

In discussing the Fermi surfaces of As, Sb and Bi we shall depart from our usual procedure of going down each column of the periodic table from the element of lowest atomic number to that of highest atomic number. We shall discuss Bi first because of the very large amount of theoretical and experimental work which has been performed on this semi-metal, right from the pioneering days of Fermi surface studies until now. There already exist several reviews which give some discussion of the electronic structure of Bi (Kahn and Frederikse 1959, Lax and Mavroides 1960, Boyle and Smith 1963, Shoenberg 1969).

7.2 *Bismuth*

Historically, Bi was one of the first materials for which the free-electron energy bands and the free-electron Fermi surface were investigated theoretically (Jones 1934, Mott and Jones 1936, see also Jones 1960). As well as being studied theoretically in the very early days of the investigations of the electronic properties of metals, Bi was also one of the first materials to be studied experimentally by some of the methods which have now become commonplace in Fermiology. As early as 1930 oscillations were observed in the magnetoresistance of Bi by Shubnikov and de Haas and in the diamagnetic susceptibility by de Haas and van Alphen (for references see Boyle and Smith (1963)). Moreover, no other material seems to have been studied more extensively since the discovery of these oscillatory effects. A comprehensive list of references up to about 1962 on experiments on the electronic structure of Bi is given in the review by Boyle and Smith (1963); we do not repeat all these here, but we shall include references to more recent work.

Fig. 38

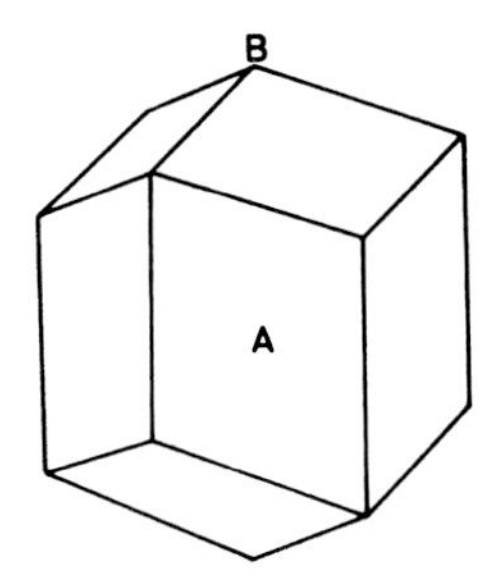

The Brillouin zone for Bi used by Jones (1934).

The Brillouin zone used by Jones (1934) was a zone in extended **k** space, see fig. 38, and corresponded to five electrons per atom; the relationship between this large zone and the conventional Brillouin zone is explained in § 42 of the book by Jones (1960). By studying the electrical resistivity as a function of concentration of Sn impurities Jones concluded that the number of electrons which spilled out of this large zone must be less than about 0·0013 per atom. There will be an equal number of holes in the large zone. Mott and Jones (1936) estimated this number to be about 10^{-4} per atom and recent values include:

$1{\cdot}6 \times 10^{-5}$ per atom	Abelès and Meiboom (1956,
$1{\cdot}4 \times 10^{-5}$ per atom	Jain and Koenig (1962)
$0{\cdot}88 \times 10^{-5}$ per atom	Zitter (1962)
$1{\cdot}1 \times 10^{-5}$ per atom	Williams (1965), Bhargava (1967)
$1{\cdot}2 \times 10^{-5}$ per atom	Noothoven Van Goor (1968).

In the free-electron approximation these pockets of electrons would occur at A in the large zone which corresponds to the points L or X in the conventional Brillouin zone. These pockets of electrons will be repeated by the three-fold rotation axis of symmetry of the crystal. The pockets of holes are at B in the large zone which can be shown (Jones 1960) to correspond to ΓT in the conventional Brillouin zone. It is the existence of only these relatively small pockets of electrons and holes in otherwise completely empty or full bands which leads to the low electrical conductivity of Bi and its description as a semi-metal. It was pointed out by Sondheimer (1952) that in metals containing a very small number of conduction electrons, such as Bi, the electronic mean free path $\bar{\lambda}$ should be abnormally long, of the order of $1\,\mu$ at room temperature. This was confirmed by measurements of $\sigma/\bar{\lambda}$ by means of the anomalous skin effect in Bi (Pippard and Chambers 1952). The temperature dependence of $\bar{\lambda}$ was investigated by Gantmakher and Leonov (1968) by measuring the amplitudes of the lines in the radio-frequency size effect. One or two 'effects' previously only observed in semiconductors have now also been observed in Bi; these include the photoelectromagnetic effect (Young 1960) and the acoustoelectric effect, in the presence of a transverse applied magnetic field (Yamada 1965, Lopez 1968).

This model of the Fermi surface was found to be consistent with the galvanomagnetic measurements of Abelès and Meiboom (1956) in low magnetic fields. Shoenberg (1939, 1952) concluded from the de Haas–van Alphen measurements that the three ellipsoidal pockets of electrons have one principal axis parallel to a two-fold axis of symmetry but are tilted away from the trigonal axis through a small angle (see also Shoenberg and Uddin 1936b, Blackman 1938, Dhillon and Shoenberg 1955). It is customary to represent each of these tilted ellipsoidal pieces of Fermi surface enclosing pockets of electrons by:

$$E_{\mathbf{k}} = \frac{\hbar^2}{2m}(\alpha_{11}k_x{}^2 + \alpha_{22}k_y{}^2 + \alpha_{33}k_z{}^2 + 2\alpha_{23}k_yk_z), \quad . \quad . \quad (7.2.1)$$

where m is the mass of a free electron and the set of orthogonal axes $Oxyz$ are oriented with Oz along the three-fold axis and Ox along one of the two-fold axes. The components of the effective mass tensor can be written in terms of the parameters α_{ij}:

$$\left.\begin{aligned} \frac{m_{11}}{m} &= \frac{1}{\alpha_{11}}, \\ \frac{m_{22}}{m} &= \frac{\alpha_{33}}{\alpha_{22}\alpha_{33} - \alpha_{23}{}^2}, \\ \frac{m_{33}}{m} &= \frac{\alpha_{22}}{\alpha_{22}\alpha_{33} - \alpha_{23}{}^2}, \\ \frac{m_{23}}{m} &= \frac{-\alpha_{23}}{\alpha_{22}\alpha_{33} - \alpha_{23}{}^2}. \end{aligned}\right\} \quad . \quad . \quad . \quad . \quad . \quad . \quad . \quad (7.2.2)$$

Most workers are agreed that the Fermi surface enclosing a region of holes consists of an ellipsoid of revolution where the axis of revolution is parallel to the trigonal axis (Pippard and Chambers 1952, Heine 1956, Aubrey and Chambers 1957). The ellipsoid of revolution for the Fermi surface of holes is commonly represented by:

$$E_{\mathbf{k}} = \frac{\hbar^2}{2m}(\beta_1(k_x^2+k_y^2)+\beta_3 k_z^2), \quad . \quad . \quad . \quad . \quad (7.2.3)$$

where

$$\beta_1 = 1/m_{11} \text{ and } \beta_3 = 1/m_{33}.$$

Cyclotron resonance investigations of Bi have been performed by many workers (Dexter and Lax 1955, Galt *et al.* 1955, Foner *et al.* 1956, Lax *et al.* 1956, Tinkham 1956, Everett 1962, Khaĭkin *et al.* 1962, Kirsch and Miller 1962, Kao 1963, Smith *et al.* 1963, Khaĭkin and Édel'man 1964, Hebel 1965, Marsh and Aubrey 1965). A large number of investigations have been performed on the galvanomagnetic effects in Bi (see the references given by Boyle and Smith (1963), also Mase and Tanuma (1959), Clark and Assenheim (1960), Gitsu and Ivanov (1960), Tanaka *et al.* (1961), Mase *et al.* (1962), Colombani and Huet (1963 a, b), Grenier *et al.* (1963), Hall and Koenig (1964), Pospelov and Kechin (1964), Alekseevskiĭ and Kostina

Table 20. Parameters for the inverse effective mass tensors of the electrons and holes in Bi†

	α_{11}	α_{22}	α_{33}	α_{23}	β_1	β_3
Anomalous skin effect[1]	160	1·56	83	6·7	1·7	0·13
de Haas–van Alphen effect[2]	420	0·8	40	4·0	—	—
de Haas–van Alphen effect[3]	—	—	—	—	20	1·43
de Haas–van Alphen effect[4]	202	1·67	70	7·0	—	—
Cyclotron resonance[5]	202	1·67	83·3	8·33	—	—
Cyclotron resonance[6]	196	1·71	84	8·6	21·2	1·46
Cyclotron resonance[7]‡	114	1·39	108	9·47	14·7	1·07
Far infra-red studies[8]	133	1·2	91	6·1	—	—
Ultrasonic attenuation[9]	178	1·1	84·5	7·2	—	—
Ultrasonic attenuation[10]	197	1·64	81·1	9·41	16·6	1·84
Ultrasonic attenuation[11]	115	6·0	75·2	19·0	—	—

† This table is an extended version of table 1 of Boyle and Smith (1963).
‡ The cyclotron resonance data of Everett (1962) and Édel'man and Khaĭkin (1965) and the computer calculations of Fal'kovskiĭ and Razina (1965) are discussed in the text.

[1] Smith (1959); [2] Shoenberg (1957); [3] Brandt (1960), Brandt *et al.* (1959); [4] Weiner (1962); [5] Aubrey and Chambers (1957); [6] Aubrey (1961); [7] Galt *et al.* (1959); [8] Boyle and Brailsford (1960); [9] Reneker (1959); [10] Mase *et al.* (1966); [11] Giura *et al.* (1967) (the data given by these authors referred to the principal axes, $\alpha_1 = 115$, $\alpha_2 = 2{\cdot}1$, $\alpha_3 = 79$, and the tilt angle $= 13°$).

(1965), Morimoto (1965), Bate and Einspruch (1965), Bogod (1967), Bogod and Eremenko (1967 a), Friedman (1967), Toureille (1967), Bogod *et al.* (1968, 1969), Hattori (1968), Otake and Koike (1968), Gitsu *et al.* (1969)). The effect of the self-magnetic field on the galvanomagnetic effects in Bi was investigated by Hattori and Tosima (1965) and qualitative agreement between experiments and theory was obtained, using existing models for the Bi band structure. Various experimental results for the effective masses of both the electrons and the holes in Bi are collected in table 20.

Cohen (1961) investigated the band structure of Bi by a perturbation calculation, using the experimentally observed symmetries of the Fermi surface ellipsoids to take into consideration the non-parabolic nature of the energy bands and the consequent non-ellipsoidal nature of the constant-energy surfaces. At that time it was not known at which points in the Brillouin zone the ellipsoids of the electron Fermi surface were situated. If the ellipsoids of the Fermi surface of electrons are assumed to be situated at the points L, as seems to be well established now, then the form of the distorted ellipsoidal Fermi surface of electrons can be represented by (see Kao 1963):

$$\frac{p_1^2}{2m_1}+\frac{p_2^2}{2m_2}+\frac{p_3^2}{2m_3}=E\left(1+\frac{E}{E_g}\right)-\left(\frac{p_2^2}{2m_2}\right)\frac{1}{E_g} \quad . \quad . \quad (7.2.4)$$

where the subscripts 1, 2 and 3 refer to the principal-axis system of the original ellipsoid; the expressions for the components of the effective mass tensor, in terms of the parameters in eqn. (7.2.4), are given in Appendix A to the paper by Kao (1963). Experimental evidence for the non-parabolic band structure and non-ellipsoidal Fermi surfaces was obtained in the study

Fig. 39

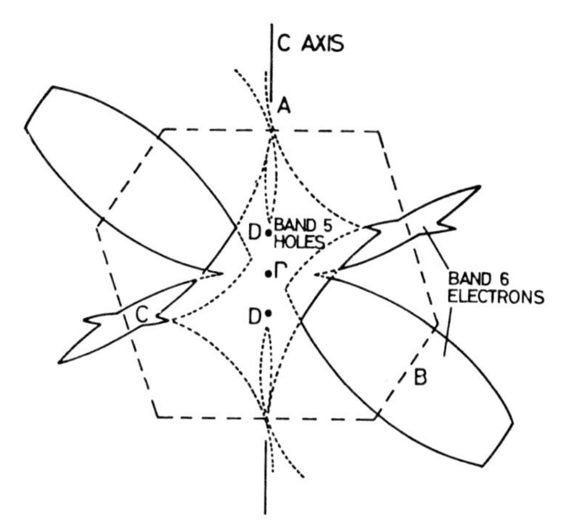

Cross sections of the Fermi surface of Bi in the free-electron approximation (Harrison 1960 c); points B and C correspond to X and L respectively in fig 37.

of the variation of the effective mass as a function of direction in the cyclotron resonance experiments of Édel'man and Khaĭkin (1965); this was supported by the associated computer analysis of various pieces of experimental data by Fal'kovskiĭ and Razina (1965).

The construction of the free-electron Fermi surface directly in the conventional Brillouin zone of Bi has been studied by Harrison (1960c) in an extension of the work on f.c.c., b.c.c. and h.c.p. metals that was described in § 2.1. The spherical free-electron Fermi surface contains ten electrons per unit cell of the crystal (that is, five electrons per atom) and this gives rise to pieces of Fermi surface in bands 2–8. When the actual crystal potential is considered several of these pieces of Fermi surface disappear and the extensive experimental evidence available (see below) suggests that for Bi the only pieces of Fermi surface to survive are regions of holes in band 5 and regions of electrons in band 6. Therefore Harrison (1960c) only gives the details for these two bands, see fig. 39. The Fermi surface of electrons that is actually observed experimentally is expected to arise from the electrons of region X or L. Experimentally it is known that the electron Fermi surface is elongated and is tipped slightly out of the plane perpendicular to the c axis. It is clear from the symmetry of the point X that the addition of a non-zero potential and spin–orbit coupling could not give rise to such segments there, while they might easily appear at L. The two other electron Fermi surfaces are obtained by the use of the three-fold rotation axis of symmetry. The Fermi surface in band 5 which surrounds a pocket of holes and which has been observed experimentally presumably arises from the complex hole surface, which is

Fig. 40

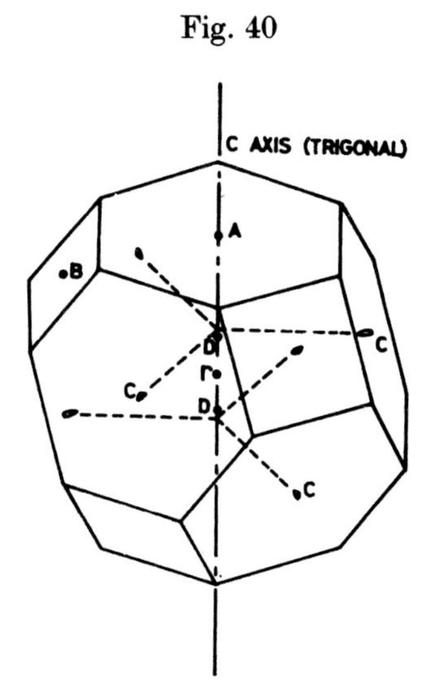

The disposition of the electron Fermi surface pockets in Bi (Harrison 1960 c); the points A, B and C correspond to T, X and L respectively in fig. 37.

shown in fig. 39, and would appear to be centred at Γ. However, according to the band structure calculations of Mase (1958, 1959 a) and Cohen *et al.* (1964) the Fermi surface of holes in band 5 would be at T rather than Γ. Harrison's arguments about the potential suggested that this piece of Fermi surface was centred somewhere between T and Γ. The final picture of the Fermi surface of electrons in band 6 deduced by Harrison (1960 c) by modifying the free-electron Fermi surface in the light of experimental evidence and the band structure calculations of Mase (1958, 1959 a) is shown in fig. 40.

Oscillations in a number of properties other than the diamagnetic susceptibility (de Haas–van Alphen effect) have been observed by many workers (see table 1 of Kahn and Frederikse (1969) and the references given by Boyle and Smith (1963)); these include the Hall effect (Overton and Berlincourt 1955, Connell and Marcus 1957), the resistivity (see the references given by Boyle and Smith (1963), also Brown (1964, 1966), Antcliffe and Bate (1967)), the thermoelectric power and thermal conductivity (Steele and Babiskin 1955), the ultrasonic velocity (Mavroides *et al.* 1962), the ultrasonic attenuation (Toxen and Tansal 1965, Mase *et al.* 1966, Inoue and Tsuji 1967), and the infra-red transmission (Boyle and Rodgers 1959). However, Engeler (1963) produced evidence that the oscillations

Table 21. Periods of oscillations in Bi†

Axis (parallel to **H**)	Period (in $1/H$) (10^{-5} G^{-1})									
	1	2	3	4	5	6	7	8	9	10
Electrons										
Trigonal	1·2	1·18	1·57	1·6	—	1·2	—	—	1·17	1·56
Binary	—	7·4	7·1	7·5	7·1	6·8	7·05	7·6	7·20	‡
	0·48	0·25	0·30	—	—	0·50	0·52	—	0·53	‡
Bisectrix	4·3	4·3	4·1	4·2	4·1	4·0	4·1	4·5	4·17	‡
	8·2	8·5	8·2	8·8	8·2	7·8	8·2	8·9	8·30	‡
Holes										
Trigonal	1·56	—	—	—	—	1·54	1·58	1·6	1·575	‡
Binary	0·46	—	—	—	—	—	—	0·50	—	‡
Bisectrix	—	—	—	—	—	—	0·48	0·50	0·45	‡

† This is an extended version of table 2 of Boyle and Smith (1963).
‡ Can be read off visually from the graphs in figs. 15–18 of Mase *et al.* (1966).

[1] Brandt (1960), Brandt and Razumeenko (1960), Brandt *et al.* (1963), Brandt and Lyubutina (1964) ; [2] Shoenberg (1952) ; [3] Steele and Babiskin (1955), Babiskin (1957) ; [4] Connell and Marcus (1957) ; [5] Kunzler and Hsu (1960) ; [6] Lerner (1962, 1963) ; [7] Brown (1964) ; [8] Eckstein and Ketterson (1965) ; [9] Bhargava (1967) ; [10] Mase *et al.* (1966).

in the infra-red transmission of Bi in a magnetic field observed by Boyle and Rodgers resulted from direct interband transitions between Landau levels rather than from quantum oscillations of the Fermi level as previously hypothesized. Quantum oscillations of the Fermi level of Bi were, however, observed by Pelikh and Eremenko (1967) via the contact potential and by Takano and Kawamura (1968) in Alfvén wave transmission through Bi. Magnetothermal oscillations have been observed in Bi (Boyle *et al.* 1960, Kunzler and Hsu 1960, Kunzler *et al.* 1962, Brown 1964, Noguchi and Tanuma 1967). The results of various de Haas–van Alphen type oscillation experiments are collected in table 21.

Overton and Berlincourt studied the effect of pressure on the period (in $1/H$) of the oscillations in the Hall coefficient; the period was found to increase by about 1·5% for a pressure of about 120 bar and was explained by assuming that the conduction band moves up and the valence band moves down under pressure, thereby reducing the Fermi surface cross-sectional areas. The de Haas–van Alphen effect in Bi under pressure up to about 1 kbar has also been investigated (Brandt and Venttsel' 1958, Brandt and Ryabenko 1959). The effect of pressure on the electrical conductivity, Hall effect, and magnetoresistance of Bi has been investigated experimentally by Sekoyan and Likhter (1960) and Vaišnys and Kirk (1967) but provided no new information about the Fermi surface; the situation is complicated by the fact that there are several phase transitions (Bundy and Strong 1962). Changes in the period of the magnetoresistance oscillations (Shubnikov–de Haas effect) in Bi as a function of uniaxial stress (35·5 bar) were observed by Bate and Einspruch (1965); the observed dependence of these changes on the magnetic field orientation suggested that the electron quasi-ellipsoids do not contract uniformly under uniaxial stress. The effect of pressure on the Shubnikov–de Haas oscillations in Bi has been studied experimentally up to 15 kbar (Itskevich and Fisher 1967 b, Itskevich *et al.* 1967); all the measured cross sections, both of the electron and of the hole ellipsoids, were found to decrease with increasing pressure. The effect of pressure on the temperature dependence of the electrical conductivity of Bi was measured for pressures up to 25 kbar at 2°K–300°K by Balla and Brandt (1964). The dependence of the carrier density on pressure was determined and it was predicted that at a pressure of about 26 kbar the overlap between the valence band and conduction band in Bi should disappear completely, so that at low temperatures Bi would become an insulator (see also Fal'kovskiĭ (1967), and Brandt *et al.* (1968)). The effect of impurities on the Fermi surface in Bi has been investigated using the de Haas–van Alphen effect (Brandt and Razumeenko 1960, Brandt and Shchekochikhina 1961, Brandt and Lyubutina 1967, Ermolaev and Kaganov 1967).

Although, as we have seen, it has been known for a long time that the Fermi surface for electrons in Bi consists of a set of equivalent ellipsoids tilted away from the trigonal axis, the actual number (that is, three or six) of these ellipsoids and their location in the Brillouin zone was much more difficult to ascertain. Similarly, although the Fermi surface for holes was known to be an ellipsoid of revolution about the trigonal axis, there was again doubt about the number of these ellipsoids (that is, one or two) and about their location in the Brillouin zone. Several authors suggested the existence of a third set of carriers (heavy holes) beyond the two sets described so far on the Jones–Shoenberg picture (see, for example, Heine 1956, Brandt and Venttsel' 1958, Kalinkina and Strelkov 1958, Brandt *et al.* 1959, Brandt 1960, Smith 1961, Lerner 1962, 1963, Sybert *et al.* 1962). This problem has been studied by Jain and Koenig (1962) who surveyed the data existing at that time on the total number of carriers and the number of carriers in each ellipsoid. They concluded that for the electrons there are just three ellipsoids which, from the band structure calculations of Mase (1958, 1959 a), are really six half-ellipsoids at L on the irregular hexagonal faces of the Brillouin zone (see fig. 40). Jain and Koenig also concluded that there is only one light-hole ellipsoid centred at T on the top hexagonal face of the Brillouin zone, see fig. 40. The analysis led to a volume for the possible heavy-hole Fermi surface that was of the same order of magnitude as the errors in the calculation. The number of such heavy holes, if they exist, would then have to be less than about 15% of the number of electrons (unless some heavy electrons also exist). Measurements of the small-field ($\lesssim 1$ G) galvanomagnetic tensor components of Bi at 4·2°K by Zitter (1962) showed that the electron concentration and the hole concentration are very nearly the same. When compared with cyclotron resonance and de Haas–van Alphen experiments, these results are consistent with a three-ellipsoid model for electrons and a single ellipsoid for light holes. Recently there have been two new band structure calculations performed for Bi, one is an *ab initio* relativistic A.P.W. calculation (Ferreira 1967, 1968) and the other is a pseudopotential calculation using only three adjustable parameters (Golin 1968). The calculation by Ferreira provided fairly clear evidence that there are pockets of holes at T and nowhere else and pockets of electrons at L and nowhere else; constant energy surfaces in bands 5 and 6 near both L and T were plotted and it was shown that the holes at T are in a T_4^- band as suggested by Smith *et al.* (1964). The effect of spin–orbit coupling on the band structure of Bi was found to be much larger than in As or Sb, as would be expected.

Although many experimental results showed that the tilt angle in Bi is about 6° (see, for example, Édel'man and Khaĭkin (1965), Fal'kovskiĭ and Razina (1965)) the sign of the tilt angle remained undetermined for some time. That is, it was not clearly established whether the vertical cross section should be represented by the ellipsoid indicated with solid lines or broken lines in fig. 41. It is convenient to establish a sign convention to describe the tilted ellipsoids and the usual convention, described by Brown *et al.* (1968) is as follows. The positive bisectrix (line bisecting the angle between two binary rotation axes) is shown in fig. 41. The positive binary axis is then chosen so that the binary (x), bisectrix (y) and trigonal (z) axes form a right-handed triad. A positive tilt angle is then usually (but

Fig. 41

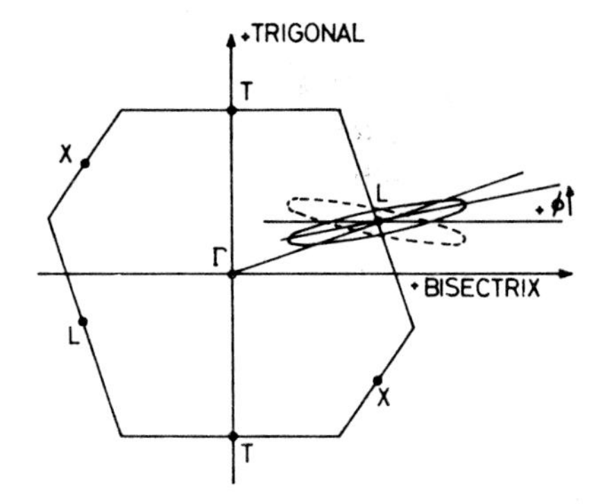

Cross section of the Fermi surface of Bi with the cross section drawn in continuous lines for a positive tilt angle and in broken lines for a negative tilt angle (Brown *et al.* 1968).

not universally) defined to involve a rotation about the binary axis so as to rotate $+y$ through the first quadrant towards $+z$. Thus the solid ellipsoid in fig. 41 has its major axis tilted by a positive small angle from the bisectrix direction. Brown *et al.* (1968) studied the various pieces of experimental evidence available and also measured the de Haas–van Alphen effect for Bi and Sb specimens which were oriented by X-rays and therefore known to be in the same orientation. They found that the sign of the tilt angle in Bi is *positive* in the convention just described and also that the signs of the tilts of the electron Fermi surfaces in Bi and Sb are opposite, independent of the actual sign convention used. Some of the ambiguities which existed previously about the sign of the tilt angle were ascribed to difficulties associated with the use of etch pits instead of X-rays by some workers when orienting their samples (see, for instance, plate I of Boyle and Smith (1963)).

Having come to the general conclusion that the Fermi surface of Bi consists of three small tilted ellipsoids of electrons and one ellipsoid of rotation of holes, and having already described the geometry and location of these ellipsoids, we conclude this section by describing briefly a number of experimental results that are closely related to the problem of determining the band structure and Fermi surface of Bi.

Values of γ for Bi have been given by various authors:

0·63 mJ mole^{-1} deg^{-2}	Blackman (1938)
(0·078 ± 0·013) mJ mole^{-1} deg^{-2}	Keesom and Pearlman (1954)
0·048 mJ mole^{-1} deg^{-2}	Ramanathan and Srinivasan (1955)
0·067 mJ mole^{-1} deg^{-2}	Kalinkina and Strelkov (1958)
0·021 mJ mole^{-1} deg^{-2}	Phillips (1960).

The band structure of Bi has been investigated by Esaki and Stiles (1965) by the use of electron tunnelling through Al_2O_3 between Bi and Al; many bands were detected, a large number of which had not been seen before with other experimental methods. Interband transitions in Bi have been studied in experiments on infra-red reflection in the presence of a magnetic field (Lax *et al.* 1960, Brown *et al.* 1960, 1963). There have been several experimental investigations of the variation of the resistivity of Bi with sample thickness (Friedman and Koenig 1960, Friedman 1967, García and Kao 1968, Vatamanyuk *et al.* 1968, Aubrey and Creasey 1969). According to the theory given by Cohen and Blount (1960) the g factor of the conduction electrons in Bi can be very large; for instance g would be expected to exceed 200 when the magnetic field is in the direction corresponding to the smallest effective mass. Observations of resonances at microwave frequencies have been made and attributed to spin resonances corresponding to these large g factors (Smith *et al.* 1960, 1963, Everett 1962). However, it was claimed that calculations of the spin magnetization by Hebel *et al.* (1965) showed that the observed resonances could not be spin resonances because their intensities were much too large relative to the observed cyclotron resonances. An alternative explanation of the observed resonances was advanced by Hebel (1965) in terms of the propagation of microwave power in a sample in which a 'slightly anomalous' skin effect condition is assumed to exist. However, Carter and Picard (1967) claimed to have observed genuine spin resonances in Bi and values of g determined from the spin splitting of magnetothermal oscillations by Smith *et al.* (1964) and McCombe and Seidel (1967) varied from 1·3 to about 220 according to carriers and orientation. Positron annihilation experiments in Bi were performed by Dekhtyar and Mikhalenkov (1960) and the effect of temperature changes on the positron annihilation results were investigated later (Dekhtyar and Mikhalenkov 1961, Faraci *et al.* 1969). Spin splitting of the Landau levels in Bi has been observed in many different experiments, by Saito (1963 b) in the de Haas–van Alphen effect, by Lerner (1963) in the Shubnikov–de Haas effect, by Boyle *et al.* (1960), Kunzler *et al.* (1962), Smith *et al.* (1964) and McCombe and Seidel (1967) in magnetothermal experiments, and by Mase *et al.* (1966) and Sakai *et al.* (1969) in the magnetoacoustic 'giant' quantum oscillations. At large magnetic fields the transmission of microwaves through Bi is essentially undamped and can be regarded as Alfvén (helicon) waves in a solid state plasma. Kirsch and Miller (1962) observed a large kink in the 9 GHz microwave absorption of Bi as a perpendicularly applied magnetic field was varied up to 1·5 kG; this was identified as a Doppler-shifted cyclotron resonance. Buchsbaum and Galt (1961) re-interpreted the cyclotron resonance results of Galt *et al.* (1959) in terms of Alfvén wave propagation in a solid. Alfvén wave transmission has been observed and studied in Bi by several workers, see for example, Brownell and Hygh (1967), Guthmann and Libchaber (1967), Édel'man (1968) and Nagata and Kawamura (1968) for references.

In the course of magnetoresistance measurements in Bi, Esaki (1962) observed a non-ohmic behaviour, that is, the curves of current, I, against voltage, V (at a given value of the magnetic field), were not linear but possessed a sharp change in slope at some critical value of the applied voltage, V. The electric field, E_c, at which this kink occurred was simply related to the applied magnetic field by $E_c = \alpha B$, where $\alpha \sim 10^{-3}$ volt cm^{-1} oersted^{-1}. This effect was explained by Esaki by assuming that a strong electron–phonon interaction (phonon emission) occurs when the cycloidal drift velocity ($v_x = cE_y/B_z$) in the crossed electric and magnetic fields (E_y and B_z) reaches the velocity of sound propagation for that particular direction; this is consistent with the experimental value just mentioned for α ($= E_c/B$). When this phonon emission once starts, the scattering time τ may drastically decrease and therefore there would be a sharp decrease in the resistance (see also Esaki and Heer (1962), Hopfield (1962), Miyake and Kubo (1962) and Toxen and Tansal (1963)).

7.3. *Arsenic and Antimony*

Neither As nor Sb has been the object of such intensive experimental and theoretical investigation as has Bi. Nevertheless, enough work has been done to establish that the Fermi surfaces of these two elements are very similar to the Fermi surface of Bi and also to establish the quantitative details of the Fermi surfaces of As and Sb. The geometry of the electron Fermi surface for As and Sb is very similar to that of Bi. The tilt angles for As and Sb have now been firmly established from de Haas–van Alphen measurements (Windmiller 1966, Priestley *et al.* 1967), and are negative in the convention spelled out by Brown *et al.* (1968), which we have already described in § 7.2 in connection with Bi.

The overall band structure of these semi-metals has been studied very little, as we have already noticed in the case of Bi. Of these three semi-metals it is As that has been studied least on the experimental side. However, it is As which has the lowest atomic number and is therefore the easiest of the three on which to perform an *ab initio* band structure calculation. The band structure of As has been calculated by Falicov and Golin (1965) using a pseudopotential method and by a separate O.P.W. calculation. The pseudopotential coefficients were selected by comparison with the known pseudopotential coefficients for Ge, which is next to As in the periodic table. On the band structure calculated by Falicov and Golin (1965) for As there would be a pocket of holes at T (in contradiction with the results of the earlier band structure calculation of Mase (1958, 1959 a)) and six pockets of electrons probably at L (as in Bi) but, just possibly, at X instead. The pockets of holes at T were predicted to be substantially warped ellipsoids.

The de Haas–van Alphen experiments of Berlincourt (1955) on As were originally interpreted in terms of an electron Fermi surface consisting of three ellipsoids as in Bi but with a larger tilt angle (– 36°). The long period oscillations observed by Berlincourt were ascribed to a single hole ellipsoidal Fermi surface similar to that in Bi, although beats were observed in these oscillations and the simple ellipsoid model did not seem entirely satisfactory for this piece of the Fermi surface. A better atomic pseudopotential was then constructed by Lin and Falicov (1966) by adding to the pseudopotential of Ge one-half of the antisymmetric pseudopotential of GaAs and renormalizing according to the volumes of the unit cells. This produced a band structure and Fermi surface for As which was in good agreement with the more recent experimental results on the de Haas–van Alphen effect (Priestley *et al.* 1967, Vanderkooy and Datars 1967, Ishizawa 1968 b), galvanomagnetic properties (Jeavons and Saunders 1968, 1969, Sybert *et al.* 1967, 1968), cyclotron resonance (Datars and Vanderkooy 1966), magnetoacoustic attenuation (Ketterson and Eckstein 1965, Shapira and Williamson 1965, Fukase and Fukuroi 1967) and magnetothermal and Shubnikov–de Haas oscillations (Noguchi and Tanuma 1967,

Table 22. The electron and hole Fermi surfaces of As (areas are in A.U.)

(*a*) Electrons		
	Theory†	Experiment‡
Area normal to binary axis	0·016	0·020
Area normal to trigonal axis	0·018	0·020
Minimum area for **H** in the trigonal–bisectrix plane	0·0055§	0·0055
Tilt angle for maximum area‖	$-8°$	$\sim -9°$
Tilt angle for minimum area‖	$\sim +80°$	$+85{\cdot}7° \pm 0{\cdot}5°$
Effective mass along the binary axis	0·11	—
Principal effective masses in binary–bisectrix plane	0·038	—
	0·94	—
(*b*) Holes		
Cross section of the cylinders	$6{\cdot}9 \times 10^{-5}$	$6{\cdot}9 \times 10^{-5}$
Tilt angle of cylinders	$-11°$	$-11°$
Area of pockets normal to binary axis	$\sim 9{\cdot}6 \times 10^{-3}$	—
Tilt angle of minimum area	$+44°$	$+36{\cdot}4° \pm 0{\cdot}5°$

† Lin and Falicov (1966).

‡ Experimental data are collated from Shapira and Williamson (1965), Priestley *et al.* (1967) and Vanderkooy and Datars (1967).

§ The Fermi energy used by Lin and Falicov (1966) was determined by fitting this area to the experimental results.

‖ Tilt angles are measured in the sense of rotation from ΓT (0°) ΓX (59° 17′). ΓL corresponds to $-72° 50'$ or equivalently 107° 10′.

(Adapted from Lin and Falicov (1966).)

Vanderkooy and Datars 1968). The electron Fermi surface was still found to consist of three ellipsoids and the parameters of this Fermi surface are given in table 22. The Fermi surface of holes obtained by Lin and Falicov (1966) was rather different from that of Bi in that it consisted of six connected pockets of holes in a 'crown' centred at T. The point in each individual pocket corresponding to the highest energy in band 5 was the point with coordinates (0·2043, 0·3758, 0·2043) and the points generated from it by the operations of the point group of the crystal. This crown is illustrated in fig. 42 and good agreement with the experimental results was obtained, see table 22. Infra-red magnetoreflectivity experiments of Maltz and Dresselhaus (1968) suggested that there might be another small

Fig. 42

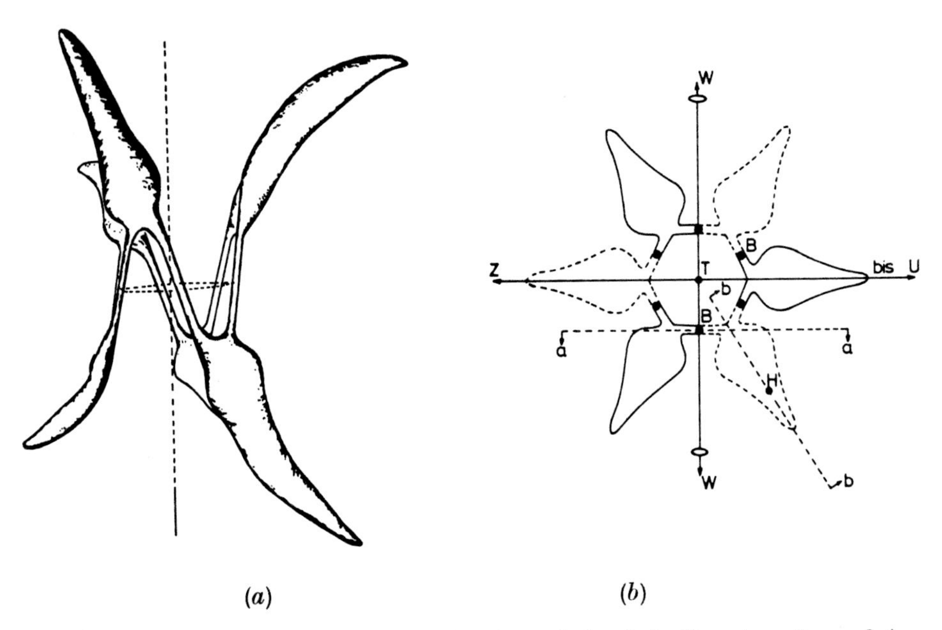

(*a*) Perspective drawing and (*b*) projection of the hole Fermi surface of As (Lin and Falicov 1966).

pocket of carriers in As, containing only about 1% of the total carriers; the measured band gaps at T did not agree with the calculation of Lin and Falicov (1966). Measurements of γ include:

$(0{\cdot}194 \pm 0{\cdot}007)$ mJ mole^{-1} deg^{-2} Culbert (1967),

$(0{\cdot}191 \pm 0{\cdot}001)$ mJ mole^{-1} deg^{-2} Taylor *et al.* (1967),

which are to be compared with calculated values of from 0·184 to 0·188 mJ mole^{-1} deg^{-2} (Taylor *et al.* 1967). Under pressure the carrier density in As is thought to decrease (Brandt and Minina 1968) as occurs in Bi (see § 7.2). The de Haas–van Alphen effect in doped As has been measured by Ishizawa (1968 b) and spin-splitting has been observed in the 'giant' quantum oscillations in the magnetoacoustic attenuation in As by Fukase (1969).

The amount of experimental work performed on Sb is intermediate between the vast amount of work which has been done on Bi and the very small amount which has been done on As. Prior to the band structure calculation of Falicov and Lin (1966) there already existed a substantial body of experimental results. These experiments included work on the de Haas–van Alphen effect (Shoenberg and Uddin 1936 a, Shoenberg 1952, Saito 1963 a, Ishizawa and Tanuma 1965 b, Windmiller and Priestley 1965, Yamaguchi and Tanuma 1965), the Shubnikov–de Haas effect (Ketterson and Eckstein 1963, Lerner and Eastman 1963, Rao *et al.* 1964), cyclotron resonance (Datars 1961, 1962 a, Datars and Dexter 1961, Everett 1964, Datars and Vanderkooy 1964), magnetoacoustic effects (Eckstein 1963, Ketterson 1963, Beckman *et al.* 1964, Eriksson *et al.* 1964, Eckstein *et al.* 1964), anomalous skin effect (Greenfield *et al.* 1959), Shubnikov–de Haas oscillations in the optical magnetoreflection coefficient (Dresselhaus and Mavroides 1964 a) and in the infra-red absorption (Nanney 1963), electron spin resonance (Smith *et al.* 1960, Datars 1962 b) and galvanomagnetic effects (Steele 1955, Freedman and Juretschke 1961, Datars and Eastman 1962, Epstein and Juretschke 1963, Rao *et al.* 1964, Harris and Corrigan 1965, Long *et al.* 1965). Spin-splitting of the Landau levels in Sb has been observed by McCombe and Seidel (1967) in the magnetothermal oscillations. The carrier density in Sb is slightly higher than in Bi and the following values have been collected by Öktü and Saunders (1967):

$1{\cdot}29 \times 10^{-3}$ per atom	Eriksson *et al.* (1964)
$1{\cdot}32 \times 10^{-3}$ per atom	Epstein and Juretschke (1963)
$1{\cdot}25 \times 10^{-3}$ per atom	Ketterson and Eckstein (1963)
$1{\cdot}55 \times 10^{-3}$ per atom	Rao *et al.* (1964)
$1{\cdot}29 \times 10^{-3}$ per atom	Öktü and Saunders (1967)
$1{\cdot}69 \times 10^{-3}$ per atom	Windmiller and Priestley (1965)

and, like Bi, the semi-metal Sb is generally assumed to be compensated. A table with a comparison of the periods obtained from several different oscillatory experiments in Sb is given by Dresselhaus and Mavroides (1964 a) The results of all these various experiments were interpreted in terms of a Fermi surface for the electrons based on the model of three tilted ellipsoids which was described by Shoenberg (1952) and which is very similar to the Fermi surfaces for electrons in As and Bi. However, the interpretation of these results for the Fermi surface of holes was less unambiguous. Some results were interpreted in terms of a Fermi surface for holes consisting of a set of three distorted ellipsoids with only a slight tilt while other results were interpreted in terms of one ellipsoid of light holes and one ellipsoid of heavy holes.

The pseudopotential approach used by Lin and Falicov (1966) for As which we have already described was also used for Sb. The pseudopotential used for Sb by Falicov and Lin (1966) was obtained by using the pseudopotential of InSb and neglecting the differences in the crystal environment of the Sb. The pseudopotential for InSb was obtained as a sum of two contributions: (i) a symmetric part assumed to be the average pseudopotential of In and Sb, namely the pseudopotential of grey Sn, and (ii) an antisymmetric part which is equal to the difference between the atomic pseudopotentials of Sb and In. The pseudopotential of InSb constructed in this way was then renormalized by the ratio of the unit cells of InSb and of Sb to give the pseudopotential which was used for Sb. The detailed band structure that was obtained is shown in fig. 3 of the paper

Fig. 43

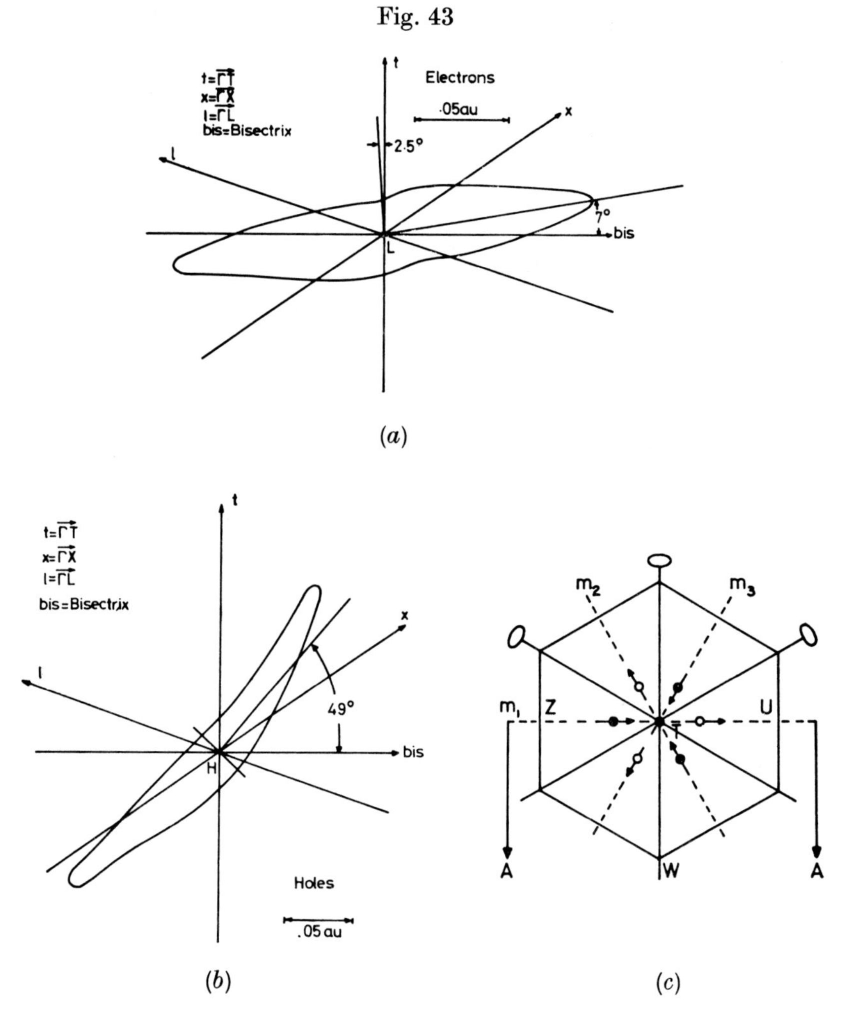

Cross sections of the calculated Fermi surface of Sb (Falicov and Lin 1966).

by Falicov and Lin (1966). The pockets of electrons in band 6 were found to consist of distorted ellipsoids at L with a tilt angle of $-7°$; a vertical section of this surface is shown in fig. 43 (*a*) and its dimensions are given in table 23. The Fermi surface of holes in Sb obtained from the calculation of Falicov and Lin (1966) is different from both of those of As and Bi. It consists of six pockets of holes arranged around the point T (again in contradiction, like As, with the results of the earlier calculation of Mase (1958, 1959 a)) and obviously related to the six pockets of holes in As but now in Sb no longer connected into a crown as they were in As. The point in each individual pocket corresponding to the highest energy in band 5 is the point H with coordinates (0·2452, 0·3929, 0·2452) and the points generated from H by the operations of the point group of the crystal.

Table 23. Dimensions of the Fermi surface of Sb

(*a*) Electrons		
	Experiment	Theory§
Tilt angle of maximum area	$-6{\cdot}0°$†	$-7°$
Tilt angle of minimum area	$+87{\cdot}7°$	$+87{\cdot}5°$
Binary effective mass m_{11}	0·093‡	0·09
Ratio of maximum to minimum area in trigonal–bisectrix plane	6·37†	~6·7
Binary cyclotron mass m_c	0·32‡	~0·2
(*b*) Holes		
Tilt angle of minimum area	53·0°†	~41°
Binary effective mass m_{11}	0·068‡	0·06
Ratio of binary area to minimum area	3·52†	~4·8
Binary cyclotron mass	0·215‡	~0·3

† Windmiller (1966).
‡ Datars and Vanderkooy (1964) except that the assignments of electrons and holes have been interchanged.
§ Falicov and Lin (1966).
(Adapted from Falicov and Lin (1966).)

These pockets are illustrated in fig. 43 (*b*) and (*c*). Subsequent to the band structure and Fermi surface calculation by Falicov and Lin (1966) some further experimental results became available. The dimensions of the electron and hole Fermi surfaces obtained from the de Haas–van Alphen results of Windmiller (1966) are compared with the results of the calculations of Falicov and Lin in table 23. The period for the holes was also observed in the magnetothermal oscillations by Noguchi and Tanuma (1967). A compilation of other experimental values for the tilt angles for Sb is given in table 8 of Öktü and Saunders (1967). The direct identification of the types of carriers as electrons or holes appropriately was performed by Ishizawa and Tanuma (1965 a) by doping pure Sb with varying amounts of Te (donor) or Sn (acceptor) and observing the changes in the sizes of the pieces of Fermi surface by means of the de Haas–van Alphen effect (see also Ishizawa 1968 a). In addition to the numerical values in table 23 a much more sensitive test was that the general features of the angular variation of the de Haas–van Alphen periods observed by Windmiller (1966) were in good agreement with those expected from the Fermi surface proposed for Sb by Falicov and Lin (1966). Acoustic cyclotron resonance measurements have been made by Korolyuk *et al.* (1968) in Sb. Magneto-acoustic geometric resonances in Sb, with the magnetic field parallel to the trigonal axis, were observed by Miller (1966) with a period of $(44 \pm 2) \times 10^{-4}\ \text{G}^{-1}$ compared with $47 \times 10^{-4}\ \text{G}^{-1}$ calculated from the model described above for the hole Fermi surface in Sb; 'giant' quantum oscillations in the magnetoacoustic attenuation were observed by Kazanskii and Korolyuk (1966).

Values of γ for Sb obtained by various workers include:

$(0{\cdot}112 \pm 0{\cdot}005)$ mJ mole^{-1} deg^{-2} Culbert (1967)

$(0{\cdot}105 \pm 0{\cdot}014)$ mJ mole^{-1} deg^{-2} McCollum and Taylor (1967)

$\left.\begin{matrix}(0{\cdot}117 \pm 0{\cdot}006)\\(0{\cdot}123 \pm 0{\cdot}010)\end{matrix}\right\}$ mJ mole^{-1} deg^{-2} Zebouni and Blewer (1967).

New measurements of the galvanomagnetic properties of Sb (Öktü and Saunders 1967) and of the Seebeck coefficient (Saunders and Öktü 1968) could be explained adequately on the basis of the Fermi surface proposed by Falicov and Lin (1966), although other galvanomagnetic measurements by Tanaka *et al.* (1968) suggested the possibility of the existence of an extra pocket of holes at T. A non-ohmic behaviour of the magneto-resistance of Sb was observed by Bogod and Krasovitskii (1968) which was similar to that observed in Bi by Esaki (1962) which we mentioned in § 7.2 (see also Datars and Eastman (1962) and Eastman and Datars (1963)). Two distinct series of optical interband transitions in Sb have been observed by Dresselhaus and Mavroides (1965) and evidence was presented for the existence of energy bands lying close to the Fermi surface consistent with the band structure calculation of Cohen *et al.* (1964). Investigations of the pressure dependence of the cross-sectional area of the Fermi surface of Sb by Minomura *et al.* (1969) using the de Haas–van Alphen effect showed that the cross-sectional area increases with pressure, which is the opposite of that observed in Bi (see § 7.2) and As, but is similar to that observed in graphite (see § 6.2.) The corresponding increase in carrier density with pressure was observed by Brandt and Minina (1968). However, according to the galvanomagnetic work of Kechin (1967), at pressures up to 10 kbars the ellipsoids in Sb become elongated but thinner so that the total number of carriers is very nearly constant (see also Kechin *et al.* (1965)). The temperature dependence of the magnetoresistance of Sb was measured by Bogod and Eremenko (1967 b).

§ 8. Conclusion

8.1. *Pseudopotentials, Electron–Electron Interactions and Electron–Phonon Interactions*

Although it may become intensely absorbing to the individual worker who is actually involved, it has to be remembered that the determination of the shape of the Fermi surface of a metal is not just an end in itself. The importance of determining the shape of the Fermi surface of a metal is to try to obtain a better understanding of all those physical properties of a metal that depend in some way on the behaviour of the conduction electrons in the metal. Some of these properties of a metal are such that measurements of their behaviour give direct information about the geometry of the Fermi surface, either in the form of the extremal areas of cross section or of caliper dimensions of the Fermi surface. Such experiments rely on the ability to construct a physical situation in which there is some interaction between the outside world and a selected group of electrons on the Fermi surface. These are the properties that we have described already in this article in connection with the various experimental determinations of detailed Fermi surface geometry and which we may describe as *microscopic* or *direct* properties; they include the various size effects and the 'quantum oscillations' in a variety of properties as a function of applied magnetic field, of which the best known are the de Haas–van Alphen effect and the Shubnikov–de Haas effect. There are also some properties which, although they do not give quantitative geometrical information such as radii or cross-sectional areas, give qualitative geometrical information about the Fermi surface. For example, observation of the magnetoresistance of a metal at high magnetic fields yields information about whether open orbits are possible in the plane normal to the magnetic field and hence whether the Fermi surface is multiply-connected in one particular plane. There is a second set of properties of a metal in which it is not possible to isolate the contributions from individual groups of electrons and the observed magnitude of any experimental number will arise from contributions from all the electrons on the Fermi surface of the metal. Although the magnitudes of these properties give no direct information about the geometry of the Fermi surface they still depend on the behaviour of the electrons in the metal and in particular on the behaviour of those electrons near the Fermi surface. We may describe these as *global* and *indirect* (Mercouroff 1967) or *macroscopic* properties. Although these properties generally give much less direct information about the geometry of the Fermi surface we have, nevertheless, also made use of some of them in this article. For example, measurement of the electronic contribution to the specific heat gives a measure of the density of states at the Fermi surface. These macroscopic properties were discussed in connection with the alkali metals in § 3.3. These properties also include mechanical, electrical, optical, thermal, and low-field or spontaneous magnetic properties and various transport properties and the theoretical expression for the magnitude of an experimental measurement of a parameter describing one of these properties will involve a surface integral of some function over the whole of the Fermi surface. It is, of course, possible to use the experimental measurements of one of these properties to test various proposed Fermi surface geometries by seeing which one gives the closest agreement, when substituted into the integral, with the experimental value. This is obviously rather tedious and is no longer used in Fermi surface determinations because of the existence of the more direct methods described earlier; however, it was used in the classic work by Pippard (1957) on the determination of the shape of the Fermi surface of Cu by measuring the surface impedance in the anomalous skin-effect régime. In general it now seems possible that the present state of knowledge of the Fermi surfaces of at least the simple metals is sufficiently detailed that it should be possible to make fairly detailed theoretical predictions of the magnitudes of these various macroscopic and indirect properties.

It is a well-known feature of *ab initio* band structure calculations that in the present state of the subject the least accurate part of the process is the construction of a realistic potential $V(\mathbf{r})$, including exchange and correlation effects, in which an electron in the individual particle approximation is assumed to move. The degree of sophistication and reliability of the numerical methods now used in the solution of the appropriate equations to determine the electronic wave functions $\psi_{\mathbf{k}}(\mathbf{r})$ and the energy eigenvalues $E_{\mathbf{k}}$ far exceeds the accuracy of the initial potential $V(\mathbf{r})$. Although the construction of a potential $V(\mathbf{r})$ *ab initio* has been performed quite carefully in a number of cases (for example, by Heine (1957 a, b, c) for Al) the results have generally been disappointing. While the gross features of the band structure are fairly well determined in this way it often happens that the uncertainties in the potential $V(\mathbf{r})$ produce uncertainties in $E_{\mathbf{k}}$ which are quite significant for just those parts of the bands that are near to the Fermi energy E_{F}. The calculation of the exchange and correlation contributions to $V(\mathbf{r})$ not only involves approximations because of the necessity to truncate the expressions but also because of uncertainties about the details of the electronic wave functions as well. We have seen in the previous sections of this article that several of the more successful band structure calculations have been made with the use of pseudopotentials rather than by trying to construct a potential $V(\mathbf{r})$ from first principles. We have already mentioned the concept of a pseudopotential in § 2.1. One of the main advantages of using a pseudopotential is that a realistic pseudopotential for any one of the simple metals can be expressed in terms of quite a small number of parameters which can be adjusted until the calculated Fermi surface provides the best agreement with a set of experimental measurements of caliper dimensions or extremal areas of cross section of the Fermi surface. For example, this has been applied to the alkali metals by Lee (1969), to Al by Ashcroft (1963 a, b), to Mg by Kimball *et al.* (1967) as well as to several other metals, as we have mentioned in various previous sections. To the purist a pseudopotential calculation mav lack some of the elegance of a completely *ab initio* band structure calculation, but it should not be despised on these grounds. One can, if one wishes, regard the pseudopotential method as a sophisticated interpolation scheme which is based on physical rather than mathematical ideas and which enables the shape of the entire Fermi surface to be determined numerically from the necessarily rather small number of experimentally determined caliper dimensions or extremal areas of cross section. Having used the various direct experimental measurements to determine the parameters that specify the pseudopotential, the effect of the conduction electrons on the macroscopic and indirect properties can then be expressed in terms of the pseudopotential; see, for example, Holland (1968) on the use of pseudopotentials in calculating Knight shifts for Li, Na, Cs, Al and Ga.

The effect of electron–phonon interactions is often an important contribution to the magnitude of one of the macroscopic or indirect properties of a metal, and one important aspect of the use of pseudopotentials is that the same pseudopotential that is used to calculate the electronic band structure of a metal can also be used to calculate the lattice vibration spectrum (phonon dispersion relations) of the metal (see, for example, Vosko *et al.* (1965) on Na, Al and Pb, Animalu *et al.* (1966 b) on the alkali metals and Al, Schneider and Stoll (1966 a, b, 1967 a) on the alkali metals and Pb, Ho and Ruoff (1967) on Na, and Stafleu and de Vroomen (1967 b) on Sn). Therefore, recently some workers have attempted to use the phonon dispersion relations, as determined experimentally by inelastic neutron scattering, to determine the pseudopotential which can then subsequently be used to calculate the electronic band structure. Conversely, other people have used the measurements of Fermi surface dimensions to determine the parameters of a pseudopotential used in a band structure calculation; this pseudopotential can then be used to calculate the phonon dispersion relations. So far this kind of work has been limited to the alkali metals and one or two others. There is considerable scope for the extension of this work to the more complicated metallic elements. The phonon dispersion curves of K have been measured by inelastic neutron scattering by Cowley *et al.* (1966) and they analysed their results in terms both of conventional Born–von Kármán models and of pseudopotentials; their pseudopotentials were found to be in close agreement with the results of the pseudopotential calculation of Animalu *et al.* (1966 b). The experimental measurements, by inelastic neutron scattering, of the phonon dispersion curves of Na (Woods *et al.* 1962) were analysed by the pseudopotential method by Cochran (1963). Similar experimental measurements on Al were made by Brockhouse *et al.* (1962). These experimental results for Na and Al were also found to be in good agreement with the theoretical phonon dispersion relations calculated by Animalu *et al.* (1966 b) using a Heine–Abarenkov type of pseudopotential. The screened pseudopotentials in K (Cowley *et al.* 1966) and in Na (Cochran 1963) were found to change sign very close to $2k_F$ so that the magnitude of the Kohn effect is expected to be very small for both Na and K, as is indeed observed experimentally. The experimental measurements of the phonon dispersion relations of Li by neutron scattering were used by Wallace (1969) to determine a pseudopotential for Li. A model pseudopotential for Pb was used by Harrison (1965) and its two parameters were adjusted to fit two selected phonon frequencies; Kohn anomalies in the phonon dispersion relations were then predicted and were in order of magnitude agreement with those observed experimentally. Pseudopotentials have also been used to calculate the phonon dispersion relations of Be (Sahni *et al.* 1966) and Mg (Roy and Venkataraman 1967).

Having determined the forms of the band structure and the phonon dispersion relations of a metal it is then important, for an understanding of many of the physical properties of a metal, to investigate the interactions among the electrons and the interactions between the electrons and the phonons. Electron–electron interactions are considered, for example, in the review by Pines (1955) and electron–phonon interactions in the book edited by Bak (1964). These interactions too can be studied in terms of the pseudopotential. The electron–phonon interaction is particularly important in connection with the ordinary electronic transport properties of normal metals (for a development of the formal theory see Ziman (1960)) and in connection with the existence of superconductivity (Bardeen *et al.* (1957)). In addition to these properties there will also be significant contributions to other properties, such as the electronic specific heat, as a result of electron–phonon interactions. There exist several discussions of pseudopotentials, in connection with both Fermi surfaces and electron–phonon interactions (see, for example, Sham and Ziman (1963), Ziman (1964), Harrison (1966)). Although the formal theory of transport coefficients or superconductivity in terms of electron–phonon interactions is highly developed and for the most part fairly well understood, the real problem that remains is to produce reliable quantitative results using realistic band structures, Fermi surfaces and phonon dispersion relations. A considerable number of calculations have been made of transport coefficients, mostly on the alkali metals. Several references are cited in § 1 of the paper by Robinson (1968); in addition to those, we mention one or two other calculations. Collins and Ziman (1961) assumed that the phonon dispersion curves for the alkali metals, which had not then been determined experimentally, obey a 'law of corresponding states' and calculated the electrical and thermal resistivities and the phonon-drag thermoelectric power. The observed values of these transport coefficients were shown to be consistent with a Fermi surface that is quite distorted in Li, becomes nearly spherical in Na and K, and again is distorted in Rb and Cs. Some work has also been done on the phonon-drag thermoelectric power in Bi (Kuznetsov and Shalyt 1967) and in Sb (Red'ko and Shalyt 1968). Darby and March (1964) used the measured phonon dispersion relations in Na (Woods *et al.* 1962) to calculate the magnitude and temperature dependence of the electrical resistivity of Na. They obtained good agreement with the experimental results of Dugdale and Gugan (1960, 1962) by assuming a spherical Fermi surface and only considering anisotropy in the phonon dispersion relations. A modified pseudopotential of the form given by Heine and Abarenkov (1964) was used by Bonsignori and Bortolani (1966) to calculate the resistivity of K and the results agreed with the experimental results of Dugdale and Gugan (1962) to within 10% over the range of 0°K to 300°K; the results of similar calculations by Bonsignori and Bortolani (1966) for Na showed a less satisfactory agreement with experiment. However, it is probably not too unfair to say that, at present, most of these calculations of transport coefficients give more information about how to construct sensible pseudopotentials than about the reliability or otherwise of the theory of transport properties (see, for example, Robinson and Dow (1968)). The phonon dispersion curves and a pseudopotential determined from normal-state data have been used by Carbotte and Dynes (1967, 1968) to calculate directly the superconducting energy gap and transition temperature for a number of simple metals. Values of the critical temperature for Al and Pb were in good agreement with experimental values, with errors of only about 10%, and it was shown that if Na and K become superconducting this can only happen at temperatures much lower than 10^{-5} °K.

We have also mentioned that electron–electron interactions and electron–phonon interactions can make appreciable contributions to other macroscopic properties of a metal, such as the electronic specific heat or the attenuation of ultrasonic waves (in the absence of a magnetic field). Again, most of the quantitative calculations have been performed for the alkali metals or for Al. The effect of electron–electron interactions on the electronic specific heat (expressed in terms of a thermal effective mass) of the alkali metals was calculated by Silverstein (1962, 1963), while the same was done for electron–phonon interactions by Animalu *et al.* (1966a), see table 24. Using the results of the de Haas–van Alphen measurements of the Fermi surfaces of K and Rb (see § 3.4) Ashcroft (1965) used pseudopotentials to determine the contributions to the total observed thermal effective mass arising from electron–electron interactions, $(\delta m/m)_{\mathrm{e-e}}$, from electron–phonon interactions, $(\delta m/m)_{\mathrm{e-p}}$, and from the band structure, $(m^*/m)_{\mathrm{BS}}$. The total effective mass $(m^*/m)_{\mathrm{total}}$ is then given by:

$$\left(\frac{m^*}{m}\right)_{\mathrm{total}} = \left(\frac{m^*}{m}\right)_{\mathrm{BS}} \left\{1 + \left(\frac{\delta m}{m}\right)_{\mathrm{e-e}} + \left(\frac{\delta m}{m}\right)_{\mathrm{e-p}}\right\}. \qquad (8.2.1)$$

Similar calculations of the total effective mass were made for Na, Al and Pb by Ashcroft and Wilkins (1965) and for In by Ashcroft and Lawrence (1968). The results are collected in table 24, from which it can be seen that the electron–phonon enhancement of the specific heat effective mass $(m^*/m)_{\mathrm{SH}}$ is fairly small for Na, K, Rb and Cs but quite large for Li, Al and In and very large for Pb. The values of $(\delta m/m)_{\mathrm{e-e}}$ used in these calculations were obtained by Ashcroft by linear extrapolation from the work of Rice (1965) on electron–electron interactions. In some later work Rice used several different treatments of the electron–electron interactions and of the electron–phonon interactions to calculate the parameters for Na and K in the Landau Fermi-liquid theory; the results are given in tables VII and VIII of Rice (1968). Experimental values of two of the parameters, B_0 and B_1, in the Landau Fermi-liquid theory were obtained by Dunifer *et al.* (1968) from the effect of spin waves on the line shapes in conduction–electron spin resonance in Na and K. We have mentioned at various points in previous sections (see, for example, § 5.2 on Al) the use of ultrasonic attenuation, in the absence of any magnetic field, as a rather direct method of investigating electron–phonon interactions and hence the origins of superconductivity.

Table 24. Electron–electron and electron–phonon contributions to the thermal effective mass for various metals

	$(m^*/m)_{BS}$	$(\delta m/m)_{e-e}$	$(\delta m/m)_{e-p}$	$(m^*/m)_{calc}$			$(m^*/m)_{exp}$
Li	1·64[1]	0·03[2]	0·34[2]	2·25†	1·96[3]	1·78[4]	2·19[5]
Na	1·00[6]	0·06[6]	0·18[6]	1·24[6]	1·15[3]	1·13[4]	1·25[5]
K	0·94†	0·11[2]	0·15[2]	1·18[2]	1·35[3]	1·24[4]	1·23[7]
Rb	1·06†	0·13[2]	0·17[2]	1·38[2]	1·56[3]	1·38[4]	1·36[5]
Cs	1·75[1]	0·15[2]	0·18[2]	2·33†	2·86[3]	1·97[4]	1·63[5]
Al	1·06[6]	−0·01[6]	0·49[6]	1·57[6]	—	—	1·60[8]
In	0·91[8]	~0·0[8]	0·60[8]	1·45[8]	—	—	1·46[9]
Pb	0·90[8]	0·00[6]	1·05[6]	1·85[8]	—	—	2·20[8]

† Calculated from eqn. (8.2.1) and the other data in this table.

[1] Ham (1962 b); [2] Ashcroft (1965); [3] Silverstein (1962, 1963); [4] Animalu *et al.* (1966 a); [5] Martin (1961 b), Sundström (1965); [6] Ashcroft and Wilkins (1965); [7] Filby and Martin (1965); [8] Ashcroft and Lawrence (1968); [9] Clement and Quinnell (1953).

Experimental evidence for the importance of electron–electron interactions in the resistivity of simple metals was obtained by Garland and Bowers (1968) who observed the presence of a T^2 term, due to electron–electron interactions, in the resistivity of Al and In at low temperatures, but not in Na or K. A pseudopotential method was used by Wallace (1968) to calculate the thermal expansion coefficient of Na and K and qualitative agreement with experimental results was obtained.

8.2. *Gaps in the Present Knowledge of the Topology of the Fermi Surfaces of s-block and p-block Metals*

As we have seen in earlier sections, the general features of the Fermi surface topology of the s-block and p-block elements are fairly well established, although there still remain some details to be clarified by further experimental or theoretical work.

Group IA. The dimensions of the Fermi surfaces of the alkali metals have been determined with considerable accuracy. One of the few remaining unsettled points seems to be the question of whether or not Overhauser's spin-density waves (or charge-density waves) actually exist (Thomas and Turner 1968), see § 3.9.

Group IIA. Similarly the dimensions of the Fermi surfaces of Be and Mg have been measured with a considerable degree of accuracy. There is still room for a considerable amount of new and accurate experimental and theoretical work on Ca, Sr and Ba. While the Fermi surface given by Vasvari (1968) for Ca is probably qualitatively correct the details still remain to be established. The Fermi surface of Sr is also probably qualitatively similar to that proposed by Vasvari (1968) for Ca, but there is still scope for both band structure calculations and experimental Fermi surface determinations, even of a fairly approximate nature, for Sr and Ba.

Group IIIB. The details of the Fermi surfaces of Al and In appear to be well established. In Ga, which it will be recalled has a very complicated structure, there remains the problem of determining how many of the smaller pieces of the Fermi surface in bands 3, 4, 6, 7, 8 and 9 in the free-electron model actually survive in the real metal. Again, for Tl the general features of the larger pieces of Fermi surface are well established but there are still some unsolved problems in relation to some of the smaller pieces of the Fermi surface.

Group IVB. The details of the Fermi surfaces of graphite and Pb are well established but for Sn, which has a complicated structure, there still remains some scope for further work on determining precisely the dimensions of some of the pieces of Fermi surface although it appears that the general features are well established, see § 6.3.

Group VB. The details of the dimensions and locations of the various small pockets of electrons and holes which make up the Fermi surface of As, Sb or Bi are well known.

Therefore, we would summarize the remaining problems in determining the details of the geometry of the Fermi surfaces of the s-block and p-block metals:

(i) a conclusive test of Overhauser's spin-density wave (charge-density wave) theory for alkali metals,

(ii) a detailed checking of the Fermi surface proposed by Vasvari (1968) for Ca,

(iii) a determination of the gross features and details of the Fermi surfaces of Sr and Ba (and Fr and Ra),

(iv) an accurate determination of the dimensions of some of the smaller pieces of the Fermi surfaces of Ga, Tl and Sn.

As we nave noted at various points in the preceding sections, several workers have investigated the effect of pressure or tension on the shape of the Fermi surface of a metal. There is clearly scope for further experiments and calculations on the effect of pressure on the Fermi surfaces of the remaining metals. In the most simple-minded approach, the effect of pressure or tension is just to change the lattice constants of the crystal. There is therefore a corresponding change in the size of the Brillouin zone; in the free-electron approximation the linear dimensions of the Fermi surface will then also be scaled by the same factor as the linear dimensions of the Brillouin zone. However, in a real metal, changing the lattice constant will also change the potential $V(\mathbf{r})$ experienced by a conduction electron in the metal, and at ultra-high pressures ionization of the core may occur. When this occurs there is an order-of-magnitude increase in the electrical resistivity and this has been observed experimentally in K, Rb and Cs by Stocks and Young (1969). Since the determination of $V(\mathbf{r})$ is the most difficult part of *ab initio* band structure calculations at present, any attempts to determine quantitatively the effect of pressure on $V(\mathbf{r})$, and thence on the Fermi surface, from first principles are bound to be very hazardous. Even in the comparatively simple case of a compensated semi-metal one finds that while very high pressure reduces the overlap between the valence and conduction bands in Ca or Bi it increases the overlap in graphite. There is also scope for studying the effect of impurities on the shape of the Fermi surfaces of various metals, if this can be shown to be useful. Doping with impurities has long been used as an aid to Fermiology studies, right from the time of the early work on Bi by Jones (1934). Positron annihilation is one of the few direct Fermi surface techniques that is suitable for use on metals in the superconducting state. We have already mentioned in § 6.4 the positron annihilation experiments of Briscoe *et al.* (1966) to investigate the Fermi surface of superconducting Pb; there is scope for extension of these experiments to other superconducting metals (see also Dekhtjar (1969)).

On the theoretical side there is considerable scope for further attempts to produce quantitatively satisfactory calculations of transport properties and other macroscopic properties of metals with complicated Fermi surfaces, while it is clear that the use of realistic pseudopotentials in quantitative calculations of superconducting properties is only just beginning. We have seen in the previous section that it is only for the alkali metals and one or two other metals that quantitative calculations have so far met with any degree of success. It is also now becoming popular to attempt to perform iterative band structure calculations that are self-consistent solutions of the Hartree–Fock equations for a metal (see, for example, Chow and Kleinman (1969) on Al) and it seems that, at last, quantitatively meaningful determinations of the binding energy of a metallic solid may emerge.

Note added in proof.—In the course of obtaining permission to reproduce copyright figures and tables several miscellaneous pieces of information emerged. The author is grateful to the workers mentioned for their comments.

(i) Further measurements have been made on the Knight shift and its temperature dependence for Mg (P. D. Dougan, S. N. Sharma and D. L. Williams, *Can. J. Phys.*, **47**, 1047 (1969)) and for Sn (S. N. Sharma, Thesis, University of British Columbia) (E. P. Jones).

(ii) New cellular band structure calculations for Ca suggest that there may be pockets of holes in band 1 at both K and W rather than just at W as stated in § 4.4 (S. L. Altmann).

(iii) A new relativistic A. P. W. band structure calculation for Ba (G. Johansen, *Solid st. Commun.*, **7**, 731 (1969)) produced a Fermi surface in the b.c.c. phase which consisted of a pocket of electrons at H in band 2 (described as a 'superegg') and a pocket of holes at P in band 1 (described as a 'tetracube'). These pieces of Fermi surface are considerably larger than the pockets in Ca described in § 4.4. A similar calculation for the hypothetical f.c.c. phase of Ba produced a band structure with only a very small overlap between band 1 and band 2 similar to the bands for f.c.c. Ca and Sr mentioned in § 4.4 (G. Johansen).

(iv) In connection with table 13 Cochran and Lyall have recently remeasured the anomalous skin depth in polycrystalline In and obtained new values of $\sigma/\bar{\lambda}$ much closer to their previous low frequency results than to the microwave measurements of Dheer quoted in table 13 (J. F. Cochran and K. R. Lyall).

(v) The identification of the carriers in graphite as electrons and holes in § 6.2 should probably be reversed; this was concluded from magnetoreflection experiments on graphite (P. R. Schroeder, M. S. Dresselhaus, and A. Javan, *Phys. Rev. Lett.*, **20,** 1292 (1968)) using polarized infra-red radiation from a gas laser (J. W. McClure).

(vi) In connection with several metals we have mentioned the use of pseudopotential band structure calculations as interpolation schemes for the parametrization of Fermi surfaces. It has also been suggested (B. Segall and F. S. Ham, *Methods in Computational Physics*, **8,** 251 (1968)) that the Green's function (or K.K.R.) method is also very suitable as a parametrization scheme for band structures and Fermi surfaces (F. S. Ham).

References

Abarenkov, I. V., and Heine, V., 1965, *Phil. Mag.*, **12,** 529.

Abelès, B., and Meiboom, S., 1956, *Phys. Rev.*, **101,** 544.

Abrikosov, A. A., and Fal'kovskiĭ, L. A., 1962, *Zh. éksp. teor. Fiz.*, **43,** 1089. English translation : *Soviet Phys. JETP*, **16,** 769.

Adams, E. N., and Holstein, T. D., 1959, *J. Phys. Chem. Solids*, **10,** 254.

Agababyan, K. Sh., Mina, R. T., and Pogosyan, V. S., 1968, *Zh. éksp. teor. Fiz.*, **54,** 721. English translation : *Soviet Phys. JETP*, **27,** 384.

Ahlers, G., 1966, *Phys. Rev.*, **145,** 419.

Aleksandrov, B. N., 1962, *Zh. éksp. teor. Fiz.*, **43,** 399. English translation : *Soviet Phys. JETP*, **16,** 286.

Alekseevskiĭ, N. E., and Egorov, V. S., 1963 a, *Zh. éksp. teor. Fiz.*, **45,** 388. English translation : *Soviet Phys. JETP*, **18,** 268 ; 1963 b, *Ibid.*, **45,** 448. English translation : *Soviet Phys. JETP*, **18,** 309 ; 1964, *Ibid.*, **46,** 1205. English translation : *Soviet Phys. JETP*, **19,** 815 ; 1968 a, *Ibid.*, **55,** 1153. English translation : *Soviet Phys. JETP*, **28,** 601 ; 1968 b, *Ibid., Pis'ma*, **8,** 301. English translation : *Soviet Phys. JETP Lett.*, **8,** 185.

Alekseevskiĭ, N. E., Egorov, V. S., and Dubrovin, A. V., 1967, *Zh. éksp. teor. Fiz., Pis'ma*, **6,** 793. English translation : *Soviet Phys. JETP Lett.*, **6,** 249.

Alekseevskiĭ, N. E., and Gaĭdukov, Yu. P., 1959 a, *Zh. éksp. teor. Fiz.*, **36,** 447. English translation : *Soviet Phys. JETP*, **9,** 311 ; 1959 b, *Ibid.*, **37,** 672. English translation : *Soviet Phys. JETP*, **10,** 481 ; 1960, *Ibid.*, **38,** 1720. English translation : *Soviet Phys. JETP*, **11,** 1242 ; 1961 a, *Ibid.*, **41,** 354. English translation : *Soviet Phys. JETP*, **14,** 256 ; 1961 b, *Ibid.*, **41,** 1079. English translation : *Soviet Phys. JETP*, **14,** 770 ; 1962, *Ibid.*, **43,** 2094. English translation : *Soviet Phys. JETP*, **16,** 1481.

Alekseevskiĭ, N. E., Gaĭdukov, Yu. P., Lifshitz, I. N., and Peschanskiĭ, V. G., 1960, *Zh. éksp. teor. Fiz.*, **39,** 1201. English translation : *Soviet Phys. JETP*, **12,** 837.

Alekseevskiĭ, N. E., and Kostina, T. I., 1965, *Zh. éksp. teor. Fiz.*, **48,** 1209. English translation : *Soviet Phys. JETP*, **21,** 807.

Alexander, E., Feller, S., Fraenkel, B. S., and Perel, J., 1965, *Nuovo Cim.*, **35,** 311.

Alig, R. C., Quinn, J. J., and Rodriguez, S., 1965, *Phys. Rev. Lett.*, **14,** 981 ; 1966, *Phys. Rev.*, **148,** 632.

Alstadheim, T., and Risnes, R., 1968, *Phil. Mag.*, **18,** 885.

Altmann, S. L., and Cracknell, A. P., 1964, *Proc. phys. Soc.*, **84,** 761.

Alty, J. L., and Stringer, J., 1969, *Phys. Stat. Sol.*, **32,** 243.

Amundsen, T., and Seeberg, P., 1969, *J. Phys.* C (*Solid St. Phys.*), **2,** 694.

Anderson, J. G., and Young, R. C., 1968, *Phys. Rev.*, **168,** 696.

Anderson, J. R., and Gold, A. V., 1965, *Phys. Rev.*, **139,** A1459.

ANDERSON, J. R., O'SULLIVAN, W. J., and SCHIRBER, J. E., 1967, *Phys. Rev.*, **153,** 721.
ANDERSON, J. R., O'SULLIVAN, W. J., SCHIRBER, J. E., and SOULE, D. E., 1967, *Phys. Rev.*, **164,** 1038.
ANDERSON, W. T., RUHLIG, M., and HEWITT, R. R., 1967, *Phys. Rev.*, **161,** 293.
ANIMALU, A. O. E., 1967, *Phys. Rev.*, **161,** 445.
ANIMALU, A. O. E., BONSIGNORI, F., and BORTOLANI, V., 1966 a, *Nuovo Cim.* B, **42,** 83 ; 1966 b, *Ibid.*, **44,** 159.
ANIMALU, A. O. E., and HEINE, V., 1965, *Phil. Mag.*, **12,** 1249.
ANTCLIFFE, G. A., and BATE, R. T., 1967, *Phys. Rev.*, **160,** 531.
APPELBAUM, J. A., 1966, *Phys. Rev.*, **144,** 435.
ARKHIPOV, R. G., KECHIN, V. V., LIKHTER, A. I., and POSPELOV, YU. A., 1963, *Zh. éksp. teor. Fiz.*, **44,** 1964. English translation : *Soviet Phys. JETP*, **17,** 1321.
ASHCROFT, N. W., 1963 a, *Physics Lett.*, **4,** 202 ; 1963 b, *Phil. Mag.*, **8,** 2055 ; 1965, *Phys. Rev.*, **140,** A935.
ASHCROFT, N. W., and LAWRENCE, W. E., 1968, *Phys. Rev.*, **175,** 938.
ASHCROFT, N. W., and WILKINS, J. W., 1965, *Physics Lett.*, **14,** 285.
AUBAUER, H. P., 1967, *Phys. Rev.*, **155,** 673.
AUBREY, J. E., 1960, *Phil. Mag.*, **5,** 1001 ; 1961, *J. Phys. Chem. Solids*, **19,** 321.
AUBREY, J. E., and CHAMBERS, R. G., 1957, *J. Phys. Chem. Solids*, **3,** 128.
AUBREY, J. E., and CREASEY, C. J., 1969, *J. Phys.* C (*Solid St. Phys.*), **2,** 824.
AZBEL', M. YA., and KANER, É. A., 1957, *Zh. éksp. teor. Fiz.*, **32,** 896. English translation : *Soviet Phys. JETP*, **5,** 730.
BABISKIN, J., 1957, *Phys. Rev.*, **107,** 981.
BADOUX, F., HEINRICH, F., and KALLMEYER, G., 1967, *Helv. phys. Acta*, **40,** 815.
BAK, T. A., 1964, *Phonons and Phonon Interactions* (New York : Benjamin).
BALCOMBE, R. J., 1963, *Proc. R. Soc.* A, **275,** 113.
BALCOMBE, R. J., GUPTILL, E. W., and JERICHO, M. H., 1964, *Physics Lett.*, **13,** 287.
BALLA, D., and BRANDT, N. B., 1964, *Zh. éksp. teor. Fiz.*, **47,** 1653. English translation : *Soviet Phys. JETP*, **20,** 1111.
BARDEEN, J., COOPER, L. N., and SCHRIEFFER, J. R., 1957, *Phys. Rev.*, **108,** 1175.
BARNAAL, D. E., BARNES, R. C., MCCART, B. R., MOHN, L. W., and TORGESON, D. R., 1967, *Phys. Rev.*, **157,** 510.
BARRETT, C. S., 1955, *J. Inst. Metals*, **84,** 43 ; 1956, *Acta crystallogr.*, **9,** 671 ; 1958, *Phys. Rev.*, **110,** 1071.
BARRIOL, J., 1960, *J. Chim. phys.*, **57,** 837.
BASSANI, F., and PARRAVICINI, G. P., 1967, *Nuovo Cim.* B, **50,** 95.
BATE, R. T., and EINSPRUCH, N. G., 1965, *Physics Lett.*, **16,** 11.
BEATTIE, A. G., and UEHLING, E. A., 1966, *Phys. Rev.*, **148,** 657.
BECKMAN, O., ERIKSSON, L., and HÖRNFELDT, S., 1964, *Solid St. Commun.*, **2,** 7.
BEHRINGER, R. E., 1958, *J. Phys. Chem. Solids*, **5,** 145.
BELL, R. E., and JØRGENSEN, M. H., 1960, *Can. J. Phys.*, **38,** 652.
BERKO, S., 1962, *Phys. Rev.*, **128,** 2166.
BERKO, S., and PLASKETT, J. S., 1958, *Phys. Rev.*, **112,** 1877.
BERLINCOURT, T. G., 1955, *Phys. Rev.*, **99,** 1716.
BERLINCOURT, T. G., and STEELE, M. C., 1955, *Phys. Rev.*, **98,** 956.
BERRE, B., and OLSEN, T., 1965, *Phys. Stat. Sol.*, **11,** 657.
BEZUGLYĬ, P. A., and GALKIN, A. A., 1957, *Zh. éksp. teor. Fiz.*, **33,** 1076. English translation : *Soviet Phys. JETP*, **6,** 831 ; 1958, *Ibid.*, **34,** 236. English translation : *Soviet Phys. JETP*, **7,** 163 ; 1959, *Ibid.*, **37,** 1480. English translation : *Soviet Phys. JETP*, **10,** 1049.

BEZUGLYĬ, P. A., GALKIN, A. A., and PUSHKIN, A. I., 1962, *Proc. 8th Int. Conf. Low Temp. Phys.*, p. 208 ; 1963, *Zh. éksp. teor. Fiz.*, **44,** 71. English translation : *Soviet Phys. JETP*, **17,** 50.

BEZUGLYĬ, P. A., GALKIN, A. A., PUSHKIN, A. I., and KHOMCHENKO, A. I., 1962, *Zh. éksp. teor. Fiz.*, **42,** 84. English translation : *Soviet Phys. JETP*, **15,** 60.

BEZUGLYĬ, P. A., GALKIN, A. A., and ZHEVAGO, S. E., 1964, *Zh. éksp. teor. Fiz.*, **47,** 825. English translation : *Soviet Phys. JETP*, **20,** 552 ; 1965, *Fizika tverd. Tela*, **7,** 480. English translation : *Soviet Phys. Solid St.*, **7,** 383.

BHARGAVA, R. N., 1967, *Phys. Rev.*, **156,** 785.

BHATTACHARYA, D. L., and STERN, E. A., 1964, *Proc. 9th Int. Conf. Low Temp. Phys.*, p. 1210.

BIENENSTOCK, A., and BROOKS, H., 1964, *Phys. Rev.*, **136,** A784.

BIONDI, M. A., and GUOBADIA, A. I., 1968, *Phys. Rev.*, **166,** 667.

BIRSS, R. R., 1964, *Symmetry and Magnetism* (Amsterdam : North-Holland).

BJÖRKMAN, G., LUNDQVIST, B. I., and SJÖLANDER, A., 1967, *Phys. Rev.*, **159,** 551.

BLACKMAN, M., 1938, *Proc. R. Soc.* A, **166,** 1.

BLANEY, T. G., 1967, *Phil. Mag.*, **15,** 707 ; 1968, *Ibid.*, **17,** 405 ; 1969, *Ibid.*, **20,** 23.

BLATT, F. J., BURMESTER, A., and LAROY, B., 1967, *Phys. Rev.*, **155,** 611.

BLISS, E. S., and RAYNE, J. A., 1966, *Physics Lett.*, **23,** 38 ; 1967, *Ibid.* A, **25,** 242.

BLOEMBERGEN, N., and ROWLAND, T. J., 1953, *Acta metall.*, **1,** 731.

BOGOD, YU. A., 1967, *Phys. Stat. Sol.*, **24,** K49.

BOGOD, YU. A., and EREMENKO, V. V., 1967 a, *Phys. Stat. Sol.*, **21,** 797 ; 1967 b, *Zh. éksp. teor. Fiz.*, **53,** 473. English translation : *Soviet Phys. JETP*, **26,** 311.

BOGOD, YU. A., EREMENKO, V. V., and CHUBOVA, L. K., 1968, *Phys. Stat. Sol.*, **28,** K155 ; 1969, *Zh. éksp. teor. Fiz.*, **56,** 32. English translation : *Soviet Phys. JETP*, **29,** 17.

BOGOD, YU. A., and KRASOVITSKII, V. B., 1968, *Zh. éksp. teor. Fiz.*, *Pis'ma*, **7,** 301. English translation : *Soviet Phys. JETP Lett.*, **7,** 235.

BONSIGNORI, F., and BORTOLANI, V., 1966, *Nuovo Cim.* B, **46,** 113.

BOROVIK, E. S., 1952, *Zh. éksp. teor. Fiz.*, **23,** 83.

BOROVIK, E. S., and VOLOTSKAYA, V. G., 1960, *Zh. éksp. teor. Fiz.*, **38,** 261. English translation : *Soviet Phys. JETP*, **11,** 189 ; 1965, *Ibid.*, **48,** 1554. English translation : *Soviet Phys. JETP*, **21,** 1041.

BOROVIK, E. S., VOLOTSKAYA, V. G., and FOGEL', N. YA., 1963, *Zh. éksp. teor. Fiz.*, **45,** 46. English translation : *Soviet Phys. JETP*, **18,** 34.

BORSA, F., and BARNES, R. G., 1966, *J. Phys. Chem. Solids*, **217,** 567.

BOUCKAERT, L. P., SMOLUCHOWSKI, R., and WIGNER, E. P., 1936, *Phys. Rev.*, **50,** 58.

BOUGHTON, R. I., and YAQUB, M., 1968, *Phys. Rev. Lett.*, **20,** 108.

BOYLE, W. S., and BRAILSFORD, A. D., 1960, *Phys. Rev.*, **120,** 1943.

BOYLE, W. S., HSU, F. S. L., and KUNZLER, J. E., 1960, *Phys. Rev. Lett.*, **4,** 178.

BOYLE, W. S., and RODGERS, K. F., 1959, *Phys. Rev. Lett.*, **2,** 338.

BOYLE, W. S., and SMITH, G. E., 1963, *Prog. Semicond.*, **7,** 1.

BRÄNDLI, G., COTTI, P., FRYER, E. M , and OLSEN, J. L., 1964, *Proc. 9th Int. Conf. Low Temp. Phys.*, p. 827.

BRANDT, G. B., and RAYNE, J. A., 1963, *Phys. Rev.*, **132,** 1512 ; 1964, *Physics Lett.*, **12,** 87.

BRANDT, N. B., 1960, *Zh. éksp. teor. Fiz.*, **38,** 1355. English translation : *Soviet Phys. JETP*, **11,** 975.

BRANDT, N. B., DOLGOLENKO, T. F., and STUPOCHENKO, N. N., 1963, *Zh. éksp. teor. Fiz.*, **45,** 1319. English translation : *Soviet Phys. JETP*, **18,** 908.
BRANDT, N. B., DUBROVSKAYA, A. E., and KYTIN, G. A., 1959, *Zh. éksp. teor. Fiz.*, **37,** 572. English translation : *Soviet Phys. JETP*, **10,** 405.
BRANDT, N. B., and LYUBUTINA, L. G., 1964, *Zh. éksp. teor. Fiz.*, **47,** 1711. English translation : *Soviet Phys. JETP*, **20,** 1150 ; 1967, *Ibid.*, **52,** 686. English translation : *Soviet Phys. JETP*, **25,** 450.
BRANDT, N. B., and MININA, N. YA., 1968, *Zh. éksp. teor. Fiz., Pis'ma*, **7,** 264. English translation : *Soviet Phys. JETP Lett.*, **7,** 205.
BRANDT, N. B., MININA, N. YA., and POSPELOV, YU. A., 1968, *Fizika tverd. Tela*, **10,** 1268. English translation : *Soviet Phys. Solid St.*, **10,** 1011.
BRANDT, N. B., and RAZUMEENKO, M. V., 1960, *Zh. éksp. teor. Fiz.*, **39,** 276. English translation : *Soviet Phys. JETP*, **12,** 198.
BRANDT, N. B., and RYABENKO, G. A., 1959, *Zh. éksp. teor. Fiz.*, **37,** 389. English translation : *Soviet Phys. JETP*, **10,** 278.
BRANDT, N. B., and SHCHEKOCHIKHINA, V. V., 1961, *Zh. éksp. teor. Fiz.*, **41,** 1412. English translation : *Soviet Phys. JETP*, **14,** 1008.
BRANDT, N. B., and VENTTSEL', V. A., 1958, *Zh. éksp. teor. Fiz.*, **35,** 1083. English translation : *Soviet Phys. JETP*, **8,** 757.
BRISCOE, C. V., BEARDSLEY, G. M., and STEWART, A. T., 1966, *Phys. Rev.*, **141,** 379.
BROCKHOUSE, B. N., ARASE, T., CAGLIOTI, G., RAO, K. R., and WOODS, A. D. B., 1962, *Phys. Rev.*, **128,** 1099.
BROCKHOUSE, B. N., RAO, K. R., and WOODS, A. D. B., 1961, *Phys. Rev. Lett.*, **7,** 93.
BROVMAN, E. G., and KAGAN, YU., 1966, *Fizika tverd. Tela*, **8,** 1402. English translation : *Soviet Phys. Solid St.*, **8,** 1120.
BROWN, E., and KRUMHANSL, J. A., 1958, *Phys. Rev.*, **109,** 30.
BROWN, R. D., 1964, *Bull. Am. phys. Soc.*, **9,** 264 ; 1966, *IBM Jl Res. Dev.*, **10,** 462.
BROWN, R. D., HARTMAN, R. L., and KOENIG, S. H., 1968, *Phys. Rev.*, **172,** 598.
BROWN, R. N., MAVROIDES, J. G., DRESSELHAUS, M. S., and LAX, B., 1960, *Phys. Rev. Lett.*, **5,** 243.
BROWN, R. N., MAVROIDES, J. G., and LAX, B., 1963, *Phys. Rev.*, **129,** 2055.
BROWNELL, D. H., and HYGH, E. G., 1967, *Phys. Rev.*, **164,** 916.
BRYANT, C. A., and KEESOM, P. H., 1960, *Phys. Rev. Lett.*, **4,** 460 ; 1961, *Phys. Rev.*, **123,** 491.
BUCHSBAUM, S. J., and GALT, J. K., 1961, *Physics Fluids*, **4,** 1514.
BUNDY, F. P., and STRONG, H. M., 1962, *Solid St. Phys.*, **13,** 81.
BURTON, J. J., and JURA, G., 1968, *Phys. Rev.*, **171,** 699.
CALLAWAY, J., 1956, *Phys. Rev.*, **103,** 1219 ; 1958 a, *Ibid.*, **112,** 322 ; 1958 b, *Ibid.*, **112,** 1061 ; 1960, *Ibid.*, **119,** 1012 ; 1961 a, *Ibid.*, **123,** 1255 ; 1961 b, *Ibid.*, **124,** 1824 ; 1964, *Energy Band Theory* (New York : Academic Press).
CALLAWAY, J., and HAASE, E. L., 1957, *Phys. Rev.*, **108,** 217.
CALLAWAY, J., and KOHN, W., 1962, *Phys. Rev.*, **127,** 1913.
CAPLIN, A. D., and SHOENBERG, D., 1965, *Physics Lett.*, **18,** 238.
CARBOTTE, J. P., and DYNES, R. C., 1967, *Physics Lett.* A, **25,** 685 ; 1968, *Phys. Rev.*, **172,** 476.
CAROLINE, D., 1969, *J. Phys.* C (*Solid St. Phys.*), **2,** 308.
CARTER, D. L., and PICARD, J. C., 1967, *Solid St. Commun.*, **5,** 719.
CASTLE, J. G., CHANDRASEKHAR, B. S., and RAYNE, J. A., 1961, *Phys. Rev. Lett.*, **6,** 409.
CHAMBERS, R. G., 1952, *Proc. R. Soc.* A, **215,** 481 ; 1956, *Can. J. Phys.*, **34,** 1395.

Chow, P. C., and Kleinman, L., 1969, *Phys. Rev.*, **178,** 1111.
Clark, D. E., and Assenheim, J. G., 1960, *Br. J. appl. Phys.*, **11,** 35.
Clement, J. R., and Quinnell, E. H., 1953, *Phys. Rev.*, **92,** 258.
Cochran, J. F., and Haering, R. R., 1968, *Solid State Physics. I. Electrons in Metals* (New York : Gordon & Breach).
Cochran, J. F., and Shiffman, C. A., 1965, *Phys. Rev.*, **140,** A1678.
Cochran, J. F., and Yaqub, M., 1965, *Phys. Rev.*, **140,** A2174.
Cochran, W., 1963, *Proc. R. Soc.* A, **276,** 308.
Cohen, M. H., 1961, *Phys. Rev.*, **121,** 387 ; 1964, *Phys. Rev. Lett.*, **12,** 664.
Cohen, M. H., and Blount, E. I., 1960, *Phil. Mag.*, **5,** 115.
Cohen, M. H., and Falicov, L. M., 1960, *Phys. Rev. Lett.*, **5,** 544.
Cohen, M. H., Falicov, L. M., and Golin, S., 1964, *IBM Jl Res. Dev.*, **8,** 215.
Cohen, M. H., Harrison, M. J., and Harrison, W. A., 1960, *Phys. Rev.*, **117,** 937.
Cohen, M. H., and Heine, V., 1958, *Adv. Phys.*, **7,** 395.
Cohen, M. H., and Phillips, J. C., 1964, *Phys. Rev. Lett.*, **12,** 662.
Collins, J. G., and Ziman, J. M., 1961, *Proc. R. Soc.* A, **264,** 60.
Collins, M. F., 1962, *Proc. phys. Soc.*, **80,** 362.
Colombani, A., and Huet, P., 1963 a, *C. r. hebd. Séanc. Acad. Sci., Paris*, **256,** 406 ; 1963 b, *Ibid.*, **256,** 2357.
Colombino, P., Fiscella, B., and Trossi, L., 1964, *Nuovo Cim.*, **31,** 950.
Condon, J. H., 1964, *Bull. Am. phys. Soc.*, **9,** 239 ; 1966, *Phys. Rev.*, **145,** 526.
Condon, J. H., and Marcus, J. A., 1964, *Phys. Rev.*, **134,** A446.
Connell, R. A., and Marcus, J. A., 1957, *Phys. Rev.*, **107,** 940.
Coon, J. B., Grenier, C. G., and Reynolds, J. M., 1967, *J. Phys. Chem. Solids*, **28,** 301.
Cooper, D. G., 1968, *The Periodic Table* (London : Butterworth).
Cornwell, J. F., 1961, *Proc. R. Soc.* A, **261,** 551 ; 1969, *Group Theory and Electronic Energy Bands in Solids* (Amsterdam : North-Holland).
Cornwell, J. F., and Wohlfarth, E. P., 1960, *Nature, Lond.*, **186,** 379.
Cotti, P., 1963, *Physics Lett.*, **4,** 114 ; 1964, *Phys. Kondens Materie*, **3,** 40.
Cotti, P., Fryer, E. M., and Olsen, J. L., 1964, *Helv. phys. Acta*, **37,** 585.
Coulson, C. A., 1947, *Nature, Lond.*, **159,** 265.
Cowley, R. A., Woods, A. D. B., and Dolling, G., 1966, *Phys. Rev.*, **150,** 487.
Cracknell, A. P., 1964, Thesis, Oxford University ; 1967, *Physics Lett.* A, **24,** 263 ; 1969, *J. Phys.* C (*Solid St. Phys.*), **2,** 1425.
Craven, J. E., and Stark, R. W., 1968, *Phys. Rev.*, **168,** 849.
Culbert, H. V., 1967, *Phys. Rev.*, **157,** 560.
Dahlquist, W. L., and Goodrich, R. G., 1967, *Phys. Rev.*, **164,** 944.
Darby, J. K., and March, N. H., 1964, *Proc. phys. Soc.*, **84,** 591.
Datars, W. R., 1961, *Can. J. Phys.*, **39,** 1922 ; 1962 a, *Ibid.*, **40,** 1784 ; 1962 b, *Phys. Rev.*, **126,** 975.
Datars, W. R., and Dexter, R. N., 1961, *Phys. Rev.*, **124,** 75.
Datars, W. R., and Eastman, P. C., 1962, *Can. J. Phys.*, **40,** 670.
Datars, W. R., and Vanderkooy, J., 1964, *IBM Jl Res. Dev.*, **8,** 247 ; 1966, *J. phys. Soc. Japan*, **21,** Suppl., 657.
Decker, D. L., Mapother, D. E., and Shaw, R. W., 1958, *Phys. Rev.*, **112,** 1888.
Dekhtjar, I. Ja., 1969, *Physics Lett.* A, **28,** 771.
Dekhtyar, I. Ya., and Mikhalenkov, V. S., 1960, *Dokl. Akad. Nauk SSSR*, **133,** 60. English translation : *Soviet Phys. Dokl.*, **5,** 739 ; 1961, *Ibid.*, **136,** 63. English translation : *Soviet Phys. Dokl.*, **6,** 31.
De Sorbo, W., and Nichols, G. E., 1958, *J. Phys. Chem. Solids*, **6,** 352.
Deutsch, T., Paul, W., and Brooks, H., 1961, *Phys. Rev.*, **124,** 753.
Devillers, M. A. C., and De Vroomen, A. R., 1969, *Physics Lett.* A, **30,** 159.

DEXTER, R. N., and LAX, B., 1955, *Phys. Rev.*, **100,** 1216.
DHEER, P. N., 1961, *Proc. R. Soc.* A, **260,** 333.
DHILLON, J. S., and SHOENBERG, D., 1955, *Phil. Trans. R. Soc.* A, **248,** 1.
DICKE, D. A., and GREEN, B. A., 1967, *Phys. Rev.*, **153,** 800.
DIXON, M., HOARE, F. E., HOLDEN, T. M., and MOODY, D. E., 1965, *Proc. R. Soc.* A, **285,** 561.
DONAGHY, J. J., and STEWART, A. T., 1967 a, *Phys. Rev.*, **164,** 391 ; 1967 b, *Ibid.*, **164,** 396.
DONAGHY, J. J., STEWART, A. T., ROCKMORE, D. M., and KUSMISS, J. H., 1964, *Proc. 9th Int. Conf. Low Temp. Phys.*, p. 835.
DRESSELHAUS, M. S., and MAVROIDES, J. G., 1964 a, *Solid St. Commun.*, **2,** 297 ; 1964 b, *Carbon*, **1,** 263 ; 1964 c, *IBM Jl Res. Dev.*, **8,** 262 ; 1965, *Phys. Rev. Lett.*, **14,** 259 ; 1966, *Carbon*, **3,** 465.
DRICKAMER, H. G., 1965, *Solid St. Phys.*, **17,** 1.
DRUYVESTEYN, W. F., 1968, *Phil. Mag.*, **18,** 11.
DUGDALE, J. S., and GUGAN, D., 1960, *Proc. R. Soc.* A, **254,** 184 ; 1962, *Ibid.*, **270,** 186.
DUNIFER, G., SCHULTZ, S., and SCHMIDT, P. H., 1968, *J. appl. Phys.*, **39,** 397.
DYSON, F. J., 1955, *Phys. Rev.*, **98,** 349.
EASTMAN, P. C., and DATARS, W. R., 1963, *Can. J. Phys.*, **41,** 161.
ECKSTEIN, Y., 1963, *Phys. Rev.*, **129,** 12.
ECKSTEIN, Y., and KETTERSON, J. B., 1965, *Phys. Rev.*, **137,** A1777.
ECKSTEIN, Y., KETTERSON, J. B., and ECKSTEIN, S. G., 1964, *Phys. Rev.*, **135,** A740.
ECKSTEIN, Y., KETTERSON, J. B., and PRIESTLEY, M. G., 1966, *Phys. Rev.*, **148,** 586.
ÉDEL'MAN, V. S., 1968, *Zh. éksp. teor. Fiz.*, **54,** 1726. English translation : *Soviet Phys. JETP*, **27,** 927.
ÉDEL'MAN, V. S., and KHAĬKIN, M. S., 1965, *Zh. éksp. teor. Fiz.*, **49,** 107. English translation : *Soviet Phys. JETP*, **22,** 77.
ELLIOTT, R. J., 1954, *Phys. Rev.*, **96,** 280.
ENGELER, W. E., 1963, *Phys. Rev.*, **129,** 1509.
EPSTEIN, S., and JURETSCHKE, H. J., 1963, *Phys. Rev.*, **129,** 1148.
ERFLING, H. D., and GRUNEISEN, E., 1942, *Annln Phys.*, **41,** 89.
ERIKSSON, L., BECKMAN, O., and HÖRNFELDT, S., 1964, *J. Phys. Chem. Solids*, **25,** 1339.
ERMOLAEV, A. M., and KAGANOV, M. I., 1967, *Zh. éksp. teor. Fiz., Pis'ma*, **6,** 984. English translation : *Soviet Phys. JETP Lett.*, **6,** 395.
ESAKI, L., 1962, *Phys. Rev. Lett.*, **8,** 4.
ESAKI, L., and HEER, J., 1962, *Proc. Int. Conf. Semicond, Exeter*, p. 603.
ESAKI, L., and STILES, P. J., 1965, *Phys. Rev. Lett.*, **14,** 902.
ESTERMANN, I., FRIEDBERG, S. A., and GOLDMAN, J. E., 1952, *Phys. Rev.*, **87,** 582.
ETIENNE-AMBERG, L., 1966, *Physics Lett.*, **22,** 257.
EVERETT, G. E., 1962, *Phys. Rev.*, **128,** 2564 ; 1964, *Bull. Am. phys. Soc.*, **9,** 383.
FALICOV, L. M., 1962, *Phil. Trans. R. Soc.* A, **255,** 55.
FALICOV, L. M., and COHEN, M. H., 1963, *Phys. Rev.*, **130,** 92.
FALICOV, L. M., and GOLIN, S., 1965, *Phys. Rev.*, **137,** A871.
FALICOV, L. M., and LIN, P. J., 1966, *Phys. Rev.*, **141,** 562.
FALICOV, L. M., PIPPARD, A. B., and SIEVERT, P. R., 1966, *Phys. Rev.*, **151,** 498.
FALICOV, L. M., and RUVALDS, J., 1968, *Phys. Rev.*, **172,** 498.
FALICOV, L. M., and STACHOWIAK, H., 1966, *Phys. Rev.*, **147,** 505.
FAL'KOVSKIĬ, L. A., 1967, *Zh. éksp. teor. Fiz.*, **53,** 2164. English translation : *Soviet Phys. JETP*, **26,** 1222.

FAL'KOVSKIĬ, L. A., and RAZINA, G. S., 1965, *Zh. éksp. teor. Fiz.*, **49,** 265, English translation : *Soviet Phys. JETP*, **22,** 187.

FARACI, G., QUERCIA, I. F., SPADONI, M., and TURRISI, E., 1969, *Nuovo Cim.* B, **60,** 228.

FATE, W. A., 1968, *Phys. Rev.*, **172,** 402.

FAULKNER, J. S., 1969, *Phys. Rev.*, **178,** 914.

FAWCETT, E., 1955, *Proc. R. Soc.* A, **232,** 519 ; 1956, *Phys. Rev.*, **103,** 1582 ; 1961, *J. Phys. Chem. Solids*, **18,** 320.

FEDER, J., and LOTHE, J., 1965, *Phil. Mag.*, **12,** 107.

FENTON, E. W., and WOODS, S. B., 1966, *Phys. Rev.*, **151,** 424.

FERREIRA, L. G., 1967, *J. Phys. Chem. Solids*, **28,** 1891 ; 1968, *Ibid.*, **29,** 357.

FILBY, J. D., and MARTIN, D. L., 1963, *Proc. R. Soc.* A, **276,** 187 ; 1965, *Ibid.*, **284,** 83.

FONER, S., ZEIGER, H. J., POWELL, R. L., WALSH, W. M., and LAX, B., 1956, *Bull. Am. phys. Soc.*, **1,** 117.

FØRSVOLL, K., and HOLWECH, I., 1962, *Physics Lett.*, **3,** 66 ; 1963, *J. appl. Phys.*, **34,** 2230 ; 1964 a, *Phil. Mag.*, **9,** 435 ; 1964 b, *Ibid.*, **10,** 181 ; 1964 c, *Ibid.*, **10,** 921.

FOSSHEIM, K., and LEIBOWITZ, J. P., 1966, *Physics Lett.*, **22,** 140.

FOSSHEIM, K., and OLSEN, T., 1964, *Phys. Stat. Sol.*, **6,** 867.

FOSTER, H. J., MEIJER, P. H. E., and MIELCZAREK, E. V., 1965, *Phys. Rev.*, **139,** A1849.

FREEDMAN, S. J., and JURETSCHKE, H. J., 1961, *Phys. Rev.*, **124,** 1379.

FRIEDBERG, S. A., ESTERMANN, I., and GOLDMAN, J. E., 1952, *Phys. Rev.*, **85,** 375.

FRIEDMAN, A. N., 1967, *Phys. Rev.*, **159,** 553.

FRIEDMAN, A. N., and KOENIG, S. H., 1960, *IBM Jl Res. Dev.*, **4,** 158.

FUKASE, T., 1969, *J. phys. Soc. Japan*, **26,** 964.

FUKASE, T., and FUKUROI, T., 1967, *J. phys. Soc. Japan*, **23,** 650.

FUKUDA, Y., 1965, *J. phys. Soc. Japan*, **20,** 353.

FUKUMOTO, A., and STRANDBERG, M. W. P., 1966, *Physics Lett.*, **23,** 200 ; 1967, *Phys. Rev.*, **155,** 685.

GAĬDUKOV, YU. P., 1965, *Zh. éksp. teor. Fiz.*, **49,** 1049. English translation : *Soviet Phys. JETP*, **22,** 730.

GALKIN, A. A., KANER, É. A., and KOROLYUK, A. P., 1960, *Zh. éksp. teor. Fiz.*, **39,** 1517. English translation : *Soviet Phys. JETP*, **12,** 1055.

GALKIN, A. A., NABEREZHNЎKH, V. P., and MEL'NIK, V. L., 1963, *Fizika tverd. Tela*, **5,** 201. English translation : *Soviet Phys. Solid St.*, **5,** 145.

GALT, J. K., YAGER, W. A., MERRITT, F. R., CETLIN, B. B., and BRAILSFORD, A. D., 1959, *Phys. Rev.*, **114,** 1396.

GALT, J. K., YAGER, W. A., MERRITT, F. R., CETLIN, B. B., and DAIL, H. W., 1955, *Phys. Rev.*, **100,** 748.

GANTMAKHER, V. F., 1962 a, *Zh. éksp. teor. Fiz.*, **42,** 1416. English translation : *Soviet Phys. JETP*, **15,** 982 ; 1962 b, *Ibid.*, **43,** 345. English translation : *Soviet Phys. JETP*, **16,** 247 ; 1963, *Ibid.*, **44,** 811. English translation : *Soviet Phys. JETP*, **17,** 549 ; 1964, *Ibid.*, **46,** 2028. English translation : *Soviet Phys. JETP*, **19,** 1366.

GANTMAKHER, V. F., and KANER, É. A., 1965, *Zh. éksp. teor. Fiz.*, **48,** 1572. English translation : *Soviet Phys. JETP*, **21,** 1053.

GANTMAKHER, V. F., and KRYLOV, I. P., 1964, *Zh. éksp. teor. Fiz.*, **47,** 2111. English translation : *Soviet Phys. JETP*, **20,** 1418 ; 1965, *Ibid.*, **49,** 1054. English translation : *Soviet Phys. JETP*, **22,** 734.

GANTMAKHER, V. F., and LEONOV, YU. S., 1968, *Zh. éksp. teor. Fiz., Pis'ma*, **8,** 264. English translation : *Soviet Phys. JETP Lett.*, **8,** 162.

GANTMAKHER, V. F., and SHARVIN, YU. V., 1960, *Zh. éksp. teor. Fiz.*, **39,** 512. English translation : *Soviet Phys. JETP*, **12,** 358 ; 1964, *Proc. 9th Int. Conf. Low Temp. Phys.*, p. 1193.
GARCIA, N., and KAO, Y. H., 1968, *Physics Lett.* A, **26,** 373.
GARCÍA-MOLINER, F., 1958, *Proc. phys. Soc.*, **72,** 996 ; 1959, *Proc. R. Soc.* A, **249,** 73.
GARLAND, J. C., and BOWERS, R., 1968, *Phys. Rev. Lett.*, **21,** 1007.
GASPARI, G. D., and DAS, T. P., 1968, *Phys. Rev.*, **167,** 660.
GIAUQUE, W. F., and MEADS, P. F., 1941, *J. Am. chem. Soc.*, **63,** 1897.
GILAT, G., 1965, *Solid St. Commun.* **3,** 101.
GITSU, D. V., BODIUL, P. P., and FEDORKO, A. S., 1969, *Phys. Stat. Sol.*, **33,** K143.
GITSU, D. V., and IVANOV, G. A., 1960, *Fizika tverd. Tela*, **2,** 1457. English translation : *Soviet Phys. Solid St.*, **2,** 1323.
GIURA, M., MARCON, R., PAPA, T., and WANDERLINGH, F., 1967, *Nuovo Cim.* B, **51,** 150.
GLASSER, M. L., and CALLAWAY, J., 1958, *Phys. Rev.*, **109,** 1541.
GMELIN, E., 1964, *C. r. hebd. Séanc. Acad. Sci., Paris*, **259,** 3459.
GODDARD, W. A., 1967 a, *Phys. Rev.*, **157,** 73 ; 1967 b, *Ibid.*, **157,** 81.
GOLD, A. V., 1958, *Phil. Trans. R. Soc.* A, **251,** 85 ; 1960, *Phil. Mag.*, **5,** 70.
GOLD, A. V., and PRIESTLEY, M. G., 1960, *Phil. Mag.*, **5,** 1089.
GOLDSMID, H. J., and CORSAN, J. M., 1964, *Physics Lett.*, **8,** 221.
GOLDSTEIN, A., and FONER, S., 1966, *Phys. Rev.*, **146,** 442.
GOLIN, S., 1968, *Phys. Rev.*, **166,** 643.
GOLOVASHKIN, A. I., and MOTULEVICH, G. P., 1963, *Zh. éksp. teor. Fiz.*, **44,** 398. English translation : *Soviet Phys. JETP*, **17,** 271.
GOODMAN, J. M., 1968, *Phys. Rev.*, **171,** 641.
GORDON, W. L., JOSEPH, A. S., and ECK, T. G., 1960, *The Fermi Surface.* Proceedings of a Conference held at Cooperstown, New York, on 22–24 August 1960, edited by W. A. Harrison and M. B. Webb (New York : Wiley), p. 84.
GOY, P., GOLDSTEIN, A., LANGENBERG, D. N., and PICARD, J. C., 1967, *Physics Lett.* A, **25,** 324.
GRAVES, R. H. W., and LENHAM, A. P., 1968, *J. opt. Soc. Am.*, **58,** 126.
GREENE, M., HOFFMAN, A., HOUGHTON, A., PEVERLEY, R., QUINN, J., and SEIDEL, G., 1966, *Physics Lett.*, **21,** 135.
GREENFIELD, A. F., SMITH, G. E., and LAWSON, A. W., 1959, *Bull. Am. phys. Soc.*, **4,** 409.
GRENIER, C. G., REYNOLDS, J. M., and SYBERT, J. R., 1963, *Phys. Rev.*, **132,** 58.
GRIMES, C. C., and KIP, A. F., 1963, *Phys. Rev.*, **132,** 1991.
GRIMES, C. C., KIP, A. F., SPONG, F., STRADLING, R. A., and PINCUS, P., 1963, *Phys. Rev. Lett.*, **11,** 455.
GRÜNEISEN, E., and ADENSTEDT, H., 1938, *Annln Phys.*, **31,** 714.
GUGAN, D., and JONES, B. K., 1963, *Helv. phys. Acta*, **36,** 7.
GUNNERSON, E. M., 1957, *Phil. Trans. R. Soc.* A, **249,** 299.
GUSTAFSON, D. R., and BARNES, G. T., 1967, *Phys. Rev. Lett.*, **18,** 3.
GUTHMANN, C., and LIBCHABER, A., 1967, *C. r. hebd. Séanc. Acad. Sci., Paris* B, **265,** 319.
HABERLAND, P. H., COCHRAN, J. F., and SHIFFMAN, C. A., 1969, *Physics Lett.* A, **30,** 476.
HABERLAND, P. H., and SHIFFMAN, C. A., 1967, *Phys. Rev. Lett.*, **19,** 1337.
HAERING, R. R., and MROZOWSKI, S., 1960, *Prog. Semicond.*, **5,** 273.
HALL, J. J., and KOENIG, S. H., 1964, *IBM Jl Res. Dev.*, **8,** 241.
HAM, F. S., 1962 a, *Phys. Rev.*, **128,** 82 ; 1962 b, *Ibid.*, **128,** 2524.

Harris, L., and Corrigan, F. R., 1965, *J. Phys. Chem. Solids*, **26,** 307.
Harrison, W. A., 1959, *Phys. Rev.*, **116,** 555 ; 1960 a, *Ibid.*, **118,** 1182 ; 1960 b, *Ibid.*, **118,** 1190; 1960 c, *J. Phys. Chem. Solids*, **17,** 171; 1963 a, *Phys. Rev.*, **129,** 2503 ; 1963 b, *Ibid.*, **129,** 2512 ; 1963 c, *Ibid.*, **131,** 2433 ; 1964, *Ibid.*, **136,** A1107 ; 1965, *Ibid.*, **139,** A179 ; 1966, *Pseudopotentials in the Theory of Metals* (New York : Benjamin).
Harrison, W. A., and Webb, M. B., 1960, *The Fermi Surface.* Proceedings of a Conference held at Cooperstown, New York, on 22–24 August 1960 (New York : Wiley).
Hattori, T., 1968, *J. phys. Soc. Japan*, **24,** 762.
Hattori, T., and Tosima, S., 1965, *J. phys. Soc. Japan*, **20,** 44.
Hays, D. A., and McLean, W. L., 1965, *Physics Lett.*, **17,** 215.
Hebel, L. C., 1965, *Phys. Rev.*, **138,** A1641.
Hebel, L. C., Blount, E. I., and Smith, G. E., 1965, *Phys. Rev.*, **138,** A1636.
Heine, V., 1956, *Proc. phys. Soc.* A, **69,** 505 ; 1957 a, *Proc. R. Soc.* A, **240,** 340 ; 1957 b, *Ibid.*, **240,** 354 ; 1957 c, *Ibid.*, **240,** 361 ; 1964, *Proc. 9th Int. Conf. Low Temp. Phys.*, p. 698 ; 1968, *J. Phys.* C (*Proc. phys. Soc.*), **1,** 222 ; 1969, *The Physics of Metals. I. Electrons*, edited by J. M Ziman (Cambridge University Press), Chap. 1.
Heine, V., and Abarenkov, I., 1964, *Phil. Mag.*, **9,** 451.
Heine, V., and Weaire, D., 1966, *Phys. Rev.*, **152,** 603.
Henry, N. F. M., and Lonsdale, K., 1965, *International Tables for X-ray Crystallography.* Vol. 1, Symmetry groups (Birmingham : Kynoch).
Herring, C., 1937, *Phys. Rev.*, **52,** 361 ; 1942, *J. Franklin Inst.*, **233,** 525.
Herring, C., and Hill, A. G., 1940, *Phys. Rev.*, **58,** 132.
Hill, R. W., and Smith, P. G., 1953, *Phil. Mag.*, **44,** 636.
Ho, P. S., and Ruoff, A. L., 1967, *Phys. Stat. Sol.*, **23,** 489.
Hodgson, J. N., 1965, *Proceedings of the International Colloquium on Optical Properties and Electronic Structure of Metals and Alloys*, Paris (1965), edited by F. Abelès (Amsterdam : North-Holland), p. 60.
Holcomb, D. F., Kaeck, J. A., and Strange, J. H., 1966. *Phys. Rev.*, **150,** 306.
Holland, B. W., 1968, *Phys. Stat. Sol.*, **28,** 121.
Holwech, I., and Risnes, R., 1968, *Phil. Mag.*, **17,** 757.
Hopfield, J. J., 1962, *Phys. Rev. Lett.*, **8,** 311.
Horowitz, M. Silvicli, A. A., Malakker S. F., and Daunt, J. G. 1952, *Phys. Rev.*, **88,** 1152.
Horton, G. K., and Schiff, H., 1959 *Proc. R. Soc.* A **250,** 248.
Howarth, D. J., and Jones H., 1952 *Proc. phys. Soc.* A **65,** 355.
Hughes, A. J., and Callaway, J., 1964 *Phys. Rev.*, **136,** A1390,
Hughes, A J., and Lettington, A. H., 1968, *Physics Lett.* A, **27,** 241.
Hughes, A. J., and Shepherd, J. P. G., 1969, *J. Phys.* C (*Solid St. Phys.*), **2,** 661.
Hum, R. H., and Perz, J. M., 1969, *Physics Lett.* A, **28,** 575.
Hume-Rothery, W., 1959, *Can. J. Phys.*, **37,** 1565.
Ignat'eva, T. A., Makarov, V. I., and Tereshina, N. S., 1968, *Zh. éksp. teor. Fiz.*, **54,** 1617. English translation : *Soviet Phys. JETP*, **27,** 865.
Inoue, M., 1962, *J. phys. Soc. Japan*, **17,** 808.
Inoue, S., and Tsuji, M., 1967, *J. phys. Soc. Japan*, **22,** 1191.
Ishizawa, Y., 1968 a, *J. phys. Soc. Japan*, **25,** 150 ; 1968 b, *Ibid.*, **25,** 160.
Ishizawa, Y., and Datars, W. R., 1969, *Physics Lett.* A, **30,** 463.
Ishizawa, Y., and Tanuma, S., 1965 a, *J. phys. Soc. Japan*, **20,** 1278 ; 1965 b, *Ibid.*, **20,** 1744.
Itskevich, E. S., and Fisher, L. M., 1967 a, *Zh. éksp. teor. Fiz., Pis'ma*, **5,** 141. English translation : *Soviet Phys. JETP Lett.*, **5,** 114 ; 1967 b, *Zh. éksp. teor. Fiz.*, **53,** 98. English translation : *Soviet Phys. JETP*, **26,** 66.

ITSKEVICH, E. S., KRECHETOVA, I. P., and FISHER, L. M., 1967, *Zh. éksp. teor. Fiz.*, **52,** 66. English translation : *Soviet Phys. JETP*, **25,** 41.

IYENGAR, P. K., VENKATARAMAN, G., VIJAYARAGHAVAN, P. R., and ROY, A. P., 1965 a, *J. Phys. Chem. Solids* (Suppl.), **1,** 223.; 1965 b, *Inelastic Scattering of Neutrons in Solids and Liquids*, Vol. 1. (Vienna : I.A.E.A.), p. 153.

JAIN, A. L., and KOENIG, S. H., 1962, *Phys. Rev.*, **127,** 442.

JAYARAMAN, A., KLEMENT, W., and KENNEDY, G. C., 1963 a, *Phys. Rev.*, **132,** 1620 ; 1963 b, *Phys. Rev. Lett.*, **10,** 387.

JEAVONS, A. P., and SAUNDERS, G. A., 1968, *Physics Lett.* A, **27,** 19 ; 1969, *Proc. R. Soc.* A, **310,** 415.

JENA, P., MAHANTI, S. D., and DAS, T. P., 1968, *Phys. Rev. Lett.*, **20,** 544.

JONES, B. K., 1964, *Phil. Mag.*, **9,** 217 ; 1969, *Phys. Rev.*, **179,** 637.

JONES, D., and LETTINGTON, A. H., 1967, *Proc. phys. Soc.*, **92,** 948.

JONES, E. P., and WILLIAMS, D. L., 1964, *Can. J. Phys.*, **42,** 1499.

JONES, H., 1934, *Proc. R. Soc.* A, **147,** 396 ; 1960, *The Theory of Brillouin Zones and Electronic States in Crystals* (Amsterdam : North-Holland).

KAGANOV, M. E., LIFSHITZ, I. M., and SINELNIKOV, K. D., 1957, *Zh. éksp. teor. Fiz.*, **32,** 605. English translation : *Soviet Phys. JETP*, **5,** 500.

KAGANOV, M. I., and SEMENENKO, A. I., 1967, *Fizika tverd. Tela*, **9,** 1129. English translation : *Soviet Phys. Solid St.*, **9,** 884.

KAHN, A. H., and FREDERIKSE, H. P. R., 1959, *Solid St. Phys.*, **9,** 257.

KALINKINA, I. N., and STRELKOV, P. G., 1958, *Zh. éksp. teor. Fiz.*, **34,** 616. English translation : *Soviet Phys. JETP*, **7,** 426.

KAMM, G. N., and BOHM, H. V., 1962, *Proc. 8th Int. Conf. Low Temp. Phys.*, p. 199 ; 1963, *Phys. Rev.*, **131,** 111.

KAO, Y. H., 1963, *Phys. Rev.*, **129,** 1122.

KARIMOV, YU. S., and SHCHEGOLEV, I. F., 1961, *Zh. éksp. teor. Fiz.*, **40,** 1289. English translation : *Soviet Phys. JETP*, **13,** 908.

KAZANSKII, V. B., and KOROLYUK, A. P., 1966, *Fizika tverd. Tela*, **8,** 3418. English translation : *Soviet Phys. Solid St.*, **8,** 2740.

KEARNEY, R. J., MACKINTOSH, A. R., and YOUNG, R. C., 1965, *Phys. Rev.*, **140,** A1671.

KECHIN, V. V., 1967, *Fizika tverd. Tela*, **9,** 3595. English translation : *Soviet Phys. Solid St.*, **9,** 2828.

KECHIN, V. V., LIKHTER, A. I., and POSPELOV, YU. A., 1965, *Zh. éksp. teor. Fiz.*, **49,** 36. English translation : *Soviet Phys. JETP*, **22,** 26.

KEESOM, P. H., and PEARLMAN, N., 1954, *Phys. Rev.*, **96,** 897 ; 1955, *Ibid.*, **99,** 1119.

KEESOM, P. H., and VAN DER HOEVEN, B. J. C., 1963, *Physics Lett.*, **3,** 360.

KETTERSON, J. B., 1963, *Phys. Rev.*, **129,** 18.

KETTERSON, J. B., and ECKSTEIN, Y., 1963, *Phys. Rev.*, **132,** 1885 ; 1965, *Ibid.*, **140,** A1355.

KETTERSON, J. B., and STARK, R. W., 1967, *Phys. Rev.*, **156,** 748.

KHAĬKIN, M. S., 1959, *Zh. éksp. teor. Fiz.*, **37,** 1473. English translation : *Soviet Phys. JETP*, **10,** 1044 ; 1960, *Ibid.*, **39,** 513. English translation : *Soviet Phys. JETP*, **12,** 359 ; 1961, *Ibid.*, **41,** 1773. English translation : *Soviet Phys. JETP*, **14,** 1260 ; 1962 a, *Ibid.*, **42,** 27. English translation : *Soviet Phys. JETP*, **15,** 18 ; 1962 b, *Ibid.*, **43,** 59. English translation : *Soviet Phys. JETP*, **16,** 42.

KHAĬKIN, M. S., and CHEREMISIN, S. M., 1968, *Zh. éksp. teor. Fiz.*, **54,** 69. English translation : *Soviet Phys. JETP*, **27,** 38.

KHAĬKIN, M. S., and ÉDEL'MAN, V. S., 1964, *Zh. éksp. teor. Fiz.*, **47,** 878. English translation : *Soviet Phys. JETP*, **20,** 587.

KHAĬKIN, M. S., and MINA, R. T., 1962, *Zh. éksp. teor. Fiz.*, **42,** 35. English translation : *Soviet Phys. JETP*, **15,** 24.

Khaĭkin, M. S., Mina, R. T., and Édel'man, V. S., 1962, *Zh. éksp. teor. Fiz.*, **43,** 2063. English translation : *Soviet Phys. JETP*, **16,** 1459.
Kimball, J. C., Stark, R. W., and Mueller, F. M., 1967, *Phys. Rev.*, **162,** 600.
Kip, A. F., Langenberg, D. N., Rosenblum, B., and Wagoner, G., 1957, *Phys. Rev.*, **108,** 494.
Kirsch, J., and Miller, P. B., 1962, *Phys. Rev. Lett.*, **9,** 421.
Kjeldaas, T., 1959, *Phys. Rev.*, **113,** 1473.
Klemens, P. G., 1954 a, *Aust. J. Phys.*, **7,** 70 ; 1954 b, *Proc. phys. Soc.* A, **67,** 194 ; 1954 c, *Aust. J. Phys.*, **7,** 64.
Klemens, P. G., Van Baarle, C., and Gorter, F. W., 1964, *Physica, 's Grav.*, **30,** 1470.
Knight, W. D., 1956, *Solid St. Phys.*, **2,** 93.
Knight, W. D., Berger, A. G., and Heine, V., 1959, *Ann. Phys.*, **8,** 173.
Koch, J. F., and Kip, A. F., 1962, *Phys. Rev. Lett.*, **8,** 473.
Koch, J. F., and Wagner, T. K., 1966, *Phys. Rev.*, **151,** 467.
Kohn, W., 1959, *Phys. Rev. Lett.*, **2,** 393.
Korolyuk, A. P., Matsakov, L. Ya., and Fal'ko, V. L., 1968, *Zh. éksp. teor. Fiz.*, **54,** 3. English translation : *Soviet Phys. JETP*, **27,** 1.
Koster, G. F., 1957, *Solid St. Phys.*, **5,** 173 ; 1962, *Phys. Rev.*, **127,** 2044.
Krylov, I. P., and Gantmakher, V. F., 1966, *Zh. éksp. teor. Fiz.*, **51,** 740. English translation : *Soviet Phys. JETP*, **24,** 492.
Kunzler, J. E., and Hsu, F. S. L., 1960, *The Fermi Surface.* Proceedings of a Conference held at Cooperstown, New York, on 22–24 August 1960, edited by W. A. Harrison and M. B. Webb (New York : Wiley), p. 88.
Kunzler, J. E., Hsu, F. S. L., and Boyle, W. S., 1962, *Phys. Rev.*, **128,** 1084.
Kuznetsov, M. E., and Shalyt, S. S., 1967, *Zh. éksp. teor. Fiz., Pis'ma*, **6,** 745. English translation : *Soviet Phys. JETP Lett.*, **6,** 217.
Lafon, E. E., and Lin, C. C., 1966, *Phys. Rev.*, **152,** 579.
Larson, C. O., and Gordon, W. L., 1967, *Phys. Rev.*, **156,** 703.
Lax, B., Button, J., Zeiger, H. J., and Roth, L. M., 1956, *Phys. Rev.*, **102,** 715.
Lax, B., and Mavroides, J. G., 1960, *Solid St. Phys.*, **11,** 261.
Lax, B., Mavroides, J. G., Zeiger, H. J., and Keyes, R. J., 1960, *Phys. Rev. Lett.*, **5,** 241.
Lazarev, B. G., Lazareva, L. S., Ignat'eva, T. A., and Makarov, V. I., 1965, *Dokl. Akad. Nauk SSSR*, **163,** 74. English translation : *Soviet Phys. Dokl.*, **10,** 620.
Lee, M. J. G., 1966, *Proc. R. Soc.* A, **295,** 440 ; 1969, *Phys. Rev.*, **178,** 953.
Lee, M. J. G., and Falicov, L. M., 1968, *Proc. R. Soc.* A, **304,** 319.
Lepage, J., Garber, M., and Blatt, F. J., 1964, *Physics Lett.*, **11,** 102.
Lerner, L. S., 1962, *Phys. Rev.*, **127,** 1480 ; 1963, *Ibid.*, **130,** 605.
Lerner, L. S., and Eastman, P. C., 1963, *Can. J. Phys.*, **41,** 1523.
Lewiner, J., 1967, *C. r. hebd. Séanc. Acad. Sci., Paris* B, **265,** 774.
Lewiner, J., and Biquard, P., 1967, *C. r. hebd. Séanc. Acad. Sci., Paris* B, **265,** 273.
Lien, W. H., and Phillips, N. E., 1964, *Phys. Rev.*, **133,** A1370.
Lifshitz, I. M., and Peschanskii, V. G., 1958, *Zh. éksp. teor. Fiz.*, **35,** 1251. English translation : *Soviet Phys. JETP*, **8,** 875.
Likhter, A. I., and Kechin, V. V., 1963, *Fizika tverd. Tela*, **5,** 3066. English translation : *Soviet Phys. Solid St.*, **5,** 2246.
Lin, P. J., and Falicov, L. M., 1966, *Phys. Rev.*, **142,** 441.
Linderberg, J., and Mäkilä, K. V., 1967, *Solid St. Commun.*, **5,** 353.
Lomer, W. M., 1962, *J. Phys. Paris*, **23,** 716.
Long, J. R., Grenier, C. G., and Reynolds, J. M., 1965, *Phys. Rev.*, **140,** A187.

LOPEZ, A. A., 1968, *Phys. Rev.*, **175,** 823.
LOUCKS, T. L., 1964, *Phys. Rev.*, **134,** A1618 ; 1965, *Phys. Rev. Lett.*, **14,** 1072.
LOUCKS, T. L., and CUTLER, P. H., 1964, *Phys. Rev.*, **133,** A819.
LÜCK, R., 1966 a, *Phys. Stat. Sol.*, **18,** 49 ; 1966 b, *Ibid.*, **18,** 59.
LUKIRSKII, A. P., and BRYTOV, I. A., 1964, *Fizika tverd. Tela*, **6,** 43. English translation : *Soviet Phys. Solid St.*, **6,** 33.
LÜTHI, B., 1959, *Phys. Rev. Lett.*, **2,** 503 ; 1960, *Helv. phys. Acta*, **33,** 161.
LÜTHI, B., and OLSEN, J. L., 1956, *Nuovo Cim.*, **3,** 840.
LYALL, K. R., and COCHRAN, J. F., 1967, *Phys. Rev.*, **159,** 517.
McCLURE, J. W., 1955, *Phys. Rev.*, **98,** 449 ; 1957, *Ibid.*, **108,** 612 ; 1958, *Ibid.*, **112,** 715 ; 1960, *Ibid.*, **119,** 606 ; 1964, *IBM Jl Res. Dev.*, **8,** 255.
McCLURE, J. W., and SPRY, W. J., 1968, *Phys. Rev.*, **165,** 809.
McCOLLUM, D. C., and TAYLOR, W. A., 1967, *Phys. Rev.*, **156,** 782.
McCOMBE, B., and SEIDEL, G., 1967, *Phys. Rev.*, **155,** 633.
McGARVEY, B. R., and GUTOWSKY, H. S., 1953, *J. chem. Phys.*, **21,** 2114.
McGRODDY, J. C., STANFORD, J. L., and STERN, E. A., 1966, *Phys. Rev.*, **141,** 437.
MACKINNON, L., 1966, *Experimental Physics at Low Temperatures. An Introductory Survey* (Detroit : Wayne State University Press).
MACKINNON, L., MYERS, A., and TAYLOR, M. T., 1959, *Proc. phys. Soc.*, **74,** 773.
MACKINNON, L., and TAYLOR, M. T., 1960, *The Fermi Surface.* Proceedings of a Conference held at Cooperstown, New York, on 22–24 August 1960, edited by W. A. Harrison and M. B. Webb (New York : Wiley), p. 251.
MACKINNON, L., TAYLOR, M. T., and DANIEL, M. R., 1962, *Phil. Mag.*, **7,** 523.
MACKINTOSH, A. R., 1963, *Proc. R. Soc.* A, **271,** 88.
MACKINTOSH, A., SPANEL, L. E., and YOUNG, R. C., 1963, *Phys. Rev. Lett.*, **10,** 434.
MACRAE, R. A., ARAKAWA, E. T., and WILLIAMS, M. W., 1967, *Phys. Rev.*, **162,** 615.
McWHAN, D. B., and JAYARAMAN, A., 1963, *Appl. Phys. Lett.*, **3,** 129.
MAHANTI, S. D., and DAS, T. P., 1969, *Phys. Rev.*, **183,** 674.
MAKAROV, V. I., and BAR'YAKHTAR, V. G., 1965, *Zh. éksp. teor. Fiz.*, **48,** 1717. English translation : *Soviet Phys. JETP*, **21,** 1151.
MAKAROV, V. I., and VOLYNSKII, I. YA., 1966, *Zh. éksp. teor. Fiz., Pis'ma*, **4,** 369. English translation : *Soviet Phys. JETP Lett.*, **4,** 249.
MALISZEWSKI, E., SOSNOWSKI, J., BLINOWSKI, K., KOZUBOWSKI, J., PADLO, L., and SLEDZIEWSKA, D., 1963, *Inelastic Scattering of Neutrons in Solids and Liquids*, Vol. 2 (Vienna : I.A.E.A.), p. 87.
MALTZ, M., and DRESSELHAUS, M. S., 1968, *Phys. Rev. Lett.*, **20,** 919.
MANNING, M. F., and KRUTTER, H. M., 1937, *Phys. Rev.*, **51,** 761.
MARSH, N. W. A., and AUBREY, J. E., 1965, *Nature, Lond.*, **205,** 894.
MARTIN, D. L., 1961 a, *Proc. phys. Soc.*, **78,** 1482 ; 1961 b, *Phys. Rev.*, **124,** 438 ; 1965, *Ibid.*, **139,** A150.
MASE, S., 1958, *J. phys. Soc. Japan*, **13,** 434 ; 1959 a, *Ibid.*, **14,** 584 ; 1959 b, *Ibid.*, **14,** 1538.
MASE, S., FUJIMORI, Y., and MORI, H., 1966, *J. phys. Soc. Japan*, **21,** 1744.
MASE, S., and TANUMA, S., 1959, *J. phys. Soc. Japan*, **14,** 1644.
MASE, S., VON MOLNAR, S., and LAWSON, A. W., 1962, *Phys. Rev.*, **127,** 1030.
MATATAGUI, E., and CARDONA, M., 1968, *Solid St. Commun.*, **6,** 313.
MATTHEISS, L. F., 1964, *Phys. Rev.*, **133,** A1399.
MATZKANIN, G. A., and SCOTT, T. A., 1966, *Phys. Rev.*, **151,** 360.
MAVROIDES, J. G., LAX, B., BUTTON, K. J., and SHAPIRA, Y., 1962, *Phys. Rev. Lett.*, **9,** 451.
MAXWELL, E., and LUTES, O. S., 1954, *Phys. Rev.*, **95,** 333.
MAYER, H., and EL NABY, M. H., 1963, *Z. Phys.*, **174,** 289.

MEADS, P. F., FORSYTHE, W. R., and GIAUQUE, W. F., 1941, *J. Am. chem. Soc.*, **63,** 1092.
MELNGAILIS, J., and DE BENEDETTI, S., 1966, *Phys. Rev.*, **145,** 400.
MELZ, P. J., 1966, *Phys. Rev.*, **152,** 540.
MERCOUROFF, W., 1967, *La Surface de Fermi des Metaux* (Paris : Masson).
MEYER, A., and YOUNG, W. H., 1965, *Phys. Rev.*, **139,** A401.
MIĄSEK, M., 1958, *Bull. Acad. pol. Sci. Sér. Sci. math. astr. phys.*, **8,** 89 ; 1963, *Phys. Rev.*, **130,** 11.
MIĄSEK, M., and SUFFCZYŃSKI, M., 1961 a, *Bull Acad. pol. Sci. Sér. Sci. math. astr. phys.*, **9,** 477 ; 1961 b, *Ibid.*, **9,** 483.
MICAH, E. T., STOCKS, G. M., and YOUNG, W. H., 1969 a, *J. Phys.* C (*Solid St. Phys.*), **2,** 1653 ; 1969 b, *Ibid.*, **2,** 1661.
MILFORD, F. J., and GAGER, W. B., 1961, *Phys. Rev.*, **121,** 716.
MILLER, B. I., 1966, *Phys. Rev.*, **151,** 519.
MILLIKEN, J. C., and YOUNG, R. C., 1966, *Phys. Rev.*, **148,** 558.
MILLS, J. J., MORANT, R. A., and WRIGHT, D. A., 1965, *Br. J. appl. Phys.*, **16,** 479.
MINA, R. T., ÉDEL'MAN, V. S., and KHAĬKIN, M. S., 1966, *Zh. éksp. teor. Fiz.*, **51,** 1363. English translation : *Soviet Phys. JETP*, **24,** 920.
MINA, R. T., and KHAĬKIN, M. S., 1963, *Zh. éksp. teor. Fiz.*, **45,** 1304. English translation : *Soviet Phys. JETP*, **18,** 896 ; 1965 a, *Ibid.*, **48,** 111. English translation : *Soviet Phys. JETP*, **21,** 75 ; 1965 b, *Zh. éksp. teor. Fiz., Pis'ma*, **1,** 34. English translation : *Soviet Phys. JETP Lett.*, **1,** 60 ; 1966, *Zh. éksp. teor. Fiz.*, **51,** 62. English translation : *Soviet Phys. JETP*, **24,** 42.
MINOMURA, S., TANUMA, S., FUJII, G., NISHIZAWA, M., and NAGANO, H., 1969, *Physics Lett.* A, **29,** 16.
MIYAKE, S. J., and KUBO, R., 1962, *Phys. Rev. Lett.*, **9,** 62.
MIZUSHIMA, S., and ENDO, T., 1968, *J. phys. Soc. Japan*, **24,** 1402.
MOORE, T. W., 1967, *Phys. Rev. Lett.*, **18,** 310 ; 1968, *Phys. Rev.*, **165,** 864.
MOORE, T. W., and SPONG, F. W., 1962, *Phys. Rev.*, **125,** 846.
MORIMOTO, T., 1965, *J. phys. Soc. Japan*, **20,** 500.
MOTT, N. F., and JONES, H., 1936, *The Theory of the Properties of Metals and Alloys* (Oxford University Press).
MÜLLER, W. E., 1966, *Solid St. Commun.*, **4,** 581.
MUNARIN, J. A., 1968, *Phys. Rev.*, **172,** 737.
MUNARIN, J. A., and MARCUS, J. A., 1965, *Proc. 9th Int. Conf. Low Temp. Phys.*, p. 743.
MUNARIN, J. A., MARCUS, J. A., and BLOOMFIELD, P. E., 1968, *Phys. Rev.*, **172,** 718.
NABEREZHNЎKH, V. P., and TOLSTOLUZHSKIĬ, V. P., 1964, *Zh. éksp. teor. Fiz.*, **46,** 18. English translation : *Soviet Phys. JETP*, **19,** 13.
NAGATA, S., and KAWAMURA, H., 1968, *J. phys. Soc. Japan*, **24,** 480.
NANNEY, C., 1963, *Phys. Rev.*, **129,** 109.
NEIGHBOR, J. E., COCHRAN, J. F., and SHIFFMAN, C. A., 1967, *Phys. Rev.*, **155,** 384.
NEIGHBOR, J. E., and SHIFFMAN, C. A., 1967, *Phys. Rev. Lett.*, **19,** 640.
NEWBOWER, R. S., and NEIGHBOR, J. E., 1967, *Phys. Rev. Lett.*, **18,** 538.
NOGUCHI, S., and TANUMA, S., 1967, *Physics Lett.* A, **24,** 710.
NOOTHOVEN VAN GOOR, J. M., 1968, *Physics Lett.* A, **26,** 490.
O'KEEFE, P. M., and GODDARD, W. A., 1969 a, *Phys. Rev. Lett.*, **23,** 300 ; 1969 b, *Phys. Rev.*, **180,** 747.
ÖKTÜ, Ö., and SAUNDERS, G. A., 1967, *Proc. phys. Soc.*, **91,** 156.
OKUMURA, K., and TEMPLETON, I. M., 1962, *Phil. Mag.*, **7,** 1239 ; 1963, *Ibid.*, **8,** 889 ; 1965, *Proc. R. Soc.* A, **287,** 89.

OLSEN, J. L., 1958, *Helv. phys. Acta*, **31,** 713.
OLSEN, T., 1960, *Phys. Rev.*, **118,** 1007 ; 1963 a, *J. Phys. Chem. Solids*, **24,** 187 ; 1963 b, *Ibid.*, **24,** 649.
O'NEAL, H. R., and PHILLIPS, N. E., 1965, *Phys. Rev.*, **137,** A748.
ORCHARD-WEBB, J. H., and COUSINS, J. E., 1968, *Physics Lett.* A, **28,** 236.
O'SULLIVAN, W. J., and SCHIRBER, J. E., 1966, *Phys. Rev.*, **151,** 484 ; 1967, *Physics Lett.* A, **25,** 124.
O'SULLIVAN, W. J., SCHIRBER, J. E., and ANDERSON, J. R., 1967, *Solid St. Commun.*, **5,** 525 ; 1968, *Physics Lett.* A, **27,** 144.
OTAKE, S., and KOIKE, S., 1968, *J. phys. Soc. Japan*, **24,** 1176.
OVERHAUSER, A. W., 1962, *Phys. Rev.*, **128,** 1437 ; 1964, *Phys. Rev. Lett.*, **13,** 190.
OVERHAUSER, A. W., and DE GRAAF, A. M., 1968, *Phys. Rev.*, **168,** 763 ; 1969, *Phys. Rev. Lett.*, **22,** 127.
OVERHAUSER, A. W., and RODRIGUEZ, S., 1966, *Phys. Rev.*, **141,** 431.
OVERTON, W. C., and BERLINCOURT, T. G., 1955, *Phys. Rev.*, **99,** 1165.
PASKIN, A., and WEISS, R. J., 1962, *Phys. Rev. Lett.*, **9,** 199.
PEERCY, P. S., WALSH, W. M., RUPP, L. W., and SCHMIDT, P. H., 1968, *Phys. Rev.*, **171,** 713.
PELIKH, L. N., and EREMENKO, V. V., 1967, *Zh. éksp. teor. Fiz.*, **52,** 885. English translation : *Soviet Phys. JETP*, **25,** 582.
PENZ, P. A., 1968, *Phys. Rev. Lett.*, **20,** 725.
PENZ, P. A., and BOWERS, R., 1967, *Solid St. Commun.*, **5,** 341 ; 1968, *Phys. Rev.*, **172,** 991.
PERZ, J. M., and DOBBS, E. R., 1967, *Proc. R. Soc.* A, **297,** 408.
PERZ, J. M., HUM, R. H., and COLERIDGE, P. T., 1969, *Physics Lett.* A, **30,** 235.
PEVERLEY, J. R., 1968, *Phys. Rev.*, **173,** 689.
PHILLIPS, N. E., 1959, *Phys. Rev.*, **114,** 676 ; 1960, *Ibid.*, **118,** 644 ; 1964, *Ibid.*, **134,** A385.
PHILLIPS, N. E., LAMBERT, M. G., and GARDNER, W. R., 1964, *Rev. mod. Phys.*, **36,** 131.
PHILLIPS, W. C., and WEISS, R. J., 1968, *Phys. Rev.*, **171,** 790.
PINCHERLE, L., 1960, *Rep. Prog. Phys.*, **23,** 355.
PINES, D., 1955, *Solid St. Phys.*, **1,** 367.
PIPPARD, A. B., 1954, *Adv. Electronics Electron Phys.*, **6,** 1 ; 1957, *Phil. Trans. R. Soc.* A, **250,** 325 ; 1960, *Rep. Prog. Phys.*, **23,** 176 ; 1965, *The Dynamics of Conduction Electrons* (London : Blackie).
PIPPARD, A. B., and CHAMBERS, R. G., 1952, *Proc. phys. Soc.* A, **65,** 955.
POSPELOV, YU. A., and KECHIN, V. V., 1963, *Fizika tverd. Tela*, **5,** 3574. English translation : *Soviet Phys. Solid St.*, **5,** 2622 ; 1964, *Ibid.*, **6,** 3206. English translation : *Soviet Phys. Solid St.*, **6,** 2565.
PRESSLEY, R. J., and BERK, H. L., 1965, *Phys. Rev.*, **140,** A1207.
PRIESTLEY, M. G., 1962, *Phil. Mag.*, **7,** 1205 ; 1963, *Proc. R. Soc.* A, **276,** 258 ; 1966, *Phys. Rev.*, **148,** 580.
PRIESTLEY, M. G., FALICOV, L. M., and WEISZ, G., 1963, *Phys. Rev.*, **131,** 617.
PRIESTLEY, M. G., WINDMILLER, L. R., KETTERSON, J. B., and ECKSTEIN, Y., 1967, *Phys. Rev.*, **154,** 671.
RAMANATHAN, K. G., and SRINIVASAN, T. M., 1955, *Phys. Rev.*, **99,** 442.
RAO, G. N., ZEBOUNI, N. H., GRENIER, C. G., and REYNOLDS, J. M., 1964, *Phys. Rev.*, **133,** A141.
RAYNE, J. A., 1962 a, *Physics Lett.*, **2,** 128 ; 1962 b, *Proc. 8th Int. Conf. Low Temp. Phys.*, p. 204 ; 1963 a, *Phys. Rev.*, **129,** 652 ; 1963 b, *Ibid.*, **131,** 653.
RAYNE, J. A., and CHANDRASEKHAR, B. S., 1962, *Phys. Rev.*, **125,** 1952.
RED'KO, N. A., and SHALYT, S. S., 1968, *Fizika tverd. Tela*, **10,** 1557. English : translation : *Soviet Phys. Solid St.*, **10,** 1233.

REED, W. A., and MARCUS, J. A., 1962, *Phys. Rev.*, **126,** 1298.
REITZ, J. R., 1955, *Solid St. Phys.*, **1,** 1.
REITZ, J. R., and OVERHAUSER, A. W., 1968, *Phys. Rev.*, **171,** 749.
RENEKER, D. H., 1959, *Phys. Rev.*, **115,** 303.
REYNOLDS, J. M., GOODRICH, R. A., and KHAN, S. A., 1966, *Phys. Rev. Lett.*, **16,** 609.
RICE, T. M., 1965, *Ann. Phys.*, **31,** 100 ; 1968, *Phys. Rev.*, **175,** 858.
ROBERTS, B. W., 1960, *Phys. Rev.*, **119,** 1889 ; 1961, *Phys. Rev. Lett.*, **6, 453**.
ROBINSON, J. E., 1968, *Nuovo Cim.* (Suppl.), **6,** 745.
ROBINSON, J. E., and DOW, J. D., 1968, *Phys. Rev.*, **171,** 815.
ROSENBERG, H. M., 1963, *Low Temperature Solid State Physics. Some Selected Topics* (Oxford University Press).
ROWE, J. M., BROCKHOUSE, B. N., and SVENSSON, E. C., 1965, *Phys. Rev. Lett.*, **14,** 554.
ROY, A. P., and VENKATARAMAN, G., 1967, *Phys. Rev.*, **156,** 769.
SAGALYN, P. L., and HOFMANN, J. A., 1962, *Phys. Rev.*, **127,** 68.
SAGAWA, T., 1966, *J. phys. Soc. Japan*, **21,** 49.
SAHA, A. R., 1965, *Proc. I.E.E.E.*, **53,** 106.
SAHNI, V. C., VENKATARAMAN, G., and ROY, A. P., 1966, *Physics Lett.*, **23,** 633.
SAITO, Y., 1963 a, *J. phys. Soc. Japan*, **18,** 452 ; 1963 b, *Ibid.*, **18,** 1845.
SAKAI, T., MATSUMOTO, Y., and MASE, S., 1969, *J. phys. Soc. Japan*, **27,** 862.
SATO, H., 1959, *J. phys. Soc. Japan*, **14,** 609.
SATO, Y., 1968, *J. phys. Soc. Japan*, **24,** 489.
SAUNDERS, G. A., and ÖKTÜ, Ö., 1968, *J. Phys. Chem. Solids*, **29,** 327.
SCHIRBER, J. E., 1963, *Phys. Rev.*, **131,** 2459.
SCHIRBER, J. E., and O'SULLIVAN, W. J., 1969, *Phys. Rev.*, **184,** 628.
SCHLOSSER, H., and MARCUS, P. M., 1963, *Phys. Rev.*, **131,** 2529.
SCHMUNK, R. E., BRUGGER, R. M., RANDOLPH, P. D., and STRONG, K. A., 1962, *Phys. Rev.*, **128,** 562.
SCHNEIDER, T., and STOLL, E., 1966 a, *Phys. Kondens. Materie*, **5,** 331 ; 1966 b, *Ibid.*, **5,** 364 ; 1967 a, *Ibid.*, **6,** 135 ; 1967 b, *Physics Lett.* A, **24,** 258.
SCHULTZ, S., and SHANABARGER, M. R., 1966, *Phys. Rev. Lett.*, **16,** 178.
SCHUMACHER, R. T., and VANDERVEN, N. S., 1966, *Phys. Rev.*, **144,** 357.
SEGALL, B., 1961, *Phys. Rev.*, **124,** 1797 ; 1963, *Ibid.*, **131,** 121.
SEIDEL, G., and KEESOM, P. H., 1958, *Phys. Rev.*, **112,** 1083.
SEKOYAN, S. S., and LIKHTER, A. I., 1960, *Fizika tverd. Tela*, **2,** 1940. English translation : *Soviet Phys. Solid St.*, **2,** 1748.
SHAM, L. J., and ZIMAN, J. M., 1963, *Solid St. Phys.*, **15,** 221.
SHAND, J. B., 1969, *Physics Lett.* A, **30,** 478.
SHAPIRA, Y., 1964 a, *Bull. Am. phys. Soc.*, **9,** 239 ; 1964 b, *Phys. Rev. Lett.*, **13,** 162.
SHAPIRA, Y., and LAX, B., 1964, *Phys. Rev. Lett.*, **12,** 166.
SHAPIRA, Y., and NEURINGER, L. J., 1967, *Phys. Rev. Lett.*, **18,** 1133.
SHAPIRA, Y., and WILLIAMSON, S. J., 1965, *Physics Lett.*, **14,** 73.
SHEPELEV, A. G., and FILIMONOV, G. D., 1966, *Zh. éksp. teor. Fiz.*, **51,** 746. English translation : *Soviet Phys. JETP*, **24,** 496.
SHEPHERD, J. P. G., and GORDON, W. L., 1968, *Phys. Rev.*, **169,** 541.
SHEPHERD, J. P. G., LARSON, C. O., ROBERTS, D., and GORDON, W. L., 1964, *Proc. 9th Int. Conf. Low Temp. Phys.*, p. 752.
SHOCKLEY, W., 1937, *Phys. Rev.*, **52,** 866.
SHOENBERG, D., 1939, *Proc. R. Soc.* A, **170,** 341 ; 1952, *Phil. Trans. R. Soc.* A, **245,** 1 ; **1953**, *Physica, 's Grav.*, **19,** 791 ; 1957, *Prog. Low Temp. Phys.*, **2,** 226 ; 1962, *Phil. Trans. R. Soc.* A, **255,** 85 ; 1964, *Proc. 9th Int. Conf. Low Temp. Phys.*, p. 680 ; 1969, *The Physics of Metals.* 1. *Electrons*, edited by J. M. Ziman (Cambridge University Press), Ch. 2.

SHOENBERG, D., and STILES, P. J., 1963, *Physics Lett.*, **4,** 274 ; 1964, *Proc. R. Soc.* A, **281,** 62.
SHOENBERG, D., and UDDIN, M. Z., 1936 a, *Proc. Camb. phil. Soc. math. phys. Sci.*, **32,** 499 ; 1936 b, *Proc. R. Soc.* A, **156,** 687.
SHYU, W. M., DAS, T. P., and GASPARI, G. D., 1966, *Phys. Rev.*, **152,** 270.
SHYU, W. M., GASPARI, G. D., and DAS, T. P., 1966, *Phys. Rev.*, **141,** 603.
SILVERSTEIN, S. D., 1962, *Phys. Rev.*, **128,** 631 ; 1963, *Ibid.*, **130,** 912.
SIMON, G., 1965, *Z. angew. Phys.*, **20,** 161.
SINGER, L. S., and WAGONER, G., 1962, *J. chem. Phys.*, **37,** 1812.
SLATER, J. C., 1965, *Quantum Theory of Molecules and Solids.* Vol. 2. *Symmetry and Energy Bands in Crystals* (New York : McGraw-Hill) ; 1967, *Ibid.*, Vol. 3, *Insulators, Semiconductors and Metals* (New York : McGraw-Hill).
SLATER, J. C., KOSTER, G. F., and WOOD, J. H., 1962, *Phys. Rev.*, **126,** 1307.
SLONCZEWSKI, J. C., and WEISS, P. R., 1958, *Phys. Rev.*, **109,** 272.
SMITH, G. E., 1959, *Phys. Rev.*, **115,** 1561 ; 1961, *J. Phys. Chem. Solids*, **20,** 168.
SMITH, G. E., BARAFF, G. A., and ROWELL, J. M., 1964, *Phys. Rev.*, **135,** A1118.
SMITH, G. E., GALT, J. K., and MERRITT, F. R., 1960, *Phys. Rev. Lett.*, **4,** 276.
SMITH, G. E., HEBEL, L. C., and BUCHSBAUM, S. J., 1963, *Phys. Rev.*, **129,** 154.
SMITHELLS, C. J., 1967, *Metals Reference Book* (London : Butterworth).
SNIDER, J. L., and NICOL, J., 1957, *Phys. Rev.*, **105,** 1242.
SNODGRASS, H. J., and BENNETT, L. H., 1963, *Phys. Rev.*, **132,** 1465.
SNOW, E. C., 1967, *Phys. Rev.*, **158,** 683.
SONDHEIMER, E. H., 1952, *Proc. phys. Soc.* A, **65,** 561.
SOULE, D. E., 1964, *IBM Jl Res. Dev.*, **8,** 268.
SOULE, D. E., MCCLURE, J. W., and SMITH, L. B., 1964, *Phys. Rev.*, **134,** A453.
SOVEN, P., 1965 a, *Phys. Rev.*, **137,** A1706 ; 1965 b, *Ibid.*, **137,** A1717.
SPARLIN, D. M., and SCHREIBER, D. S., 1964, *Proc. 9th Int. Conf. Low Temp. Phys.*, p. 823.
SPONG, F. W., and KIP, A. F., 1965, *Phys. Rev.*, **137,** A431.
SPRY, W. J., and SCHERER, P. M., 1960, *Phys. Rev.*, **120,** 826.
SQUIRES, G. L., 1966, *Proc. phys. Soc.*, **88,** 919.
STAFLEU, M. D., and DE VROOMEN, A. R., 1966, *Physics Lett.*, **23,** 179 ; 1967 a, *Phys. Stat. Sol.*, **23,** 675 ; 1967 b, *Ibid.*, **23,** 683.
STAGER, R. A., and DRICKAMER, H. G., 1963 a, *Phys. Rev.*, **131,** 2524 ; 1963 b, *Ibid.*, **132,** 124.
STANFORD, J. L., and STERN, E. A., 1966, *Phys. Rev.*, **144,** 534.
STARK, R. W., 1967, *Phys. Rev.*, **162,** 589.
STARK, R. W., ECK, T. G., and GORDON, W. L., 1964, *Phys. Rev.*, **133,** A443.
STARK, R. W., ECK, T. G., GORDON, W. L., and MOAZED, F., 1962, *Phys. Rev. Lett.*, **8,** 360.
STEDMAN, R., ALMQVIST, L., NILSSON, G., and RAUNIO, G., 1967, *Phys. Rev.*, **163,** 567.
STEDMAN, R., and NILSSON, G., 1965, *Phys. Rev. Lett.*, **15,** 634 ; 1966, *Phys. Rev.*, **145,** 492.
STEELE, M. C., 1955, *Phys. Rev.*, **99,** 1751.
STEELE, M. C., and BABISKIN, J., 1955, *Phys. Rev.*, **98,** 359.
STERN, E. A., 1961, *Phys. Rev.*, **121,** 397 ; 1963, *Phys. Rev. Lett.*, **10,** 91.
STEVENSON, R., 1967, *Can. J. Phys.*, **45,** 4115.
STEWART, A. T., 1961, *Phys. Rev.*, **123,** 1587 ; 1964, *Ibid.*, **133,** A1651.
STEWART, A. T., SHAND, J. B., DONAGHY, J. J., and KUSMISS, J. H., 1962, *Phys. Rev.*, **128,** 118.
STOCKS, G. M., and YOUNG, W. H., 1969, *J. Phys.* C (*Solid St. Phys.*), **2,** 680.
STROUD, D., and EHRENREICH, H., 1968, *Phys. Rev.*, **171,** 399.
SUFFCZYŃSKI, M., 1961, *Bull. Acad. pol. Sci. Sér. Sci. math. astr. phys.*, **9,** 489.

SUGIHARA, K., and ONO, S., 1966, *J. phys. Soc. Japan*, **21,** 631.

SULLIVAN, P., and SEIDEL, G., 1967, *Physics Lett.* A, **25,** 229.

SUNDSTRÖM, L. J., 1965, *Phil. Mag.*, **11,** 657.

SURMA, M., 1968, *Physics Lett.* A, **26,** 562.

SURMA, M., FURDYNA, J. K., and PRADDAUDE, H. C., 1964, *Phys. Rev. Lett.*, **13,** 710.

SYBERT, J. P., GRENIER, C. G., and REYNOLDS, J. M., 1962, *Bull. Am. phys. Soc.*, **7,** 74.

SYBERT, J. R., MACKEY, H. J., and HATHCOX, K. L., 1968, *Phys. Rev.*, **166,** 710.

SYBERT, J. R., MACKEY, H. J., and MILLER, R. E., 1967, *Physics Lett.* A, **24,** 655.

TAKANO, S., and KAWAMURA, H., 1968, *Physics Lett.* A, **26,** 187.

TAKEYA, K., and YAZAWA, K., 1964, *J. phys. Soc. Japan*, **19,** 138.

TANAKA, K., SURI, K., and JAIN, A. L., 1968, *Phys. Rev.*, **170,** 664.

TANAKA, K., TANUMA, S., and KUKUORI, F., 1961, *Sci. Rep. Res. Insts Tôhoku Univ.* A, **13,** 67.

TAYLOR, M. T., 1964, *Phys. Rev. Lett.*, **12,** 497 ; 1965, *Phys. Rev.*, **137,** A1145.

TAYLOR, W. A., MCCOLLUM, D. C., PASSENHEIM, B. C., and WHITE, H. W., 1967, *Phys. Rev.*, **161,** 652.

TERRELL, J. H., 1964, *Physics Lett.*, **8,** 149 ; 1966, *Phys. Rev.*, **149,** 526.

THOMAS, R. L., and BOHM, H. V., 1966, *Phys. Rev. Lett.*, **16,** 587.

THOMAS, R. L., and TURNER, G., 1968, *Phys. Rev.*, **176,** 768.

THORSEN, A. C., and BERLINCOURT, T. G., 1961 a, *Bull. Am. phys. Soc.*, **6,** 511 ; 1961 b, *Phys. Rev. Lett.*, **6,** 617.

TINKHAM, M., 1956, *Phys. Rev.*, **101,** 902.

TORGESON, D. R., and BARNES, R. G., 1962, *Phys. Rev. Lett.*, **9,** 255.

TOUREILLE, A., 1967, *C. r. hebd. Séanc. Acad. Sci., Paris*, B, **265,** 778.

TOWNES, C. H., HERRING, C., and KNIGHT, W. D., 1950, *Phys. Rev.*, **77,** 852.

TOXEN, A. M., and TANSAL, S., 1963, *Phys. Rev. Lett.*, **10,** 481 ; 1965, *Phys. Rev.*, **137,** A211.

TRIPP, J. H., EVERETT, P. M., GORDON, W. L., and STARK, R. W., 1969, *Phys. Rev.*, **180,** 669.

TRIPP, J. H., GORDON, W. L., EVERETT, P. M., and STARK, R. W., 1967, *Physics Lett.* A, **26,** 98.

TRIVISONNO, J., SAID, M. S., and PAUER, L. A., 1966, *Phys. Rev.*, **147,** 518.

TSOI, V. S., and GANTMAKHER, V. F., 1969, *Zh. éksp. teor. Fiz.*, **56,** 1232. English translation : *Soviet Phys. JETP*, **29,** 663.

VAIŠNYS, J. R., and KIRK, R. S., 1967 *J. appl. Phys.*, **38,** 4335.

VALIČ, M. I., SHARMA, S. N., and WILLIAMS, D. L., 1968, *Physics Lett.* A, **26,** 528.

VAN DER HOEVEN, B. J. C., and KEESOM, P. H., 1963, *Phys. Rev.*, **130,** 1318 ; 1964, *Ibid.*, **135,** A631 ; 1965, *Ibid.*, **137,** A103.

VANDERKOOY, J., and DATARS, W. R., 1967, *Phys. Rev.*, **156,** 671 ; 1968, *Can. J. Phys.*, **46,** 1935.

VAN DER LUGT, W., and KNOL, J. S., 1967, *Phys. Stat. Sol.*, **23,** K83.

VANDERVEN, N. S., 1968, *Phys. Rev.*, **168,** 787.

VAN NIEUWSTADT, H. M. M., and DE VROOMEN, A. R., 1967, *Physics Lett.* A, **24,** 367.

VASVARI, B., 1968, *Rev. mod. Phys.*, **40,** 776.

VASVARI, B., ANIMALU, A. O. E., and HEINE, V., 1967, *Phys. Rev.*, **154,** 535.

VASVARI, B., and HEINE, V., 1967, *Phil. Mag.*, **15,** 731.

VATAMANYUK, V. I., KULYUPIN, YU. A., and SARBEI, O. G., 1968, *Zh. éksp. teor. Fiz., Pis'ma*, **7,** 23. English translation : *Soviet Phys. JETP Lett.*, **7,** 15.

VERKIN, B. I., DMITRENKO, I. M., and MIKHAILOV, I. F., 1955, *Dokl. Akad. Nauk SSSR*, **101,** 233.

VERKIN, B. I., LAZAREV, B. G., and RUDENKO, N. S., 1950 a, *Zh. éksp. teor. Fiz.*, **20,** 93 ; 1950 b, *Ibid.*, **20,** 995.

VOLOTSKAYA, V. G., 1963 a, *Zh. éksp. teor. Fiz.*, **44,** 80. English translation : *Soviet Phys. JETP*, **17,** 56 ; 1963 b, *Ibid.*, **45,** 49. English translation : *Soviet Phys. JETP*, **18,** 36.

VOL'SKIĬ, E. P., 1964, *Zh. éksp. teor. Fiz.*, **46,** 123. English translation : *Soviet Phys. JETP*, **19,** 89.

VON DER LAGE, F. C., and BETHE, H. A., 1947, *Phys. Rev.*, **71,** 612.

VON GUTFELD, R. J., and NETHERCOT, A. H., 1967, *Phys. Rev. Lett.*, **18,** 855.

VOSKO, W. H., TAYLOR, R., and KEECH, G. H., 1965, *Can. J. Phys.*, **43,** 1187

WAGNER, T. K., and KOCH, J. F., 1968, *Phys. Rev.*, **165,** 885.

WALLACE, D. C., 1968, *Phys. Rev.*, **176,** 832 ; 1969, *Ibid.*, **178,** 900.

WALLACE, P. R., 1947, *Phys. Rev.*, **71,** 622 ; 1960, *Solid St. Phys.*, **10,** 1.

WALSH, W. M., RUPP, L. W., and SCHMIDT, P. H., 1966, *Phys. Rev.*, **142,** 414.

WANG, E. Y., and MCCARTHY, K. A., 1969, *Phys. Rev.*, **183,** 653.

WATTS, B. R., 1963, *Physics Lett.*, **3,** 284 ; 1964, *Proc. R. Soc.* A, **282,** 521.

WAYNE, R. C., and COTTS, R. M., 1963, *J. chem. Phys.*, **39,** 1337.

WEAIRE, D., 1968, *J. Phys.* C (*Proc. phys. Soc.*), **1,** 210.

WEDDELING, F. K., 1965, *J. appl. Phys.*, **36,** 328.

WEINER, D., 1962, *Phys. Rev.*, **125,** 1226.

WEISZ, G., 1966, *Phys. Rev.*, **149,** 504.

WIGNER, E., and SEITZ, F., 1933, *Phys. Rev.*, **43,** 804 ; 1934, *Ibid.*, **46,** 509.

WILLIAMS, G. A., 1965, *Phys. Rev.*, **139,** A771.

WILLIAMSON, S. J., FONER, S., and DRESSELHAUS, M. S., 1964, *Proc. 9th Int. Conf. Low Temp. Phys.*, p. 771 ; 1965, *Phys. Rev.*, **140,** A1429 ; 1966, *Carbon*, **4,** 29.

WILLIAMSON, S. J., SURMA, M., PRADDAUDE, H. C., PATTEN, R. A., and FURDYNA, J. K., 1966, *Solid St. Commun.*, **4,** 37.

WINDMILLER, L. R., 1966, *Phys. Rev.*, **149,** 472.

WINDMILLER, L. R., and PRIESTLEY, M. G., 1965, *Solid St. Commun.*, **3,** 199.

WOLCOTT, N. M., 1956, *Bull. Am. phys. Soc.*, **1,** 289.

WOOD, J. H., 1966, *Phys. Rev.*, **146,** 432.

WOODS, A. D. B., BROCKHOUSE, B. N., MARCH, R. H., STEWART, A. T., and BOWERS, R., 1962, *Phys. Rev.*, **128,** 1112.

WOOLLAM, J. A., 1968, *Physics Lett.* A, **27,** 246.

WOOLLAM, J. A., and SCHROEDER, P. A., 1968, *Phys. Rev. Lett.*, **21,** 81.

WOOTEN, F., HUEN, T., and STUART, R. N., 1965, *Proceedings of the International Colloquium on Optical Properties and Electronic Structure of Metals and Alloys*, Paris (1965), edited by F. Abelès (Amsterdam : North-Holland), p. 332.

YAHIA, J., and MARCUS, J. A., 1959, *Phys. Rev.*, **113,** 137.

YAMADA, T., 1965, *J. phys. Soc. Japan*, **20,** 1424.

YAMAGUCHI, Y., and TANUMA, S., 1965, *Physics Lett.*, **16,** 237.

YAQUB, M., and COCHRAN, J. F., 1965, *Phys. Rev.*, **137,** A1182.

YASINSKY, J. B., and ERGUN, S., 1965, *Carbon*, **2,** 355.

YNTEMA, G. B., 1953, *Phys. Rev.*, **91,** 1388.

YOUNG, R. C., 1962, *Phil. Mag.*, **7,** 2065 ; 1965, *Phys. Rev. Lett.*, **15,** 262 ; 1966, *Phys. Rev.*, **152,** 659 ; 1967, *Ibid.*, **163,** 676 ; 1968 a, *Physics Lett.* A, **27,** 539 ; 1968 b, *Phil. Mag.*, **18,** 201.

YOUNG, T., 1960, *Phys. Rev.*, **117,** 1244.

ZEBOUNI, N. J., and BLEWER, R. S., 1967, *Physics Lett.* A, **24,** 106.

ZIMAN, J. M., 1960, *Electrons and Phonons* (Oxford University Press) ; 1964, *Adv. Phys.*, **13,** 89 ; 1969, *The Physics of Metals.* 1. *Electrons* (Cambridge University Press).

ZITTER, R. N., 1962, *Phys. Rev.*, **127,** 1471.

Addendum

In recent de Haas–van Alphen measurements on Ba (K. A. McEwen, *Physics Lett.* A, **30,** 77 (1969)) three periods were observed and they were satisfactorily assigned to various parts of the ‘ tetracube ’ predicted from the band structure calculations (G. Johansen, *Solid St. Commun.*, **7,** 731 (1969)).

Further de Haas–van Alphen measurements on Tl (F. A. Capocci, P. M. Holtham, D. Parsons and M. G. Priestley, *J. Phys.* C (*Solid St. Phys.*), **3,** 2081 (1970)) showed that, in addition to the ‘ cookie ’ in band 3 and the multiply-connected region of electrons in band 4, there are two further small closed pieces of Fermi surface in Tl. One set of small pockets with the symmetry of mmm(D_{2h}) could be either holes in band 3 at M or electrons in band 4 along ΓM, but it was not feasible to distinguish between these two possibilities. The second set of pockets were dumb-bell shaped and were thought to be centred at H and in band 5 ; this result differs from the predictions of Soven (see § 5.5 of part I) which predicted two roughly ellipsoidal pockets of holes at H, one in band 5 and the other in band 6.

THE FERMI SURFACE

PART II

d- block and f- block metals

§ 1. Introduction

In part I of this article we described the theoretical and experimental work which has been performed on the determination of the shapes of the Fermi surfaces of the simple metals. In this connection we defined the term 'simple metal' as meaning a metal belonging to the s-block or p-block of the periodic table. We have seen that the general features of the Fermi surfaces of all the simple metals are now well established and that the dimensions of a large number of these Fermi surfaces have been determined with considerable precision. It is now possible to use the known Fermi surfaces of many of the simple metals in calculations aimed at giving quantitative explanations of their electronic properties. We have also seen that for many of these metals the free-electron model gives a very good first approximation to the shape of the Fermi surface and it is, in general, quite obvious for any given simple metal how many conduction electrons per atom there are. The considerable success which has attended the use of pseudopotential methods in connection with the simple metals is a consequence of the relevance of the nearly-free electron model. This is because the net scattering experienced by a conduction electron in any one of these metals is quite weak although this does not imply that the potential $V(\mathbf{r})$ itself is weak.

In part II we now turn to the consideration of the electronic band structures and the Fermi surfaces of the d-block and f-block metals, which include the transition metals and the rare-earth metals. Some authors would regard the elements of groups IB (Cu, Ag and Au) and IIB (Zn, Cd and Hg), which we include in part II, as 'simple' metals; indeed, as we shall see in §§ 2 and 3, there is some justification for this in the fact that it is still possible to see some relationship between the Fermi surfaces of these metals and the appropriate free-electron Fermi surfaces. On the other hand, some authors (for example, Brewer 1963, 1964, 1967, Engel 1967) advocate regarding the noble metals Cu. Ag and Au as transition metals. The Fermi surfaces of the remaining metals, which are either transition metals or rare-earth metals, show very substantial departures from the predictions of the free-electron model, as we shall see in later sections.

In addition to the topics that were discussed in § 1 of part I, and which we shall not repeat here, there are one or two other general points that arise in connection with the d-block and f-block metals. These include the problem of determining the number of conduction electrons per atom and the complications caused by the existence of ferromagnetic and antiferromagnetic ordering in a number of transition and rare-earth metals.

It has been a commonly accepted view for a very long time (Mott 1935, Mott and Jones 1936) that in the metals of the first transition series (Sc to Ni or Cu) the atomic 3d electrons are in a very narrow band, and can therefore be regarded as fairly well localized, while the 4s electrons are in a very broad band and are, therefore, not very strongly localized; this is often illustrated by the schematic density of states in fig. 1. However, the idea that an electron in a metal can be meaningfully described as an s electron or a d electron is only a very crude approximation and should be treated with considerable caution (for extensive discussions of electrons in transition metals see, for example, Hume-Rothery and Coles (1954), Mott and Stevens (1957), Lomer and Marshall (1958), Herring (1960).

Fig. 1

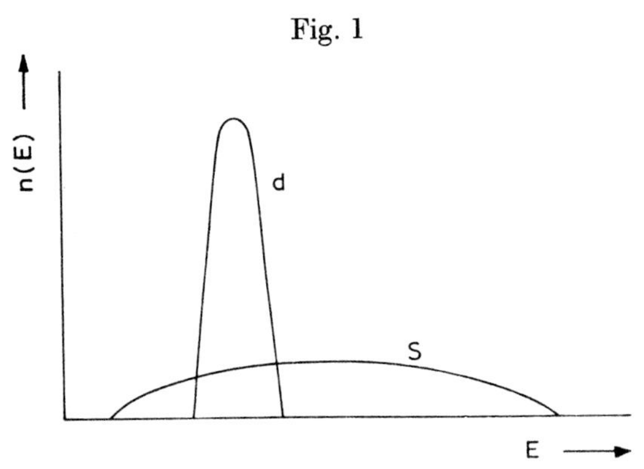

Schematic density of states for d and s bands in a transition metal.

Mott (1964), Phillips and Mueller (1967) and Friedel (1969)). Nevertheless, in so far as it is meaningful to describe the outer electrons of an atom in a metal as d or s electrons, this feature of narrow d-like bands and broad s-like bands can still be seen in some of the recent and quite reliable band-structure calculations on these metals (see, for example, the energy bands and densities of states calculated by Snow and Waber (1969) for a large number of cubic transition metals using the A.P.W. method). It might then be argued that, because the d band is narrow and the d electrons are therefore quite highly localized, it is only the s electrons which should be regarded as the conduction electrons. However, the actual relative widths of the bands are much less important than the fact that the d band and the s band overlap ; this is closely related to the chemical feature that the transition elements are notorious for exhibiting variable valence. The number of conduction electrons per atom in a metal of the first transition series is, therefore, given by the total number of 3d and 4s electrons in a free atom of that element. Similarly, in the second or third transition series the number of conduction electrons per atom is given by the total number of atomic 4d and 5s electrons or 5d and 6s electrons respectively. In the past separate band structure calculations have sometimes been performed for the d electrons taking no account of the hybridization with s orbitals or other atomic orbitals at all ; predictions of the shape of the Fermi surface of a transition metal based on the results of such calculations should be viewed with considerable suspicion (Hume-Rothery 1963). In the rare-earth metals, on the other hand, the situation is rather different. Variable valence is a much less prominent feature of the chemistry of the rare-earth elements than of the transition elements. The electronic structures of the atoms of the elements in the first rare-earth series take the general form . . . $4f^n 5s^2 5p^6 5d^1 6s^2$ and the 4f electrons are sufficiently deep within the atoms that, with only one or two exceptions, the valence of these elements is fairly well fixed as three (Sidgwick 1950). Correspondingly, in the metallic form the bands derived from the atomic 4f electrons are often claimed to be well separated from the bands derived from the atomic 5d and 6s electrons (Gupta and Loucks 1969). It is therefore, possible to consider the electronic band structure associated with the atomic 5d and 6s electrons quite separately from the 4f bands. There will then generally be three conduction electrons per atom in each rare-earth metal. The electronic properties of the rare-earth metals can therefore be expected to be quite similar to those of the metals of group IIIA. It is, in fact, very difficult to calculate directly the separation between the 4f bands and the bands derived from the 5d and 6s electrons because of their sensitivity to the choice of potential $V(\mathbf{r})$ for the metal. Optical work by Müller (1966) on Ba and Eu, which only differ by a half full shell of 4f electrons, suggested that the 4f bands were not close to the conduction bands in rare-earth metals. However, what is more definitely established and is relatively insensitive to the choice of $V(\mathbf{r})$ is the fact that the 4f bands are very narrow. For example, in their relativistic A.P.W.

calculations on Yb, Johansen and Mackintosh (1970) obtained two extremely narrow sets of 4f bands, one set consisted of four two-fold degenerate bands corresponding to $j=7/2$ and with a width of about $2{\cdot}5\times10^{-3}$ Ry while the other set consisted of three two-fold degenerate bands corresponding to $j=5/2$ and with a width of about $1{\cdot}2\times10^{-4}$ Ry. Slightly larger widths were calculated for the 4f bands in Ce by Mukhopadhyay and Majumdar (1969). The separation between the $j=5/2$ and $j=7/2$ sets of bands was, however, quite large, namely $\sim0{\cdot}1$ Ry. Because the 4f bands are narrow they will not contribute appreciably to the electronic properties of the metal, except that when they are partially filled they will produce a large magnetic moment. For the actinide series of metals it is the 5f, 6d, and 7s shells which are important, and the 6d and 7s electrons which can be regarded as the conduction electrons.

In part I we generally ignored the difference between the bands for spin-up and spin-down electrons, although for a few of the heavy metals we did discuss the importance of spin–orbit coupling. However, if a metal undergoes a transition from a paramagnetic phase to a spontaneously ordered ferromagnetic or antiferromagnetic phase there will be a very drastic change in the Fermi surface of that metal. To a first approximation the introduction of ferromagnetic ordering does not alter the shape of either the spin-up or spin-down bands. However, because of the magnetic ordering there will now be an intense spontaneous magnetic field, $\mathbf{B}_{\text{int}}$, within a specimen of the metal and this will cause a separation of $g\beta|\mathbf{B}_{\text{int}}|$ between the spin-up and spin-down bands, see fig. 2. There will now no longer be equal numbers of occupied spin-up and spin-down bands and for both sets of bands the position of the Fermi level will be moved relative to the Fermi level of the paramagnetic metal. For the more complicated forms of antiferromagnetic ordering which occur in Cr and in many of the rare-earth metals the effect of the magnetic ordering on the band structure and Fermi surface is also more complicated, see §§ 7.4 and 10.2. The origins of the magnetic phenomena are actually slightly different for the transition metals and the rare-earth metals. For the

Fig. 2

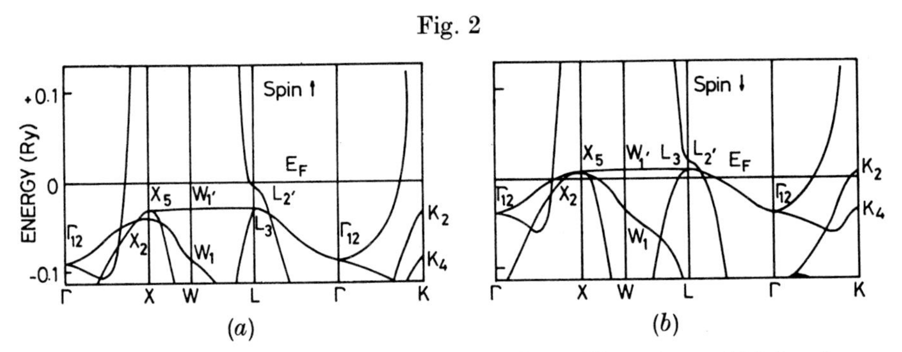

Schematic band structure for ferromagnetic Ni. The spin-up (↑) bands are assumed to have lower energy (Tsui and Stark 1966 b).

transition metals the magnetic moments are those of the conduction electrons themselves whereas for the rare-earth metals the magnetic moments arise principally from the incomplete shell of 4f electrons which we do not regard as conduction electrons. In the transition metals it is the exchange interactions among the conduction electrons which are responsible for the magnetic ordering, whereas in the rare-earth metals it is the exchange interactions between the 4f electrons in the various ion cores, via the conduction electrons, which are responsible for the magnetic ordering (Zener 1951 a, b, c, 1952, Pauling 1953, Ruderman and Kittel 1954, Kasuya 1956, Yosida 1957, Bambakidis 1970).

The appearance of a spontaneous magnetic field $\mathbf{B}_{\text{int}}$, or, for a ferromagnetic metal, of a spontaneous magnetic moment $\mathbf{M}$ when the metal becomes magnetically ordered, causes a considerable reduction in the symmetry and consequently many degeneracies in the band structure of the non-magnetic metal become lifted. The presence of a net magnetic moment, $\mathbf{M}$, in a ferromagnetic metal destroys the operation of time-reversal, θ, as a symmetry operation of the crystal. However, although the operation of time-reversal is not present on its own it may still occur in combination with some point-group or space-group operation as a symmetry operation of a domain of the magnetic metal. That is, one needs to consider the magnetic space group of the crystal, which is, of course, a subgroup of the (grey) space group of the paramagnetic metal. In recent years a considerable amount of work has been done on the theory of magnetic groups and their corepresentations (for various reviews see, for example, Birss 1964, Opechowski and Guccione 1965, Cracknell 1967, 1969 b, Bradley and Davies 1968). It was predicted by Falicov and Ruvalds (1968) and Cracknell (1969 a, 1970) that for a ferromagnetic f.c.c. or b.c.c. metal magnetized parallel to [001], [111], or [110] the energy bands are non-degenerate (except for possible accidental degeneracies) all over the appropriate Brillouin zone. For ferromagnetic h.c.p. metals magnetized parallel to [0001] the sticking together of the bands all over the top hexagonal face AHL of the Brillouin zone in the paramagnetic metal is completely removed except at the points A and L and along the line R. The energy bands along the line R, which are four-fold degenerate in the paramagnetic state including time-reversal degeneracy, become split into two two-fold degenerate bands. For ferromagnetic h.c.p. metals magnetized parallel to $[10\bar{1}0]$ or $[11\bar{2}0]$ the two-fold degeneracy, due to time-reversal symmetry, all over AHL is lifted in general, but still survives at the points A and L and along the lines R, S, and S′ (Falicov and Ruvalds 1968, Cracknell 1969 a, 1970).

In addition to the complications mentioned in the preceding paragraphs there have been considerable technical problems associated with the determination of the Fermi surfaces of the transition and rare-earth metals. On the experimental side it has, in the past, been difficult to obtain pure single-crystal specimens of many of these metals, while on the theoretical side the lack of a clear-cut division between conduction electrons and core electrons makes it very difficult to calculate realistic crystal potentials $V(\mathbf{r})$ *ab initio* in these metals. Because of the uncertainties in $V(\mathbf{r})$ it is therefore quite difficult to calculate reliable band structures for the transition metals and rare-earth metals. Consequently, for the transition metals the use of the *rigid-band model* has been quite popular in the past; this model is based on the assumption that, for metals with a common crystal structure, as one moves through any one of the series of transition metals the shapes of the electronic energy bands are not altered, apart from scaling due to differences in the lattice constants, and all that happens is that the number of conduction electrons per atom is changed and the position of the Fermi level changes accordingly. As we shall see in several later sections, this model has been used quite successfully to predict the shapes of the Fermi surfaces of a number of transition metals. The rigid-band model can also be extended to describe dilute alloys by allowing departures from the integer values for the electron/atom ratio.

It is probably true to say that the general features of the Fermi surfaces of all the simple metals, as well as the detailed dimensions of many of them, are now well established. A start has also been made on using the known Fermi surfaces of many of the simple metals in attempting to give quantitative explanations of many of the electronic properties of these metals, see § 8 of part I. However, for the transition metals and rare-earth metals the work is less far advanced and there still remains a considerable amount of rather dull routine work to be done to complete the mapping of the Fermi surfaces of these metals. For this reason we shall not include tables of dimensions of the Fermi surfaces of d-block and f-block metals in part II except when the dimensions are well established. In quoting experimental values of γ, the temperature coefficient of the electronic specific heat, we shall usually quote the mean values (given by Gschneidner (1964) or Heiniger *et al.* (1966)) together with any significant results of later and more accurate measurements. As in part I we shall not quote experimental values of χ_p, the Pauli spin paramagnetic susceptibility, for comparison with the results of band-structure calculations of the density of states, because of the difficulty of isolating the band-structure contribution from the other contributions which are present in the measured susceptibility of a metal. In the interest of progressing steadily through the periodic table we shall not use the historical approach which is probably still the easiest way to describe the present incomplete knowledge of the Fermi surfaces of the transition metals in groups IIIA to VIII. We shall see as we progress through the periodic table that, because of the close similarities among the various transition metals, the rigid-band model has been used quite extensively. Consequently those transition metals which, for historical reasons only, have been studied most extensively have tended to be used as a basis for constructing a generalized transition metal band structure for all those transition metals with the same crystal structure. For any transition metal for which the shape of the Fermi surface had not been studied experimentally it was then possible, by using the generalized transition metal band structure for that crystal structure together with the known or assumed number of conduction electrons per atom, to deduce the Fermi energy and thence the shape of the Fermi surface. Because of the many complicated magnetic ordering patterns exhibited by the rare-earth metals it seemed better to leave them until after the simpler ferromagnetic metals of group VIII had been discussed, rather than to discuss them after the elements Sc and Y of group IIIA which might have been more logical.

§ 2. d-BLOCK : GROUP IB : THE NOBLE METALS

Cu	Copper	2.8.(8.10) 1
Ag	Silver	2.8.18.(8, 10) 1
Au	Gold	2.8.18.32.(8, 10) 1

2.1. *Structures*

Cu, Ag and Au all have the f.c.c. structure with lattice constants (Smithells 1967) of $a = 3{\cdot}61$ Å for Cu, $a = 4{\cdot}08$ Å for Ag, and $a = 4{\cdot}07$ Å for Au. In the electronic structure of each of these elements a completed d shell has been formed at the expense of one s electron.

2.2. *Copper*

The electronic structure of metallic Cu is characterized by a full d band with one electron per atom in the conduction band. It is to be expected, by comparison with the electronic structure of a free Cu atom, that the d band is not far below the conduction band. The free-electron Fermi surface for a metal with the f.c.c. structure and one conduction electron per atom is shown in fig. 2 (*a*) of part I. This Fermi surface consists of a sphere with volume equal to half the volume of the Brillouin zone ; it is completely contained within the Brillouin zone and does not touch the surface of the Brillouin zone. However, from the relationship between the electrical conductivity and the thermal conductivity of Cu, Klemens (1954) argued that the Fermi surface of Cu must be sufficiently distorted from sphericity so as to touch the Brillouin zone boundary ; the most likely boundary point is L, at the centre of the hexagonal face of the Brillouin zone, since it is nearest to Γ. The sign of the thermoelectric power (Jones 1955), the magnitude of the Hall effect and the non-saturation of the magnetoresistance (Chambers 1956) also indicated that the Fermi surface of Cu is multiply-connected. The details of the shape of the Fermi surface of Cu were established quite early in the history of Fermiology by Pippard (1957) using the anomalous skin effect. This was before any of the other effects which give direct measurements of Fermi surface geometry, such as the de Haas–van Alphen effect, had been observed in Cu. Because the anomalous skin effect does not enable one to make direct deductions about features of the shape of the Fermi surface, Pippard had to guess a form for the Fermi surface of Cu and calculate the anisotropy of the anomalous skin resistance for comparison with the experimental

Fig. 3

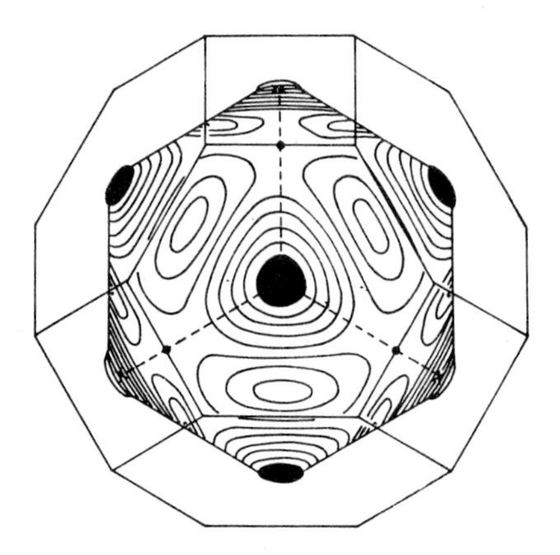

The Fermi surface of Cu (Pippard 1957).

measurements. Adjustments were then made to the assumed Fermi surface until the best fit was obtained. Although this procedure was successful for Cu it is unlikely that it would lead to similar success for a metal with a more complicated Fermi surface. Previous work on Cu was quite sparse and was discussed in detail by Pippard (1957) ; this included anomalous skin effect work on polycrystalline Cu (Chambers 1952), measurements of the electronic contribution to the specific heat and one or two band-structure calculations (Jones 1937, Howarth 1953, 1955). The Fermi surface of Cu deduced by Pippard is illustrated in fig. 3 and the contours are mapped in fig. 17 of Pippard (1957). This Fermi surface can be fitted to 1% accuracy by a simple analytical function satisfying the correct boundary and symmetry conditions

$$E_{\mathbf{k}} = \alpha[(-3 + \cos \tfrac{1}{2}ak_x \cos \tfrac{1}{2}ak_y + \cos \tfrac{1}{2}ak_y \cos \tfrac{1}{2}ak_z + \cos \tfrac{1}{2}ak_z \cos \tfrac{1}{2}ak_x) + r(-3 + \cos ak_x + \cos ak_y + \cos ak_z)] \tag{2.2.1}$$

with $r = 0{\cdot}0995$ and $E_{\mathrm{F}}/\alpha = 3{\cdot}6301$ (García-Moliner 1958).

The de Haas–van Alphen effect in Cu was first observed by Shoenberg (1959) and further measurements confirmed the general features of the Fermi surface determined by Pippard (Shoenberg 1960 a, b) as also did the experiments of Alekseevskiĭ and Gaĭdukov (1959 a, b) on the non-saturation of the magnetoresistance of Cu in certain directions. A simple pseudopotential plane wave calculation of the band structure of Cu was performed by Cornwell (1961) and the parameters were adjusted to give the best fits to these de Haas–van Alphen results. A detailed set of de Haas–van Alphen measurements on Cu were made by Shoenberg (1962) who followed the angular variations of the principal frequencies ; a photograph of a solid model of the Fermi surface of Cu is given in fig. 1 of the paper by Shoenberg (1962). In addition to the obvious orbits around

Fig. 4

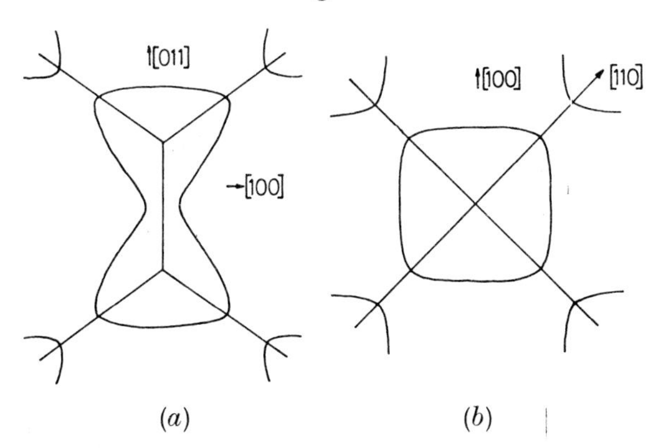

(*a*) The ' dog's bone ' and (*b*) the ' four-cornered rosette ' orbits in Cu (Shoenberg 1962).

the belly and around the necks several other orbits are also possible, two of which, the ' dog's bone ' and ' four-cornered rosette ', are shown schematically in fig. 4. An interpolation scheme to represent the Fermi surface of Cu analytically was devised by Roaf (1962) based on the expression

$$\begin{aligned} E_{\mathbf{k}} = -\alpha\{3 &- \cos \tfrac{1}{2}ak_y \cos \tfrac{1}{2}ak_z - \cos \tfrac{1}{2}ak_z \cos \tfrac{1}{2}ak_x - \cos \tfrac{1}{2}ak_x \cos \tfrac{1}{2}ak_y \\ &+ C_{200}(3 - \cos ak_x - \cos ak_y - \cos ak_z) \\ &+ C_{211}(3 - \cos ak_x \cos \tfrac{1}{2}ak_y \cos \tfrac{1}{2}ak_z - \ldots) \\ &+ C_{220}(3 - \cos ak_y \cos ak_z - \ldots) \\ &+ C_{310}(6 - \cos \tfrac{3}{2}ak_x \cos \tfrac{1}{2}ak_y - \cos \tfrac{3}{2}ak_x \cos \tfrac{1}{2}ak_z - \ldots) \\ &+ C_{321}(6 - \cos \tfrac{3}{2}ak_x \cos ak_y \cos \tfrac{1}{2}ak_z - \ldots)\} \end{aligned} \qquad (2.2.2)$$

(see also Hume-Rothery and Roaf (1961)). Values of the coefficients C_{200}, C_{211}, etc. were obtained by devising the best fit to the de Haas–van Alphen frequencies determined by Shoenberg (1962). Values of the computed radius vectors and cross-sectional areas are given in tables 4 and 5, respectively, of Roaf (1962). The Fermi surface constructed in this way was used to calculate the surface impedance in the anomalous skin effect and this gave almost as good agreement with the experimental results as did the Fermi surface actually proposed by Pippard (1957) although there were some differences between the detailed dimensions of the two Fermi surfaces.

Band structure calculations for Cu were performed by Burdick (1961, 1963) and Mattheiss (1964) using the A.P.W. method and Segall (1961, 1962) using the Green's function method. The calculations performed by Burdick and by Segall produced band structures for Cu that were in close agreement with each other ($\sim$0·01 Ry usually), partly through having used the same potential ; the dimensions of their calculated Fermi surfaces were in good semiquantitative agreement ($\sim 10\%$) with the experimental results already described for the anomalous skin effect and de Haas–van Alphen effect as well as with magnetoacoustic geometric resonance measurements of some of the caliper dimensions (Morse 1960, Morse *et al.* 1961). Further magnetoacoustic measurements of caliper dimensions of the Fermi surface of Cu were made by Bohm and Easterling ; the results are compared with the de Haas–van Alphen data and the results of the band structure calculations of Segall and of Burdick in table VI of Bohm and Easterling (1962). Subsequently, both more refined experiments and more accurate calculations for Cu have been performed and the agreement between experiment and calculations has become very close indeed. The shape of the Fermi surface of Cu can now be regarded as known with sufficient accuracy to enable it to be used in testing theories of various electronic properties of metals, for example, electron–phonon interactions via ultrasonic attenuation (Kolouch and McCarthy 1965, MacFarlane *et al.* 1965, 1967, Boyd and Gavenda 1966, Alig and Rodriguez 1967, Hall 1967, MacFarlane and Rayne 1967, Mitchell and Yates 1967, Jericho and Simpson 1968) or for testing different methods for constructing potentials for use in band structure calculations (Bross and Junginger 1963, Lebègue 1963, Wakoh 1965, Ballinger and Marshall 1967, Faulkner *et al.* 1967, Altmann *et al.* 1968, Chatterjee and Sen 1968, Hubbard and Dalton 1968, Jacobs 1968 a, b, Butler *et al.* 1969, Fong and Cohen 1970, Moriarty 1970, O'Sullivan *et al.* 1970†).

The more recent band structure calculations for Cu include a Green's function calculation (Faulkner *et al.* 1967) as well as self-consistent A.P.W. calculations (Snow and Waber 1967, Greisen 1968, Snow 1968 a). Janak (1969) used the method of Gilat and Raubenheimer (1966) to re-calculate the density of states for Cu from the band structure of Snow and Waber (1967). Some idea of the measure of agreement between the various calculations and experiments can be obtained from the various values of the Fermi surface radii collected by Kamm (1970), see table 1. The more recent de Haas–van Alphen measurements in Cu include measurements by Joseph and Thorsen (1964 a), Joseph *et al.* (1966), Halse (1969) and O'Sullivan and Schirber (1969). Zornberg and Mueller (1966) showed how the method of expansion in cubic harmonics, introduced for closed Fermi surfaces by Mueller and Priestley (1966) (see also Mueller (1966)), can be modified to handle multiply-connected Fermi surfaces. The method was applied to Cu using the de Haas–van Alphen data of Joseph *et al.* (1966), while the method devised by Roaf (1962) was applied to the new experimental data by Halse (1969). The coefficients in the cubic harmonic expansion are given in table I of Zornberg and Mueller (1966) and some of the principal Fermi surface dimensions are given in table II of that paper. The volume contained within the Fermi surface calculated

† See note added in proof.

Table 1. Comparisons of Cu Fermi surface radial dimensions in symmetry directions, relative to the free-electron radius

	R_{110}	R_{100}	R_{neck}
Experimental			
Bohm and Easterling (1962)	$0{\cdot}95_6$	$1{\cdot}03_6$	$0{\cdot}19_5$
Roaf (1962)	0·943	1·076	0·200
Zornberg and Mueller (1966)	0·958	1·060	0·190
Jan and Templeton (1967)			0·1886
O'Sullivan and Schirber (1967 a)			0·1886
Halse (1969)	0·951	1·059	
Kamm (1970)	$0{\cdot}95_5$	$1{\cdot}05_5$	$0{\cdot}18_9$
Theoretical			
Faulkner *et al.* (1967)			
$V_I\ l_{max}=2$	0·9553	1·0501	0·178
$V_I\ l_{max}=4$	0·9544	1·0530	0·1808
V_{II}	0·9426	1·0658	0·2146
Snow and Waber (1967)			
$S=1$†	0·939	0·998	0·161
$S=\frac{2}{3}$†	0·836	1·056	0·279

†S = coefficient of Slater exchange term.
(Adapted from Kamm (1970).)

from this parametrization was equivalent to $(1{\cdot}006 \pm 0{\cdot}003)$ electrons per atom; it was suggested by Zornberg and Mueller that the areas of cross section obtained by Joseph *et al.* were about 0·4% too large, corresponding to an error of +0·2% in the linear dimensions of the Fermi surface. The de Haas–van Alphen measurements of Jan and Templeton (1967) and O'Sullivan and Schirber (1967 a) were in general agreement with this and the cross-sectional areas of the Fermi surface of Cu are now known with an accuracy of about 0·2% (see, for example, table I of Jan and Templeton (1967)). Radius vectors of the Fermi surface of Cu at 5° intervals of θ and ϕ are given in table 8 of Halse (1969).

2.3. *Silver and Gold*

In the free-electron model the Fermi surfaces of Ag and Au, which are monovalent f.c.c. metals like Cu, would be expected to be spheres which are completely contained within the Brillouin zone, see fig. 2 (*a*) of part I. However, on account of the general similarity between Cu, Ag and Au we might expect the Fermi surfaces of Ag and Au to be sufficiently distorted so as to touch the hexagonal face of the Brillouin zone, as happens in Cu.

The existence of open orbits was demonstrated by various work on the transverse magnetoresistance (Alekseevskiĭ and Gaĭdukov 1958 on Au, 1959 b on Ag and Au, Gaĭdukov 1959 on Au) so that it appeared that the

Fermi surface proposed by Pippard (1957) for Cu also applied to Ag and Au as well (Priestley 1960). This was further supported for both Ag and Au by the preliminary de Haas–van Alphen results of Shoenberg (1960 a, b) and some measurements of caliper dimensions using magnetoacoustic geometric resonances (Morse 1960, Morse *et al.* 1960, 1961) and for Ag alone by some further magnetoresistance measurements (Alekseevskiĭ and Gaĭdukov 1962 a). A large number of caliper dimensions of the Fermi surface were obtained from further magnetoacoustic geometric resonance measurements (Bohm and Easterling 1962 on Ag and Au, Easterling and Bohm 1962 on Ag). The detailed measurement of the angular variation of the de Haas–van Alphen frequencies by Shoenberg (1962) and the associated determination by Roaf (1962) of the coefficients in an analytical representation of the Fermi surface for Cu described in § 2.2 was also performed at the same time on Ag and Au. The computed radius vectors and cross-sectional areas are given in tables 4 and 5 of Roaf (1962). When the Fermi surface computed in this way was used to calculate the surface impedance for Ag in the anomalous skin effect the agreement with the experimental results of Morton (1960) was worse than for the Fermi surface actually suggested by Morton ; however, since this latter proposed Fermi surface did not touch the Brillouin zone boundary at all it now seems that the experimental anomalous skin effect data itself was not very accurate for Ag.

In further de Haas–van Alphen measurements (Joseph and Thorsen 1964 b on Ag and Au, 1965 on Ag, Joseph *et al.* 1965 on Au, Halse 1969 on Ag and Au) the angular variations of the belly, neck, rosette and dog's bone periods in Ag and Au were determined with an accuracy of about 0·1% and with an estimated error of about 0·5% in the absolute values of the periods. For details see the tables and graphs of Joseph and Thorsen (1965) for Ag and Joseph *et al.* (1965) for Au ; these measurements have been used as a basis for new analytical representations of the Fermi surfaces of Ag and Au (Halse 1969). In their earlier work on Ag Joseph and Thorsen (1964 b) observed a new long de Haas–van Alphen period and it was tentatively suggested that this might arise from a small pocket of electrons in band 2 in Ag ; however, their later work (Joseph and Thorsen 1965) together with a magnetoresistance study of Ag by Fink (1964) suggested that there is in fact no small pocket of electrons in Ag and that the observed long de Haas–van Alphen period arose as beats between two ordinary de Haas–van Alphen periods that are close together in frequency. Accurate measurements of the de Haas–van Alphen neck period and the ratio of the belly and neck periods were made by Jan and Templeton (1967) for Ag and Au.

Since the shapes of the Fermi surfaces of Ag and Au had been determined quite accurately before many band-structure calculations on these metals had been performed, these band-structure calculations tend to be used to give guidance about methods for constructing suitable *ab initio* potentials or suitable pseudopotentials rather than for providing any new information about the Fermi surface topology (Glasser 1962, Chatterjee and Sen 1964, 1966, 1967, 1968, Chatterjee and Chakraborty 1967, Jacobs 1968 a, b, Lewis and Lee 1968, Ballinger and Marshall 1969, Bhatnagar 1969, Christensen 1969, Kupratakuln and Fletcher 1969, Pant and Joshi 1969, Christensen and Seraphin 1970, Moriarty 1970, O'Sullivan *et al.* 1970, Ramchandani 1970, Schlosser 1970, Sommers and Amar 1970). Even in the self-consistent A.P.W. calculation by Snow (1968 b) for Ag the exchange contribution to the potential was adjusted to obtain the best fit between the calculated and experimental Fermi surface dimensions, see table 2.

Table 2. Values for the neck and belly radii of the Fermi surface of Ag (in units of free-electron sphere radii)

	R_{neck}	R_{100}	R_{110}	R_{100}/R_{110}
Experimental				
Shoenberg (1960 a)	0·14			
Bohm and Easterling (1962)	0·142	1·072	0·955	1·13
Morse†		1·006	0·975	1·03
Theoretical				
Snow (1968 b)				
S=1‡	(Does not exist)	1·007	0·975	1·03
S=$\frac{5}{6}$‡	0·095	1·015	0·940	1·08

† Quoted by Bohm and Easterling (1962).
‡ S=coefficient of Slater exchange term.
(Snow 1968 b.)

Although the general shapes of the Fermi surfaces of Cu, Ag and Au are similar, it is interesting to see how the magnitude of the distortion of the spherical free-electron Fermi surface varies from one to another of these metals. This can be seen from, for instance, the comparison of various Fermi surface dimensions for Cu, Ag and Au given in table VI of Bohm and Easterling (1962) or table 8 of Halse (1969). The Fermi surface of Ag is least distorted from the free-electron sphere and consequently one would expect the neck radius to be smaller in Ag than in Cu or Au, as is indeed observed, for example,

Cu	Ag	Au	
0·200	0·137	0·177	Shoenberg (1962)
0·195	0·142	0·180	Bohm and Easterling (1962)

where the neck radii are given in units of the radius of the spherical free-electron Fermi surface. The relative distortions of the Fermi surfaces of

Fig. 5

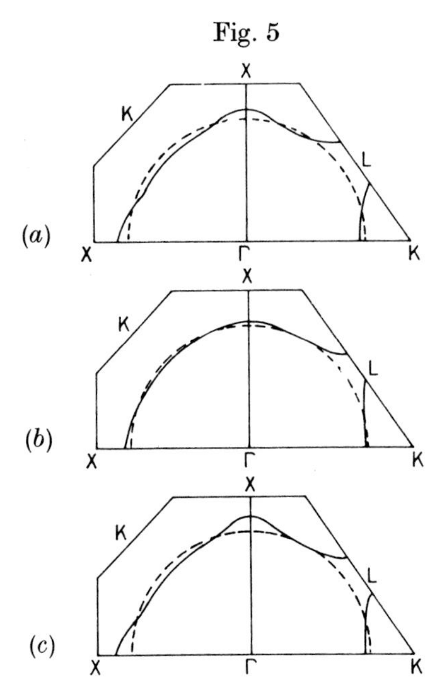

Illustration of the relative distortions of the Fermi surfaces of (*a*) Cu, (*b*) Ag, and (*c*) Au (Roaf 1962).

Cu, Ag and Au from the free-electron model are illustrated in fig. 5. Roaf's interpolation scheme has been applied to the recent de Haas–van Alphen data on Ag and Au and radius vectors of the Fermi surfaces at 5° intervals in θ and ϕ are given in table 8 of Halse (1969).

2.4. *Miscellaneous Work Related to the Electronic Structures of the Noble Metals*

In addition to the principal results which have established both the general features and the fine details of the Fermi surfaces of the noble metals and which we have described in §§ 2.2 and 2.3, useful supporting evidence has come from measurements of a number of other properties. These include cyclotron resonance (Fawcett 1956 on Cu, Kip *et al.* 1961 on Cu, Koch *et al.* 1964 on Cu, Langenberg and Marcus 1964 on Au, Howard 1965 on Ag, Häussler and Welles 1966 on Cu, Smith 1967 on Cu, Henningsen 1969 b on Ag, Hui 1969 on Cu), magnetoresistance and Hall effect measurements (Lautz and Tittes 1958 on Cu and Au, De Launay *et al.* 1959 on Cu, Lüthi 1959 on Cu, Ag and Au, Franken and Van den Berg 1960 on Cu and Ag, Priestley 1960 on Cu, Ag and Au, Jongenburger 1961 on Cu, Ag and Au, Kunzler and Klauder 1961 on Cu, Funes and Coleman 1963 on Cu, Motulevich and Shubin 1964 on Au, Larson *et al.* 1965 on Ag, Klauder *et al.* 1966 on Cu, Chopra and Bahl 1967 on Cu, Ag and Au, Lück and Saeger 1967 on Cu, Neubert 1967, 1969 on Cu and Ag, Alderson *et al.* 1968 on Cu, Ag and Au, Saeger 1968 on Cu, van Witzenburg and Laubitz 1968 on Cu, Ag and Au, Powell *et al.* 1969 on Cu), the thermal magnetoresistance (Van Witzenburg and Laubitz 1968 on Cu, Ag and Au), the Righi–Leduc effect or thermal Hall effect (Lipson 1964 on Cu), the Knight shift (Tompa 1966 on Cu), positron annihilation (De Benedetti *et al.* 1950 on Au, Lang *et al.* 1955 on Cu, Ag and Au, Stewart 1957 on Cu, Ag and Au, Berko and Plaskett 1958 on Cu, Średniawa 1960 on Cu and Au, Fujiwara 1965 on Cu, Chen Wang 1966 on Cu, Fujiwara and Sueoka 1966, 1967 on Cu, Chen *et al.* 1967 on Ag, Sueoka 1967, 1969 on Cu, Berko *et al.* 1968 on Cu, Williams *et al.* 1968 on Cu, Čížek and Adam 1969 on Cu, Mijnarends 1969 on Cu), the anomalous skin effect (Johnson and Johnson 1965 on Cu), thermal conductivity and thermopower (van Baarle *et al.* 1966 on Ag), radio-frequency size effect (Antoniewicz *et al.* 1968 on Cu, Henningsen 1969 a on Ag, Perrin *et al.* 1970 on Cu†), quantum oscillations in the velocity and attenuation of ultrasound (Alers and Swim 1963 on Au, Deaton and Gavenda 1963 on Cu, Beattie 1969 on Cu), various kinds of optical and x-ray work (Suffczyński 1959 on Cu, Shkliarevskiĭ and Padalka 1959 a on Cu and Au, 1959 b on Cu, Ag and Au, Padalka and Shkliarevskiĭ 1961 on Ag and Au, 1962 on Cu, Berglund and Spicer 1964 a, b on Cu and Ag, Lettington 1964 on Cu, Motulevich and Shubin 1964 on Au, Nikiforov and Blokhin 1964 on Cu, Spicer and Berglund 1964 on Cu, Beaglehole 1965 on Cu and Au, Yarovaya and Shkliarevskiĭ 1965 on Ag, McGroddy *et al.* 1965 on Ag and Au, McAlister *et al.* 1965 on Ag–Au alloys, Spicer 1965 on Cu and Ag, Beaglehole 1966 on Cu, Lenham and Treherne 1966 on Cu, Ag and Au, 1967 on Cu, Shkliarevskiĭ and Yarovaya, 1966 on Au, Edelmann and Ulmer 1967 on Cu, Ag and Au, Gerhardt *et al.* 1967 on Cu, Mueller and Phillips 1967 on Cu, Nilsson *et al.* 1967 on Au, Theye 1967 on Cu and Au, Bennett *et al.* 1968 on Ag, Biondi and Guobadia 1968 on Cu, Eastman and Krolikowski 1968 on Cu, Dresselhaus 1969 on Cu, Krolikowski and Spicer 1969 on Cu, Nilsson *et al.* 1969 on Au), and ion-neutralization spectroscopy measurements of the density of states (Hagstrum and Becker 1967). A band structure calculation by Davis (1968) of the Knight shift in Cu gave a value of 0·43%, which is to be compared with experimental values ranging from 0·36 to 0·52%. A value of 1·64% was obtained by Narath (1967) for the Knight shift in Au. Other miscellaneous experiments and calculations related to the electronic structure of the noble metals include measurements of the longitudinal magnetoresistance, which shows little sensitivity to the shape of the Fermi surface but can be used to study variations of λ over the Fermi surface (Powell 1964 on Cu, Strom-Olsen 1967 on Cu and Ag, Clark and Powell 1968 on Cu). Recent calculated values of γ, the temperature coefficient of the electronic specific heat, of Cu include 0·631 mJ mole^{-1} deg^{-2} and 0·720 mJ mole^{-1} deg^{-2} by Faulkner *et al.* (1967) using two

† See note added in proof.

different potentials. These two values are respectively 9% lower and 4% higher than the mean experimental value of 0·6943 mJ mole^{-1} deg^{-2} deduced by Osborne *et al.* (1967) from results given in 15 different papers; other recent experimental values of γ for Cu include (0·696 ± 0·002) mJ mole^{-1} deg^{-2} (Gmelin and Gobrecht 1967), (0·6915 ± 0·0034) mJ mole^{-1} deg^{-2} (Martin 1968 b) and 0·69327 and 0·69677 mJ mole^{-1} deg^{-2} (Boerstoel *et al.* 1968). The closeness of the calculations of Faulkner *et al.* to the experimental values suggests that the contribution to γ from electron–phonon interactions in Cu is quite small, although Zornberg and Mueller (1966) had suggested an electron–phonon enhancement of γ as high as 25%. A mean experimental value of γ of (0·659 ± 0·027) mJ mole^{-1} deg^{-2} for Ag is quoted by Gschneidner (1964); other recent experimental values are slightly lower and include:

(0·650 ± 0·002) mJ mole^{-1} deg^{-2}	Dixon *et al.* (1965)
(0·646 ± 0·005) mJ mole^{-1} deg^{-2}	Green and Culbert (1965)
(0·645 ± 0·007) mJ mole^{-1} deg^{-2}	Isaacs (1965)
0·6450 mJ mole^{-1} deg^{-2}	Ahlers (1967)
(0·6409 ± 0·0037) mJ mole^{-1} deg^{-2}	Martin (1968 b).

Lewis and Lee (1968) suggest that the enhancement of the band-structure value of γ due to electron–phonon interactions is about 10%. Values of γ for Au include

(0·748 ± 0·013) mJ mole^{-1} deg^{-2}	Gschneidner (1964)
(0·729 ± 0·018) mJ mole^{-1} deg^{-2}	Isaacs (1965)
(0·728 ± 0·018) mJ mole^{-1} deg^{-2}	Martin (1966)
(0·730 ± 0·007) mJ mole^{-1} deg^{-2}	Will and Green (1966)
(0·6911 ± 0·0042) mJ mole^{-1} deg^{-2}	Martin (1968 b).

The last value of γ for Au (Martin 1968 b), which is based on data that include measurements at lower temperatures than were used in most of the previous work, is several per cent lower than all the previous values. The phonon dispersion relations obtained from inelastic neutron scattering measurements by various workers were used by Krebs and Hölzl (1967) to determine a pseudopotential for Cu. Calculations of the thermoelectric power of Cu, making use of the known phonon dispersion relations, electronic band structure, and Fermi surface have been made by a number of workers (Klemens 1954, Abarenkov and Vedernikov 1966, Hasegawa and Kasuya 1968, 1970, Williams and Davis 1968); similar calculations of the electrical conductivity and the Hall coefficient were also made by Hasegawa and Kasuya (1970). The early work on the macroscopic properties related to the electronic structures of the noble metals was reviewed by Cohen and Heine (1958). An extensive discussion of most of the transport properties of the noble metals in terms of their known Fermi surfaces has been given by Ziman (1961). Fuchs (1936 a, b) extended the work of Wigner and Seitz (1933, 1934) on the cohesive energy of Na to the calculation of the cohesive energy and elastic constants of Cu. An expression for the ground-state energy of a metal, obtained by using many-body theory and a pseudopotential constructed from spectroscopic data, has been used to calculate the cohesive energy, lattice constant, and compressibility of Cu (Gubanov and Nikulin 1966, Nikulin 1966 a, Nikulin and Trzhaskovskaya 1968). The good agreement of these calculations with experiments indicates that a quantitative explanation of cohesion in terms of the metallic bonding of the conduction electrons can now be obtained, at least for noble metals.

The effect of pressure on the dimensions of the Fermi surfaces of the noble metals has been investigated by a number of workers. Unsuccessful attempts to observe changes in the Fermi surface dimensions, other than due to scaling as the size of the Brillouin zone changes, were made by Caroline and Schirber (1963) (on Cu and Ag) and Tan (1965) (on Cu) by magnetoresistance measurements. Davis *et al.* (1968) calculated the band structure of Cu as a function of lattice parameter ; from these results they obtained values for the pressure derivatives of several of the cross-sectional areas of the Fermi surface which were in quite good agreement with the results of de Haas–van Alphen measurements on Cu under pressure (Templeton 1966 (who measured Ag and Au as well), O'Sullivan and Schirber 1968). Collins (1966) used the results of the measurements of the pressure-dependence of the de Haas–van Alphen periods (Templeton 1966) to calculate the electronic contributions to the thermal expansions of Cu, Ag and Au at very low temperatures. Measurements of the pressure dependence of the optical constants and the band structure of Cu were made by Gerhardt (1968). The pressure dependence of the Fermi surface of Cu calculated by Jan (1968) produced pressure derivatives of the neck radius which were only about 50% of the other calculated and experimental values just mentioned, although when the calculations were repeated for Ag and Au this calculation gave the correct relative variation in the sequence Cu, Ag, Au. Shoenberg and Watts (1964, 1967) determined the changes in Fermi surface cross-sectional areas in Cu, Ag and Au under tension using the de Haas–van Alphen effect. It should also be possible to investigate the change in the Fermi surface due to elastic deformation by utilizing the fact that a contact potential will develop between two specimens of the same metal if one is stressed and the other is unstressed ; this has been studied for the noble metals by Lukhvich (1969). Chollet and Templeton (1968) determined the variation of the neck cross section and of the $\langle 111 \rangle$ belly/neck ratio as functions of impurity concentration in a number of dilute ($< 0{\cdot}1\%$) alloys of Zn, Cd, Al, Ni and Pd in Cu using the de Haas–van Alphen effect ; the results for non-transition-metal additives showed good agreement with the rigid-band model. Under conditions of very high de Haas–van Alphen magnetization relative to the de Haas–van Alphen period a state of lower energy may be able to be achieved by having domains of equal and opposite magnetization. At first only indirect evidence for these domains was available (Condon 1966), but they were later observed rather more directly in Ag as causing a splitting in the ^{109}Ag nuclear magnetic resonance line (Condon and Walstedt 1968). The problem of the magnetic interaction of two periodic terms in the de Haas–van Alphen effect which was discussed by Shoenberg (1968) was investigated experimentally for Ag by Shoenberg and Templeton (1968).

§ 3. d-BLOCK : GROUP IIB : ZINC, CADMIUM AND MERCURY

Zn	Zinc	2.8.(8.10)2
Cd	Cadmium	2.8.18.(8, 10)2
Hg	Mercury	2.8.18.32.(8, 10)2

3.1. *Structures*

Zn and Cd have the h.c.p. structure with $a = 2{\cdot}66$ Å, $c = 4{\cdot}94$ Å and $a = 2{\cdot}973$ Å, $c = 5{\cdot}605$ Å, respectively (Smithells 1967). These values correspond to c/a ratios which depart considerably from the ideal value of $\sqrt{(8/3)}$ which means that the structures are considerably more distorted than that of Mg for which the value of c/a is very close to the ideal value. Hg, which is of course a liquid at room temperature, crystallizes at about -39°C to form a trigonal crystal that belongs to the space group R$\bar{3}$m($D_{3d}{}^5$). The structure can be characterized by various unit cells of which the simplest has a single Hg atom at (0, 0, 0) with $a = 2{\cdot}9863$ Å and $\alpha = 70° \; 44{\cdot}6'$ at 5°K (Barrett 1957) ; this can be regarded as obtained by distortion of an ideal f.c.c. structure for which the trigonal lattice parameter would be $\frac{1}{2}a_0\sqrt{2}$, where a_0 is the cubic lattice parameter and $\alpha = 60°$. Other descriptions of the unit cell are given by Smithells (1967) (structure A10).

3.2. *Zinc and Cadmium*

The free-electron Fermi surface for a divalent h.c.p. metal is shown in fig. 2 (*c*) of part I. We also saw in part I that this Fermi surface was a very good approximation to the Fermi surface of Mg but did not approximate so closely to the Fermi surface of Be. Because of the similarity between their structures it might have been more appropriate to discuss the Fermi surfaces of Zn and Cd together with those of Be and Mg in part I. In many h.c.p. metals the effect of the departure of c/a from the ideal value of $\sqrt{(8/3)}$ is not very important because, although it alters the relative dimensions of the various pieces of Fermi surface it does not usually cause any alterations of the general features. However, for Cd, c/a is sufficiently greater than $\sqrt{(8/3)}$ that the vertical needle or cigar-shaped pockets of electrons in bands 3 and 4 at K do not exist at all even in the free-electron model for Cd. The critical value of c/a for the disappearance of these needles is $1{\cdot}8607$ (Harrison 1960) which is substantially smaller than the

Fig. 6

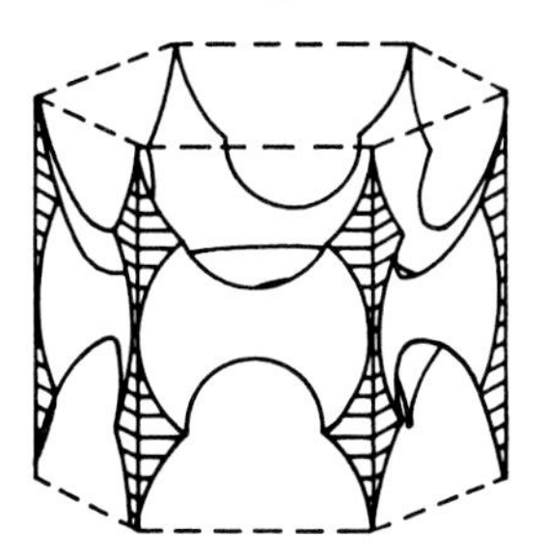

The free-electron Fermi surface of Cd in bands 1 and 2 (Harrison 1960).

actual value of c/a for Cd (of about 1·886). Simultaneously the connectivity of the arms of the monster in bands 1 and 2 becomes altered, see fig. 6. For Zn c/a is smaller than in Cd and is estimated by Harrison (1960) to be 1·8246 at the low temperatures usually used in Fermi surface studies ; this is smaller than the critical value of 1·8607 so that in the free-electron model these needles survive for Zn.

There are really two Fermi surface topologies for each of the metals Zn and Cd ; at low magnetic fields spin–orbit coupling is sufficiently large to make the double-zone scheme invalid, whereas for large magnetic fields magnetic breakdown can be expected to occur and thereby effectively restore the validity of the double-zone scheme for large magnetic fields (Cohen and Falicov 1960, 1961). The early experimental work on the electronic structures of Zn and Cd has been reviewed by Harrison (1960). This included Shubnikov–de Haas (Lazarev *et al.* 1939), cyclotron resonance (Galt *et al.* 1959) and galvanomagnetic (Borovik 1950, 1956, Lüthi 1959) measurements as well as many de Haas–van Alphen measurements which indicated that the needles in band 3 exist for Zn but not for Cd and in which for both metals several other oscillations were observed (Marcus 1947, 1950, Dingle and Shoenberg 1950, Verkin *et al.* 1950 a, b, Shoenberg 1952, Berlincourt 1954, Berlincourt and Steele 1954, Dhillon and Shoenberg 1955, Verkin and Dmitrenko 1955, 1958, Dmitrenko *et al.* 1957). The temperature dependence of the magnetic susceptibility of Zn and Cd was measured by Marcus (1949). It is interesting to note that at high pressures c/a for Cd decreases and eventually reaches a value at which, in the free-electron model, these needle-shaped pockets of electrons would be expected to appear. Evidence for the appearance of these needles in Cd has been obtained in magnetoresistance measurements on Cd at pressures up to 15 kbar (Gaĭdukov and Itskevich 1963, Itskevich and Voronovskiĭ 1966). Verkin *et al.* (1968) were able to observe the disappearance of the needles in pure Zn when the temperature was raised as well as the appearance of the needles in Cd when doped with Mg ; at the disappearance or appearance of the needles a maximum occurs in the susceptibility. The principal effect of pressure on Zn is also to decrease the axial ratio c/a and this, in the free-electron approximation, has the effect of increasing the horizontal cross-sectional area of the needles in band 3; this increase in cross-sectional area has been observed experimentally by various different methods (Balain *et al.* 1960, Gaĭdukov and Itskevich 1963, O'Sullivan and Schirber 1965) although a decrease in cross-sectional area with increasing pressure had previously been reported (Dmitrenko *et al.* 1958, Verkin and Dmitrenko 1958, 1959). An O.P.W. calculation of the band structure and Fermi surface of Zn was performed by Harrison (1962, 1963) using a potential obtained from the Hartree–Fock calculations of Piper (1961) for Zn. The calculated Fermi surface was adjusted to fit the considerable number of measured de Haas–van Alphen periods (of Joseph *et al.* 1961 and Joseph and Gordon (1962)). These experimental results included measurements of periods on the needles in band 3, and the horizontal arms and diagonal arms of the monster in band 2. The volume of the needles in Zn is very small and corresponds to about 5×10^{-6} electron per atom. These experimental results also included two periods that could only be explained satisfactorily by assuming that the spin–orbit coupling in Zn is sufficiently large as to render the double-zone scheme invalid, at least in relatively low magnetic fields, see fig. 7. Fawcett (1961 b) measured the surface conductance of polycrystalline Zn and Cd under anomalous skin effect

Fig. 7

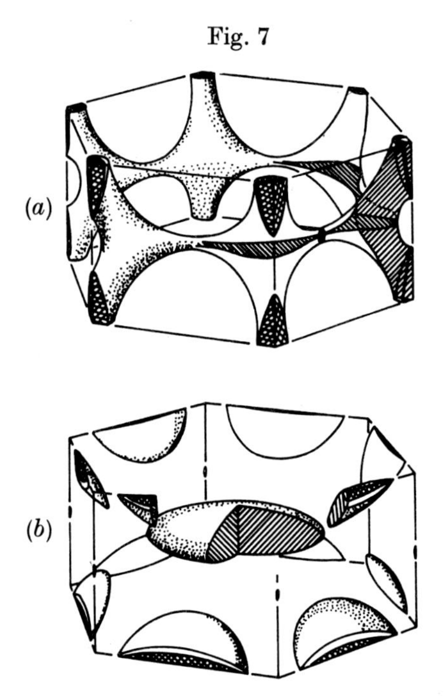

The Fermi surface of Zn: (*a*) holes in bands 1 and 2 (the caps in band 1 are shown cross-hatched), (*b*) electrons in bands 3 and 4 (the cigars in band 4 are shown cross-hatched) (Gibbons and Falicov 1963).

conditions to determine the total areas of their Fermi surfaces ; values of 41% and 39% of the free-electron values were obtained for Zn and Cd respectively.

Mattheiss (1964) performed an A.P.W. band-structure calculation for Zn, along the line ΓK only, and a substantial amount of further experimental work on both Zn and Cd was performed before any further complete band-structure calculations became available. This included work on magnetoacoustic geometric resonances (Galkin and Korolyuk 1960 on Zn, Daniel and Mackinnon 1962, 1963 on Cd, Gavenda and Deaton 1962 on Cd, Mackinnon and Daniel 1962 a on Cd, 1962 b on Zn and Cd, Mackinnon *et al.* 1962 on Cd, Gibbons and Falicov 1963 on Zn and Cd, Deaton and Gavenda 1964 on Zn and Cd), quantum oscillations in the ultrasonic attenuation (Gibbons 1961 on Zn, Korolyuk and Prushchak 1961 on Zn, Gibbons and Falicov 1963 on Zn and Cd, Bohm and Mackinnon 1964 on Zn, Bosnell and Myers 1964 on Cd, Fenton and Woods 1966 on Zn, Myers and Bosnell 1966 on Zn), the de Haas–van Alphen oscillations (Hedgecock and Muir 1963 on Zn, Grassie 1964 on Cd, Higgins *et al.* 1964, 1965 on Zn, Lawson and Gordon 1964 on Zn, Thorsen *et al.* 1964 on Zn, Tsui and Stark 1966 a on Cd, Venttsel' *et al.* 1966 on Zn), galvanomagnetic properties (Alekseevskiĭ and Gaĭdukov 1960 on Zn, 1962 b on Zn and Cd, Renton 1960 on Zn and Cd, Fawcett 1961 a on Zn and Cd, Stark 1962, 1964 on Zn, Reed and Brennert 1963 on Zn, Schirber 1964 on Zn, Zaĭtsev *et al.* 1965 on Cd, Falicov *et al.* 1966 on Zn, Tsui and Stark 1967 on Cd), cyclotron resonance (Galt and Merritt 1960 on Zn, Galt *et al.* 1961 on Cd, Shaw and Eck 1964 on Cd, Galt *et al.* 1965 on Cd, Naberezhnȳkh and Mel'nik 1965 a, b on Zn, Shaw, Eck and Zych 1966 on Cd, Shaw, Sampath and Eck 1966 on Zn, Henningsen 1967 on Zn) and the radio-frequency size effect (Naberezhnȳkh and Mar'yakhin 1967 on Cd†). A large number of linear dimensions of the Fermi surfaces of Zn and Cd are given in tables 1 and 2 respectively of Gibbons and Falicov (1963). Magnetic breakdown effects have been observed in many of the de Haas–van Alphen, magnetoacoustic and galvanomagnetic experiments, for example, between the needles and the monster in Zn or between the monster in band 2 and the pockets of holes in band 1 in both Zn and Cd (see, for example, Cohen and Falicov 1960, Stark 1962, 1964, Lawson and Gordon 1964, Gaĭdukov and Krechetova 1965 a, b, 1966, Tsui and Stark 1966 a, Venttsel' *et al.* 1966). Extensive discussions of the behaviour of the galvanomagnetic properties of Zn in terms of the details of its known Fermi surface and allowing for magnetic breakdown have been given by Pippard (1964), Falicov *et al.* (1966), and Van Dyke *et al.* (1970). Several of the principal dimensions of the Fermi surface of Zn are listed by Pippard (1964). Theoretical expressions for the galvanomagnetic tensor in weak magnetic fields, the electronic specific heat and the thermoelectric power of Cd were obtained by Tsuji and Kunimune (1963)

† See note added in proof.

Fig. 8

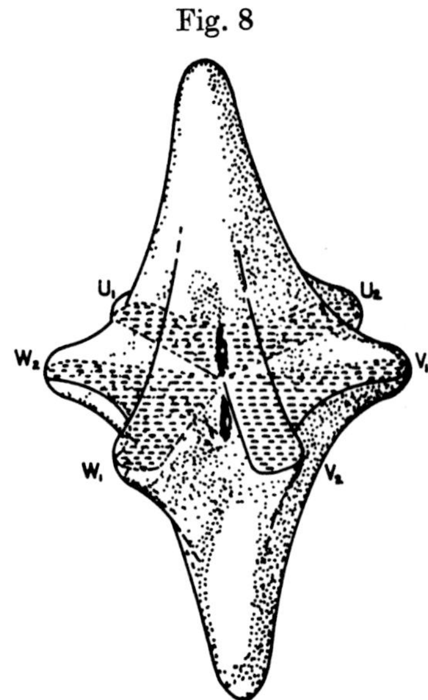

The Fermi surface of Cd ; the region of holes in band 1 formed by the dismembering of the free-electron ' monster ' (Galt *et al.* 1965).

using the lens in band 3 and a disconnected monster in band 2, parametrized by comparison with experimental measurements. From many of the various experiments mentioned it became clear that the needle-shaped pockets of electrons that appear at K in band 3 for Zn do not exist in Cd and that the monster in band 2 is disconnected by the ' pinching off ' of the $\langle 11\bar{2}0 \rangle$ waists (Fawcett 1961 a, Alekseevskiĭ and Gaĭdukov 1962 b, Gibbons and Falicov 1964, Grassie 1964), see fig. 8. These results, together with some of the caliper dimensions of the lens in band 3 determined from magnetoacoustic geometric resonances (see tables 1 and 2 and fig. 7 of Daniel and Mackinnon (1963) for the detailed dimensions), and one of the cross-sectional areas of the monster in band 2 determined from de Haas–van Alphen measurements (Joseph *et al.* 1961), were used to determine the parameters in a pseudopotential band structure calculation for Cd by Katsuki and Tsuji (1965). There have been other constructions of pseudopotentials for Cd by Mar'yakhin and Svechkarev (1967, 1969).

Stark and Falicov (1967) performed some new band structure calculations for Zn and Cd using an empirical non-local pseudopotential interpolation scheme including spin–orbit coupling. Their results agreed with the existing de Haas–van Alphen results to within a few per cent (Joseph *et al.* 1961 on Zn and Cd, Joseph and Gordon 1962 on Zn, Grassie 1964 on Cd, Thorsen *et al.* 1964 on Zn, Higgins *et al.* 1965 on Zn, Tsui and Stark 1966 a on Cd) and were claimed to be consistent with other Fermi surface measurements. The error was expressed as ΔE_{F}, the shift in the Fermi energy that would be necessary to bring the calculated frequency into exact agreement with experiment ; the values of ΔE_{F} were of the order of $1{\cdot}4 \times 10^{-3}$ Ry for Zn and 6×10^{-4} Ry for Cd. Among the conclusions deduced from these calculations was the result that the ‘ butterflies ’ (or ‘ stars ’) and the horizontal cigars around the points L in bands 3 and 4 respectively do not exist. Stark and Falicov (1967) claimed, without actually giving any details, to be able to re-assign to other sheets of the Fermi surface all the various experimental results which had previously been assigned to the butterflies. However, in further extensive work on the de Haas–van Alphen effect in Zn (Venttsel’ *et al.* 1967, Venttsel’ 1968) and on magnetoacoustic quantum oscillations in Zn and Cd by Fletcher *et al.* (1969) prominent sets of oscillations were observed, which these authors were unable to explain except in terms of the butterflies at L in band 3. These oscillations had previously been observed, although less prominently, in the de Haas–van Alphen work of Higgins *et al.* (1965) and Venttsel’ *et al.* (1966). Naberezhnȳkh *et al.* (1967) obtained caliper dimensions, from their radio-frequency size effect work on Cd, which they ascribed to the butterflies in band 3 and the cigars in band 4, although their data was insufficient to enable these shapes to be determined completely. In other radio-frequency size effect results that became available for Cd after the calculation of Stark and Falicov (1967), while giving many detailed caliper dimensions for the pockets of holes in band 1, the monster in band 2 and the lens in band 3 (see tables 3 and 4), no caliper dimensions were obtained that could be assigned to the butterflies or cigars in Cd (Goodrich and Jones 1967, Jones *et al.* 1968). Therefore, the existence of the butterflies and cigars in bands 3 and 4 must still be regarded as controversial. Two important cross sections of the Fermi surfaces in bands 1 and 2 are shown in figs. 14 and 15 of Jones *et al.* (1968).

A perusal of the Fermi surface data for Be, Mg, Zn and Cd (see also part I) which all have the h.c.p. structure reveals a similar behaviour,

Table 3. Caliper dimensions of the lens in band 3 of the Fermi surface of Cd (Å^{-1})

n, **q**† parallel to	$[11\bar{2}0]$		$[10\bar{1}0]$		$[0001]$	
H parallel to	$[0001]$	$[10\bar{1}0]$	$[0001]$	$[11\bar{2}0]$	$[10\bar{1}0]$	$[11\bar{2}0]$
Deaton (1962)		0·550	1·556	0·550		
Daniel and Mackinnon (1963)	1·458	0·54	1·428	0·498		
Gibbons and Falicov (1963)	1·560	0·550		0·540		
Gavenda and Chang (1969)	1·510	0·542	1·516	0·540		
Naberezhnȳkh *et al.* (1967)	1·46	0·53	1·47	0·54		
Jones *et al.* (1968)	1·563	0·599	1·516	0·553	1·560	1·560
Free electron values	1·658	0·540	1·658	0·540	1·658	1·658
Stark and Falicov (1967)	1·508	0·541	1·508	0·541	1·508	1·508

† Ultrasonic attenuation obtains the same calipers as the radio-frequency size effect with **q**, the direction of sound propagation, in the same direction as **n**.

(Jones *et al.* 1968.)

Table 4. Dimensions of the Fermi surface of Cd (Å^{-1})

	Experiment	Theory
Lens		
From Γ in ΓA direction	0·276	0·271
From Γ in ΓK direction	0·780	0·754
From Γ in ΓM direction	0·780	0·754
Band 2 holes		
From K in KΓ direction	0·494	0·491
From H in HA direction	0·112	0·114
From H in HL direction	0·192	0·180
Band 1 holes		
From H in HA direction	0·108	0·110
From H in HL direction	0·178	0·168
From H in HK direction		0·505
Between band 1 holes and band 2 holes in AHL plane		
Along HA direction	0·004	0·004
Along HL direction	0·014	0·013
Between arms of the band 2 hole surfaces in ΓMK plane		
Parallel to MK direction		0·054

(Jones *et al.* 1968.)

relative to the free-electron model, to that exhibited by the alkali metals. That is, the Fermi surface of the metal in the second row of the periodic table, namely Mg, is closest to the h.c.p. free-electron Fermi surface ; with increasing atomic number beyond Mg the distortion from the free-electron Fermi surface increases, see table 5 for a comparison of the relative distortions from the free-electron model in Zn and Cd. As with the alkali metals, the Fermi surface of the lightest metal, in this case Be, is also substantially distorted from the free-electron model.

We now turn our attention to work on various other electronic properties that are related, at least indirectly, to the Fermi surface topology of Zn and Cd. Experimental values of γ, the temperature coefficient of the electronic contribution to the specific heat, include for Zn :

$(0{\cdot}643 \pm 0{\cdot}012)$ mJ mole^{-1} deg^{-2} Gschneidner (1964)

$0{\cdot}653$ mJ mole^{-1} deg^{-2} Martin (1968 a)

$(0{\cdot}6359 \pm 0{\cdot}0046)$ mJ mole^{-1} deg^{-2} Martin (1969) ;

and for Cd :

$(0{\cdot}674 \pm 0{\cdot}036)$ mJ mole^{-1} deg^{-2} Gschneidner (1964).

Table 5. Comparison of lens data for Zn and Cd

H parallel to	Exp	m^*/m_0† F.E.	Exp/F.E.	Extremal area Exp/F.E.
Zn [0001]	1·20	1·00	1·20	0·97§
[0001]	1·44‡	1·00	1·44	
Basal plane	0·55	0·39	1·41	0·97‖
Cd [0001]	1·23	1·00	1·23	0·85¶
[0001]	1·29‡	1·00	1·29	
Basal plane	0·53	0·405	1·31	0·95¶

† Free electron.
‡ From limiting-point resonances.
§ Priestley and Mondino (1964).
‖ Higgins *et al.* (1965).
¶ Grassie (1964).

(Shaw, Eck and Zych 1966.)

The pseudopotentials for Zn and Cd constructed by Stark and Falicov (1967) were used by Allen *et al.* (1968) to calculate the electron–phonon mass enhancement λ for Zn and Cd. One set of values of λ was obtained using the pseudopotential determined by Stark and Falicov (1967) to calculate the band structure value of γ, and hence λ by comparison with the experimental value of γ. Values of λ were also calculated directly using the same pseudopotential. Finally, values of λ were determined from the experimental values of T_c, the superconducting transition temperature, and Θ, the Debye temperature. These various sets of λ are :

Zn	Cd	
0·43	0·36	from γ (Allen *et al.* 1968)
0·42	0·40	direct calculation (Allen *et al.* 1968)
0·38	0·38	experiment (McMillan 1968)
0·48	0·45	experiment (Garland *et al.* 1968).

This quite remarkable agreement led Allen *et al.* (1968) to believe that they had really obtained good pseudopotentials for Zn and Cd, particularly since when the direct calculation of λ was repeated using the Heine–Abarenkov model potential (Animalu and Heine 1965) substantially worse values of λ (0·27 for Zn and 0·10 for Cd) were obtained.

Magnetothermal oscillations have been observed in Zn with a period corresponding to orbits around the horizontal arms of the band 2 Fermi surface (Quadros and Oliveira 1968). Quantum oscillations in the Fermi level, which appear as oscillations in the contact potential, were observed in Zn by Whitten and Piccini (1966) (similar observations were previously made for Pb (Caplin and Shoenberg 1965) and for Bi (Verkin *et al.* 1964)). Positron annihilation experiments have been performed on Zn by Lang *et al.* (1955), Stewart (1957), Dekhtyar and Mikhalenkov (1961), and Kusmiss and Swanson (1968) and on Cd by Lang *et al.* (1955), Stewart (1957), and Faraci and Turrisi (1968). de Haas–van Alphen type oscillations, or quantum oscillations, were observed in a number of galvanomagnetic and thermomagnetic properties of Zn by Bergeron *et al.* (1960) and the temperature dependence of the Knight shift in Cd has been measured by a number of workers (Schone 1964, Borsa and Barnes 1966, Sharma and Williams 1967). Further cyclotron resonance experiments on Zn have been performed by Naberezhnӯkh and Dan'shin (1969) and Sabo (1970), and further galvanomagnetic experiments on Cd by Katyal and Gerritsen (1969 a, b) and Stringer *et al.* (1970). There have been several measurements of the de Haas–van Alphen effect in Zn doped with small amounts of various other metals (Hedgecock and Muir 1963, Higgins *et al.* 1964, Higgins and Marcus 1966, 1967). The magnetic breakdown field for orbits on the needles in Zn determined by Higgins and Marcus (1967) varied from 2·2 kG to 10 kG according to orientation. This magnetic breakdown between the needles and the monster which is observed in pure Zn vanishes in Zn–Cu alloys (Higgins *et al.* 1964). We have already mentioned the effect of pressure on the needles in band 3 in Zn. The pressure dependence of the connectivities and of the cross-sectional areas of other pieces of the Fermi surface have also been investigated (Schirber 1965 a, b, O'Sullivan and Schirber 1966 a). A considerable amount of other work has also been done on Zn and Cd under pressure (see the review by Drickamer (1965)). The pressure dependence of the electrical resistance of Zn and of Cd is very similar to that of Mg which has an almost ideal h.c.p. c/a ratio (Lynch and Drickamer 1965).

In § 7.2 of part I we encountered, in connection with Bi, the fact that very small pockets of electrons have very large g factors associated with them. We should therefore expect large values of g for the needle-shaped pockets of electrons in the Fermi surface of Zn in band 3, as has indeed been observed from the spin splitting of various quantum oscillations (Stark 1962, 1964, Bennett and Falicov 1964, Lawson and Gordon 1964, Myers and Bosnell 1965, 1966, O'Sullivan and Schirber 1966 b, 1967 b). According to O'Sullivan and Schirber (1966 b) the value of the g factor is 170.

The temperature dependence of the amplitude of the radio-frequency size effect line in Cd, for the lens-shaped Fermi surface of electrons in band 3, was investigated by Naberezhnӯkh and Tsymbal (1967). If λ increases, then the amplitude of the signal increases. Electron–electron collisions were negligible at the temperatures used (1·6–5·1°K) and λ for collisions with impurities or defects is temperature independent, so that the temperature dependence of the electron–phonon contribution to λ

can be determined. For Cd, Naberezhnӯkh and Tsymbal concluded that the electron–phonon contribution to λ was proportional to T^{-3} to a good approximation. However, in magnetoacoustic investigations of the temperature dependence of the mean free paths the temperature dependence of λ was found to exhibit considerable angular anisotropy, with considerable departures from a T^{-3} behaviour (Deaton 1963 on Cd, 1965 on Zn and Cd). Size effects have been observed in galvanomagnetic measurements on Zn (Soffer 1968) and Cd (Hambourger *et al.* 1967, Mackey *et al.* 1967).

Measurements have been made of the optical constants in the visible and near infra-red regions of the spectrum (Graves and Lenham 1968 a on Zn and Cd, 1968 b on Cd) and attempts were made to relate the interband contributions to the band-structure predictions of Harrison (1966). Lettington (1965) used measurements of the real and imaginary parts of the dielectric constant for both the o and e rays as the experimental basis of a pseudopotential calculation for Zn ; the resulting calculated Fermi surface was in reasonable agreement with the known Fermi surface of Zn. Similar experiments were also performed by Motulevich and Shubin (1969), also on Zn.

3.3. *Mercury*

Since the structure of Hg can be regarded as a distorted f.c.c. crystal, the Brillouin zone, see fig. 9, can also be regarded as a distorted f.c.c. Brillouin zone. The labels used for the special points of symmetry are an adaptation of the labels used in the f.c.c. Brillouin zone. In the free-electron model (Mott and Jones 1936) the spherical Fermi surface intersects the six large hexagonal faces of the Brillouin zone but does not reach the remaining faces, see fig. 10. The free-electron Fermi surface therefore consists of a set of lens-shaped pockets of electrons in band 2 at the large hexagonal faces, together with a complicated multiply-connected Fermi surface of holes in band 1. It should be clear that this can be obtained by a trigonal distortion of the free-electron Fermi surface for a divalent f.c.c. metal.

Apart from a few early experiments on Hg crystals of unknown orientation (Pippard (1947), on the anomalous skin effect, Verkin *et al.* (1951) and Shoenberg (1952) on the de Haas–van Alphen effect, and Stewart (1957) and Gustafson *et al.* (1963) on positron annihilation) the determination of the Fermi surface of Hg has been performed only recently. From the comparison of their positron annihilation results for solid and liquid Hg, Gustafson *et al.* (1963) concluded that the Fermi surface of Hg in extended **k** space did not depart very significantly from a sphere. de Haas–van Alphen measurements on oriented single crystals by Brandt and Rayne (1965) were able to be interpreted in terms of orbits on the multiply-connected Fermi surface of holes in band 1, provided the free-electron sphere was sufficiently distorted so as to touch the rectangular faces of the Brillouin zone at the points X in fig. 9. The results for these orbits on

Fig. 9

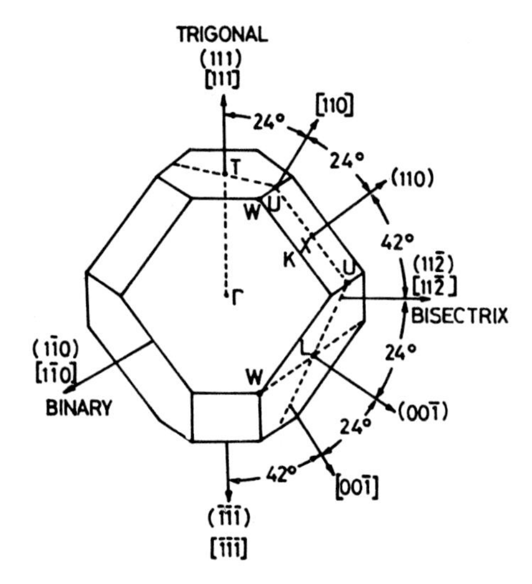

The Brillouin zone of Hg (Keeton and Loucks 1966 b).

Fig. 10

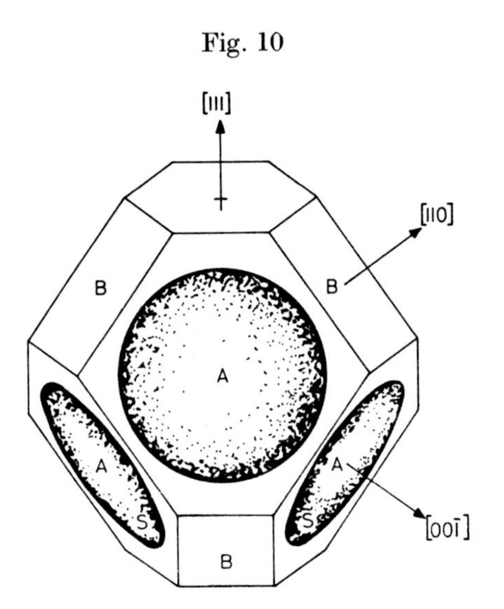

The Fermi surface of Hg in the free-electron model : holes in band 1 (Brandt and Rayne 1965).

Fig. 11

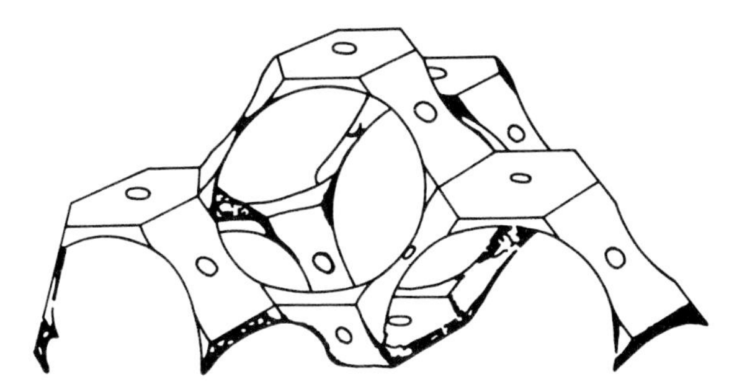

Schematic representation of the multiply-connected region of holes in band 1 in Hg (Keeton and Loucks 1966 b).

the Fermi surface of holes in band 1 are given in table II of Brandt and Rayne (1966). This contact of the Fermi surface in band 1 with the rectangular faces of the Brillouin zone was reproduced in the relativistic A.P.W. band structure calculations of Keeton and Loucks (1966 b). The Fermi surface of holes in band 1 calculated by Keeton and Loucks differed from the free-electron model in one or two other details ; occupied regions were found around the points T and K as well as around X, see fig. 11. In addition to two sets of de Haas–van Alphen periods associated with the multiply-connected Fermi surface of holes in band 1, Brandt and Rayne (1966) also observed a period associated with the lens-shaped pocket of electrons in band 2. Cyclotron masses associated with these orbits were measured by Dixon and Datars (1968). Brandt and Rayne used their experimental results to determine the parameters in a pseudopotential band structure calculation. The experimental and calculated cross-sectional areas are given in table II of Brandt and Rayne (1966). Confirmation of the existence of the open orbits that would be expected on the multiply-connected Fermi surface in band 1 was obtained from magnetoresistance and other measurements on Hg (Dixon and Datars 1965, 1968, Datars and Dixon 1967, Moss and Datars 1967, Dishman and Rayne 1966, 1968). Caliper dimensions of the Fermi surface of Hg in both band 1 and band 2 have been determined from magnetoacoustic geometric resonances and the results for the principal dimensions were compared with the results of previous work (Bellessa *et al.* 1969, Bogle *et al.* 1969†).

Other work related to the electronic structure of Hg is scarce. Optical reflectivity measurements were performed on both liquid and solid Hg by Müller and Thompson (1969) and interband transitions which were present in the solid phase, but not in the liquid phase, were able to be identified in terms of the band structure calculated by Keeton and Loucks (1966 b). The electronic specific heat appears not to have been calculated ; the experimental value of γ is $(1{\cdot}79 \pm 0{\cdot}02)$ mJ mole^{-1} deg^{-2}

† See note added in proof.

(Van der Hoeven and Keesom 1964 b). Knight shift values for Hg have been given by Knight *et al.* (1959) and Weinert and Schumacher (1968). The electrical resistivity of Hg has been measured by Andrew (1949) and Aleksandrov *et al.* (1967). It is believed that the overlap between band 1 and band 2 in Hg increases with increasing pressure (Mott 1966).

§ 4. d-BLOCK : GROUP IIIA : SCANDIUM AND YTTRIUM†

Sc Scandium 2.8.(8, 1) 2

Y Yttrium 2.8.18.(8, 1) 2

With this group of metals we enter the transition series of metals, that is, the metals which have a full 4s, 5s, or 6s shell and now begin to fill up the next lower d shell, that is, the 3d, 4d, or 5d shell. In contrast with nearly all the metals that we have considered so far, both in part I and in §§ 2 and 3 of part II, there has only been a very small amount of experimental and theoretical work performed on the Fermi surfaces of many of the transition metals. This is because of the difficulties that are often involved in obtaining sufficiently pure samples. An extensive review of early work on electrons in transition metals was given by Mott (1964), at which stage hardly any direct measurements of the dimensions of transition-metal Fermi surfaces had been made. All the three common metal structures, f.c.c., b.c.c. and h.c.p., are found among the transition metals but they do not occur randomly ; there is a steady structural trend from the group IIIA metals Sc and Y through to the noble metals Cu, Ag and Au. The band structure contribution to the total energy of a transition metal is thought to be the contribution which is primarily responsible for determining which structure is actually stable for any given transition metal. Recent discussions of proposed explanations of the structures of the transition metals have been given by Brewer (1968), Deegan (1968), and Pettifor (1970).

4.1. *Structures*

Sc and Y have the h.c.p. structure with $a = 3{\cdot}308$ Å, $c = 5{\cdot}267$ Å for Sc and $a = 3{\cdot}63$ Å, $c = 5{\cdot}75$ Å for Y (Smithells 1967). The axial ratios c/a are therefore 1·59 for Sc and 1·58 for Y which depart quite substantially from the ideal value of $\sqrt{(8/3)}$. Sc also exists in an f.c.c. phase with $a = 4.53$ Å (Smithells 1967). The free-electron Fermi surface for an h.c.p. metal of valence 3 is shown in fig. 2 (*c*) of part I. While the departure of c/a from the ideal value of $\sqrt{(8/3)}$ slightly alters the detailed dimensions of the free-electron Fermi surface it does not lead to any drastic changes such as the complete disappearance of some parts of the Fermi surface as happens in Cd and nearly happens in Zn ; in fact, since c/a is smaller than the ideal value, the horizontal cross sections of the pockets of electrons in bands 5

† The rare-earth metals are described later, see § 10.

and 6 can be expected to increase slightly, simply as a result of the change in the c/a ratio.

4.2. *Scandium and Yttrium*

The band structure, Fermi surface and density of states of Sc have been calculated by Altmann and Bradley (1967) using the cellular method and by Fleming and Loucks (1968) using the A.P.W. method. With the band structure and Fermi energy calculated by Altmann and Bradley bands 1 and 2 are full all over the Brillouin zone, bands 3 and 4 are empty at Γ, A and L, but full at M, K and H, and bands 5 and 6 are empty all over the Brillouin zone (see fig. 3 of Altmann and Bradley (1967)). The Fermi surface is therefore entirely in bands 3 and 4 and has a very complicated multiply-connected structure, see fig. 12. This complicated Fermi surface

Fig. 12

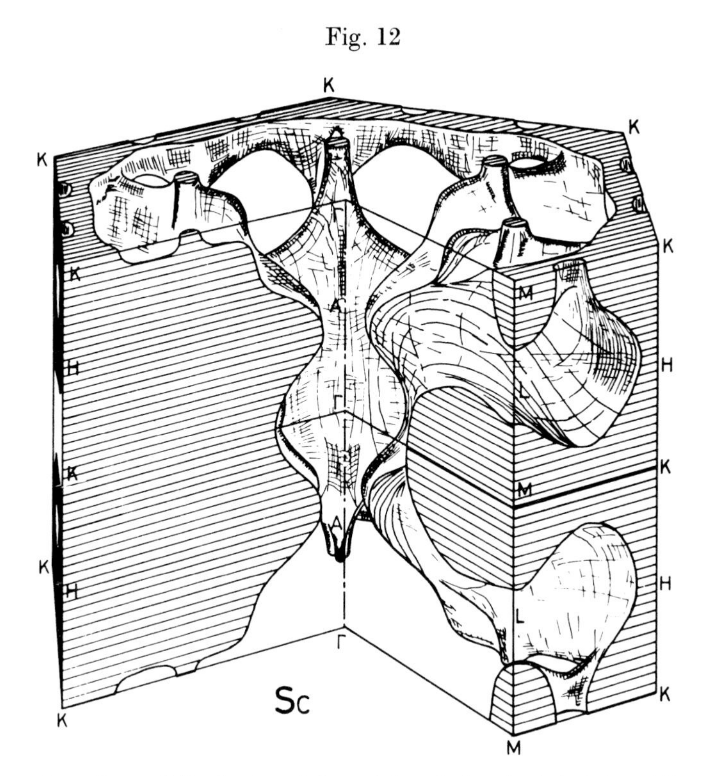

The Fermi surface of Sc in band 3 and band 4 calculated by Altmann and Bradley (1967) (electrons shaded).

of electrons in bands 3 and 4 obviously bears some relationship to the free-electron Fermi surface although it is very substantially distorted from the free-electron Fermi surface. As a result of this distortion holes appear at Γ, while narrow filaments parallel to k_z arise as a result of extra contacts between the top surface of the Brillouin zone and the region occupied by electrons at a point along ΓK and the region of holes at a point along MK. However, these contacts involve bands that are very close to the Fermi level and one or other of these connectivities could easily be destroyed by quite a small change in the Fermi energy. The Fermi surface of Sc calculated by Fleming and Loucks (1968) was different in a number of details from that calculated by Altmann and Bradley (1967). Their calculated

Fig. 13

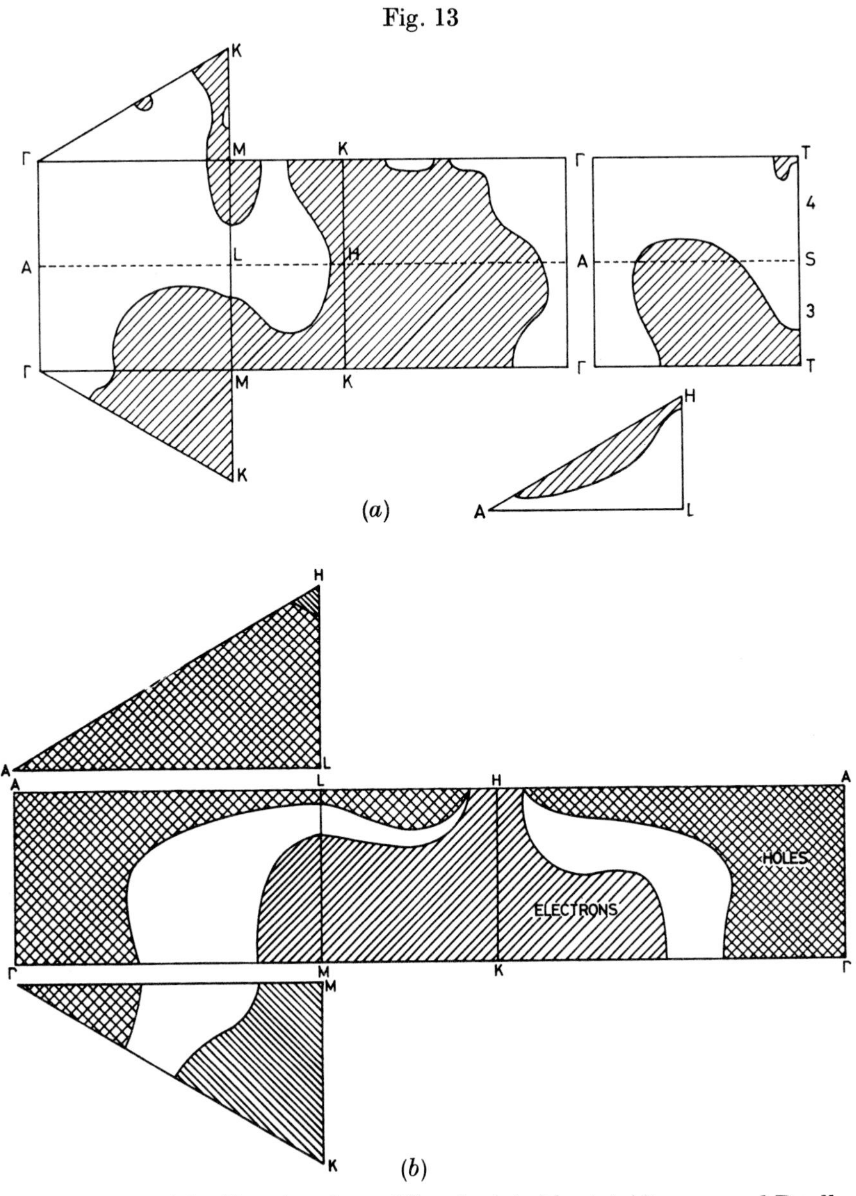

Cross sections of the Fermi surface of Sc calculated by (*a*) Altmann and Bradley (1967) (electrons shaded) and (*b*) Fleming and Loucks (1968).

Fermi surface in bands 3 and 4 also consists of a multiply-connected Fermi surface separating regions of holes which pass through the points Γ, A and L and regions of electrons which pass through the points K, M and H, see fig. 13 (*b*). However, many of the detailed connectivities and dimensions of the Fermi surface for Sc calculated by Fleming and Loucks

Fig. 14

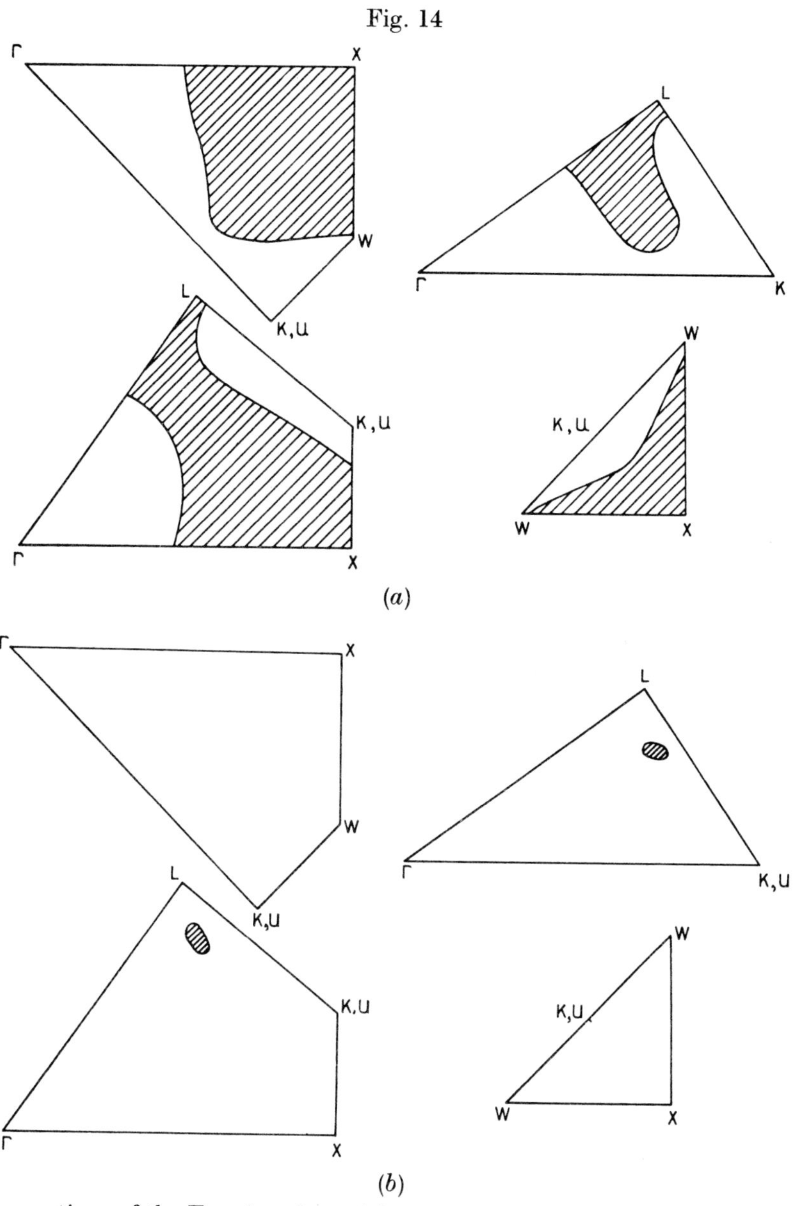

Cross sections of the Fermi surface of f.c.c. Sc : (*a*) electrons in band 2, and (*b*) electrons in band 3 (Cracknell 1964).

(1968) were different from those obtained by Altmann and Bradley. The principal topological differences were that the narrow filaments of electrons that cross ΓK and of holes that cross MK vanish ; these differences are most clearly seen from the sections shown in fig. 13. The quite substantial differences between the results of the cellular and A.P.W. calculations is almost certainly due to differences between the potentials used, rather than to any inherent faults of either method of calculating band structures (Altmann *et al.* 1968).

The only band structure calculation for the f.c.c. phase of Sc of which the author is aware is a cellular calculation by Cracknell (1964) using the same potential that was used by Altmann and Bradley (1967). The Fermi surface was found to be contained almost entirely in band 2 with just a very small pocket of electrons in band 3. In band 2 there was a multiply-connected region occupied by electrons consisting of tubes passing through the points X and L and almost reaching to W. However, the points K and U appeared not to be occupied by electrons in band 2. Sections of this Fermi surface are shown in fig. 14. Subsequent experience with this particular cellular f.c.c. programme suggests that in fact band 2 is probably occupied at K as well, so that the region of electrons would then pass through K (and U) and thereby come to resemble the free-electron Fermi surface in band 2 of an f.c.c. metal with valence 3 shown in fig. 2 (*a*) of part I. The pocket of electrons in band 3 would then vanish completely.

The most direct experimental evidence available so far about the shape of the Fermi surface of Sc comes from the magnetoresistance measurements of Bogod and Eremenko (1965) ; the $\mathbf{B}^2$ law was obeyed up to the highest magnetic fields used (96 kG), from which it was concluded that the Fermi surface of Sc is multiply-connected in some complicated fashion. However, these measurements cannot be used to distinguish between the two calculated Fermi surfaces that we have described. Detailed experimental investigations of the Fermi surface of Sc are now needed to determine the dimensions of the Fermi surface and to establish the connectivities. It appears that the only other experimental work related to the Fermi surface of Sc has been concerned with the indirect or macroscopic properties of the metal. The calculated and experimental values of γ include

calculated :

3·1 mJ mole^{-1} deg^{-2}	Altmann and Bradley (1967)
5·44 mJ mole^{-1} deg^{-2}	Fleming and Loucks (1968) ;

experimental :

(10·8 ± 0·5) mJ mole^{-1} deg^{-2}	Gschneidner (1964)
(10·9 ± 0·1) mJ mole^{-1} deg^{-2}	Lynam *et al.* (1964)
(10·66 ± 0·10) mJ mole^{-1} deg^{-2}	Flotow and Osborne (1967).

It therefore seems as though the thermal effective mass due to electron–phonon enhancement is between about 2 and 4. The experimental values of γ are quite close to those of a number of rare-earth metals, see § 10.3. Several measurements of the magnetic susceptibility of Sc have been made (for references see Gardner and Penfold (1965), also Kobayashi (1966), and Cullen and Callen (1968)). However, it is not possible to make very meaningful comparisons with calculated density of states curves because of the usual difficulty of isolating the spin paramagnetic susceptibility from the diamagnetic and orbital contributions which are also present in the total measured susceptibility. Other measurements which have been made of properties of Sc that depend indirectly on the shape of the Fermi surface include the electrical resistivity (Colvin and Arajs 1963, Volkenshteĭn and Galoshina 1963), Hall effect (Volkenshteĭn and Galoshina 1963), Knight shift (Blumberg *et al.* 1960, Barnes *et al.* 1965), positron annihilation (Rodda and Stewart 1963), thermoelectric power (Vedernikov 1969), and photoemission (Eastman 1969 a).

For Y similar cellular and A.P.W. band structure calculations have been performed to those that we have just described for Sc. The general features of the calculated Fermi surfaces for Y were found to be very similar to those calculated for Sc, that is, bands 1 and 2 are completely full and bands 5 and 6 are completely empty all over the Brillouin zone. As in Sc, the Fermi surface is entirely contained in bands 3 and 4 and separates a multiply-connected region of holes enclosing Γ, A and L and a multiply-connected region of electrons enclosing K, H and M. The Fermi surface of Y calculated by Loucks (1966 b) was almost identical with that obtained by Fleming and Loucks (1968) for Sc, see the sections given in fig. 15. Several de Haas–van Alphen frequencies predicted on the basis of these calculations are given in table III of Loucks (1966 b). The details of the dimensions and connectivities of the Fermi surface calculated by Altmann and Bradley (1967) for Y using the cellular method were rather different

Fig. 15

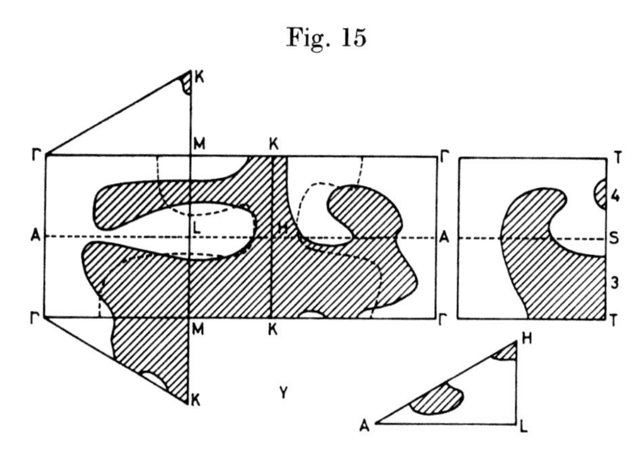

Cross sections of the Fermi surface of Y calculated by Altmann and Bradley (1967) using the cellular method (shaded areas represent electrons) ; the broken lines are the contours obtained by Loucks (1966 b) using the A.P.W. method.

Fig. 16

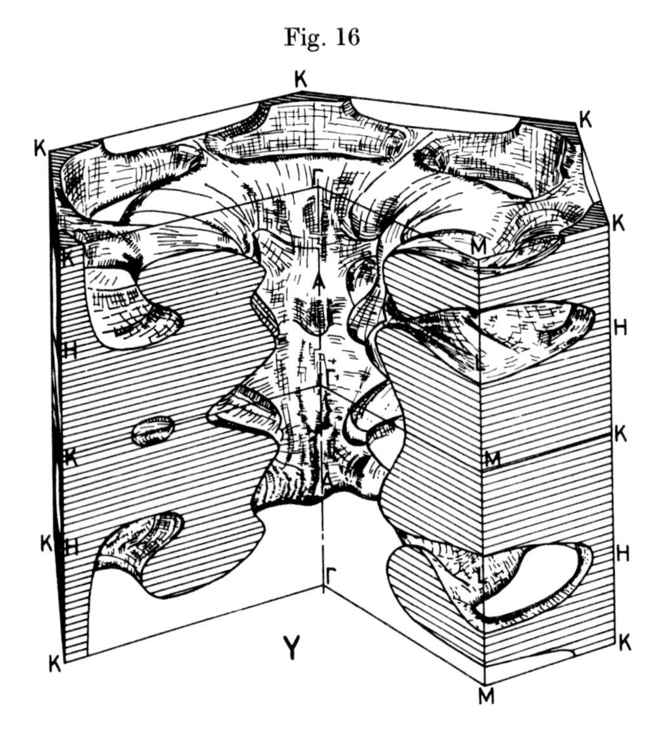

The Fermi surface of Y in bands 3 and 4 calculated by Altmann and Bradley (1967) (electrons shaded).

from those for the Fermi surface calculated by Loucks ; in particular band 4 was found to be empty at M, see figs. 15 and 16.

Once again, as with Sc, there have been no direct experimental measurements of the dimensions of the Fermi surface of Y except for some slight evidence of Kohn anomalies (Sinha *et al.* 1970) ; the only other experimental results available are concerned with indirect or macroscopic properties of the metal. Values of γ, the temperature coefficient of the electronic contribution to the specific heat, for Y are similar to those for Sc and include

calculated :

4·69 mJ mole^{-1} deg^{-2} Loucks (1966 b)
2·5 mJ mole^{-1} deg^{-2} Altmann and Bradley (1967) ;

experimental :

(10·1 ± 0·1) mJ mole^{-1} deg^{-2} Gschneidner (1964).

Various positron annihilation measurements have been performed on Y (Rodda and Stewart 1963, Sedov 1963, Williams *et al.* 1966, Williams and Mackintosh 1968). The results of the A.P.W. band structure calculation of Loucks (1966 b) were used to calculate the angular distribution of positron annihilation radiation from Y in the *c* direction (Gupta and Loucks 1968) ; the main features of the experimental results were reproduced by the theory (see fig. 7 of Williams and Mackintosh (1968)). However, this is only a very insensitive test of the details of the band structure calculation. Other experimental measurements on Y included measurements of the electrical conductivity (Tamarin *et al.* 1968), magnetic susceptibility (Chechernikov *et al.* 1964, Gardner *et al.* 1965, Gardner and Penfold 1968, Volkenshteĭn *et al.* 1969), Knight shift (Barnes *et al.* 1965, Silhouette 1968), Hall effect (Kevane *et al.* 1953, Lee and Legvold 1967), thermal conductivity (Tamarin *et al.* 1968), thermoelectric power (Vedernikov 1969) and photoemission (Eastman 1969 a).

There is clearly scope for some detailed direct experimental work on measuring the Fermi surfaces of Sc and Y with the object of testing the topologies predicted by the various band structure calculations that we have described. It is perhaps a little surprising that although the pockets of electrons in bands 5 and 6 in the free-electron model are not particularly small, in both the cellular and A.P.W. calculations for each of the metals Sc and Y bands 5 and 6 were found to be empty all over the Brillouin zone. It would be useful to have some direct experimental evidence on this point and would give some valuable indication of whether the free-electron model is likely to be useful in giving even a first approximation to the shape of the Fermi surface of a transition metal. If, after all, some pieces of Fermi surface in bands 5 and 6 of Sc and Y were found experimentally it would be useful evidence in favour of the continuance of the consideration of the free-electron model even for transition metals.

§ 5. d-BLOCK : GROUP IVA : TITANIUM, ZIRCONIUM AND HAFNIUM

Ti	Titanium	2.8.(8, 2) 2
Zr	Zirconium	2.8.18.(8, 2) 2
Hf	Hafnium	2.8.18.32.(8, 2) 2

5.1. *Structures*

Ti, Zr and Hf all have the h.c.p. structure at ordinary temperatures and undergo phase transitions to the b.c.c. structure at high temperatures (900°C for Ti, 840°C for Zr, and 1310°C for Hf). The values of the lattice constants are $a = 2{\cdot}95$ Å, $c = 4{\cdot}68$ Å for Ti, $a = 3{\cdot}23$ Å, $c = 5{\cdot}14$ Å for Zr, and $a = 3{\cdot}1969$ Å, $c = 5{\cdot}0583$ Å for Hf, corresponding to c/a ratios of 1·59 for both Ti and Zr and 1·58 for Hf (Taylor and Kagle 1963, Smithells 1967).

5.2. *Titanium*

There has been relatively little work done on the experimental or theoretical determination of the Fermi surface of Ti. Mattheiss (1964) performed an A.P.W. calculation for Ti along the line ΓK while Nikiforov and Sacheniko (1963) performed an O.P.W. calculation for the high

Fig. 17

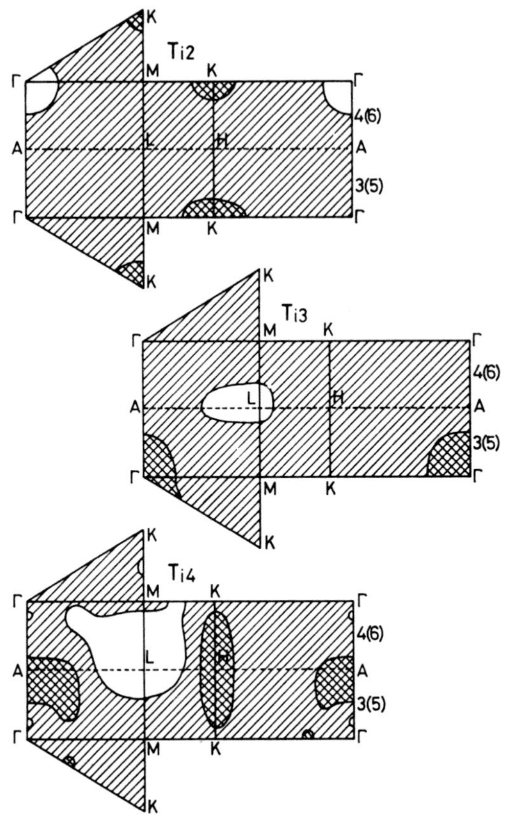

Cross sections of the Fermi surface of Ti obtained by Altmann and Bradley (1967) using three different potentials (shaded areas represent electrons in bands 3 and 4 and cross-hatched areas represent electrons in bands 5 and 6).

temperature cubic phase of Ti. Complete cellular band structure calculations for Ti were performed by Altmann and Bradley (1967) for four different choices of the potential, one of which was that used previously by Mattheiss (1964). Quite small changes in the potential used can cause considerable changes in the details of the band structure (see figs. 4–6 of Altmann and Bradley (1967)) and quite drastic changes in the calculated Fermi surface, see fig. 17. Since their field Ti3 gave the closest agreement with the experimental bandwidth this was taken to give the most reliable calculated Fermi surface. From fig. 17 it can be seen that for this field the Fermi surface would consist of a closed pocket of holes around L in bands 3 and 4 and a closed pocket of electrons around Γ in bands 5 and 6. These are similar to the major features of the Fermi surface calculated by Altmann and Bradley (1964, 1967) for Zr but they bear no resemblance to the free-electron Fermi surface of an h.c.p. metal with valence 4, see

Fig. 18

(*a*)

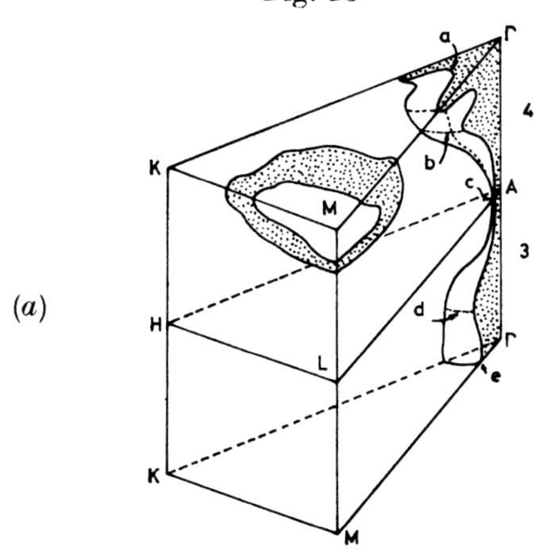

(*b*)

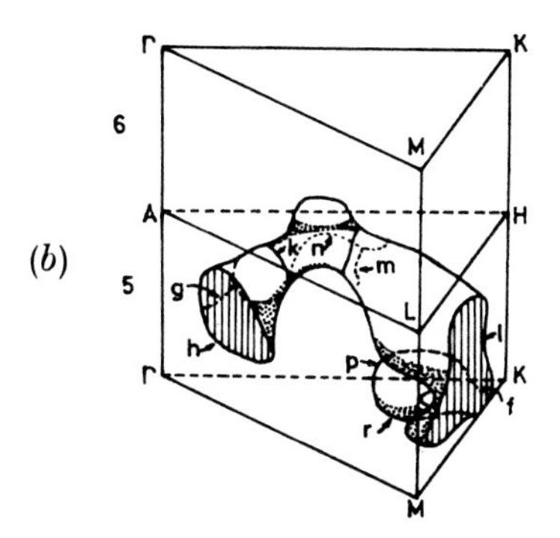

The Fermi surface of Ti calculated by Hygh and Welch (1970), (*a*) holes in bands 3 and 4, and (*b*) electrons in bands 5 and 6.

§ 5.3. A new A.P.W. calculation for Ti performed by Hygh and Welch (1970) produced a Fermi surface which is illustrated in fig. 18 and which is substantially different, both from the results of Altmann and Bradley (1964, 1967) and from what might have been expected by extrapolation from the A.P.W. calculations for Zr performed by Loucks (1967) (see § 5.3). It is clearly time that some more experimental Fermiological results for Ti were produced.

Only indirect experimental results are available for Ti. The temperature coefficient of the electronic specific heat exhibits similar electron–phonon enhancement to that observed in Sc and Y; the value of γ of 1·6 mJ mole^{-1} deg^{-2} calculated by Altmann and Bradley is to be compared with the mean experimental value of $(3{\cdot}41 \pm 0{\cdot}10)$ mJ mole^{-1} deg^{-2} quoted by Gschneidner (1964). Other work includes measurements of the thermal and electrical conductivity (Kemp *et al.* 1956), magnetic susceptibility (Kojime *et al.* 1961), Hall effect (Berlincourt 1959, Scovil 1966), magnetoresistance (Berlincourt 1959), thermoelectric power (Vedernikov 1969), positron annihilation (Stewart 1957, Wesołowski *et al.* 1963), photoemission (Eastman 1969 a) and the density of states by Compton scattering (Weiss 1970).

There is, therefore, still considerable doubt attached to the theoretical predictions of the Fermi surface of Ti as well as an almost complete lack of direct experimental evidence on this metal.

5.3. *Zirconium*

Slightly more work has been done on the determination of the Fermi surface of Zr than has been done for Ti. The de Haas–van Alphen effect was observed in Zr by Thorsen and Joseph (1963) and five periods were observed ranging in value from $2{\cdot}0$ to $3{\cdot}4 \times 10^{-8}$ G^{-1}. Of these five periods only one (α) could be followed over all angles and was therefore assumed to be associated with a closed piece of the Fermi surface. The remaining periods were present for magnetic fields in the [0001] direction and for limited inclinations to [0001] but eventually disappeared at large inclinations. The free-electron Fermi surface for an h.c.p. metal with valence 4 is not given with the others in fig. 2 (*c*) of part I ; however, its general features can be seen by extrapolation of the case of a metal of valence 3 in fig. 2 (*c*) of part I. Bands 1 and 2 are full all over the Brillouin zone and the regions occupied by electrons in bands 3, 4, 5 and 6 can be expected to be somewhat expanded. This free-electron Fermi surface is shown, in the single zone scheme, in fig. 19. Thorsen and Joseph ascribed their α

Fig. 19

(*a*) (*b*)

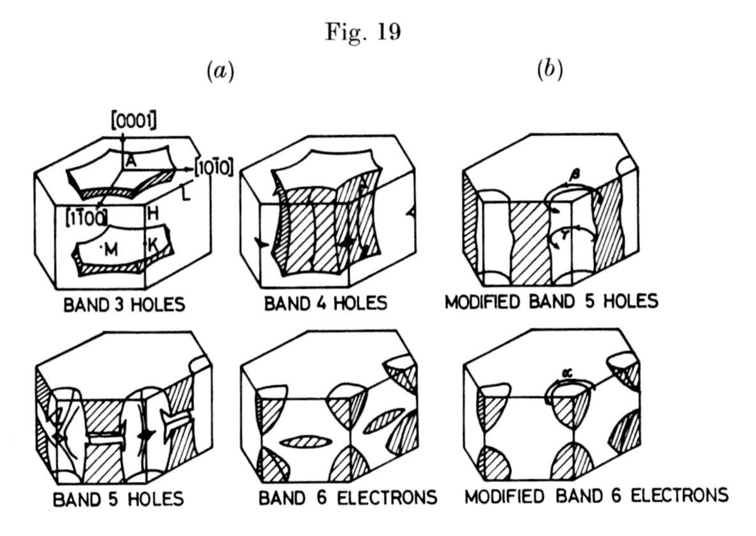

(*a*) Sketches of the free-electron Fermi surface for a h.c.p. metal of valence 4, and (*b*) modifications of the Fermi surface in bands 5 and 6 proposed by Thorsen and Joseph (1963) to account for the observed de Haas–van Alphen results in Zr.

period to the closed Fermi surface in band 6 of the free-electron model and attempted to explain two of their other periods in terms of a distorted version of the free-electron Fermi surface in band 5, see fig. 19.

There have been a number of cellular band structure calculations for Zr (Altmann 1958, Altmann and Bradley 1962, 1964, 1967) as well as an A.P.W. calculation (Loucks 1967). According to the cellular calculation of Altmann and Bradley (1964) bands 1 and 2 are full all over the Brillouin zone, in bands 3 and 4 in the double-zone scheme there are a number of closed regions of holes, see fig. 20 and in bands 5 and 6 in the double-zone scheme there is a region occupied by electrons which is almost isolated except for very small necks across the vertical faces of the Brillouin zone, see fig. 21. A small change in the Fermi surface in bands 5 and 6 was found in later calculations (Altmann and Bradley 1967) in which band 5

Fig. 20

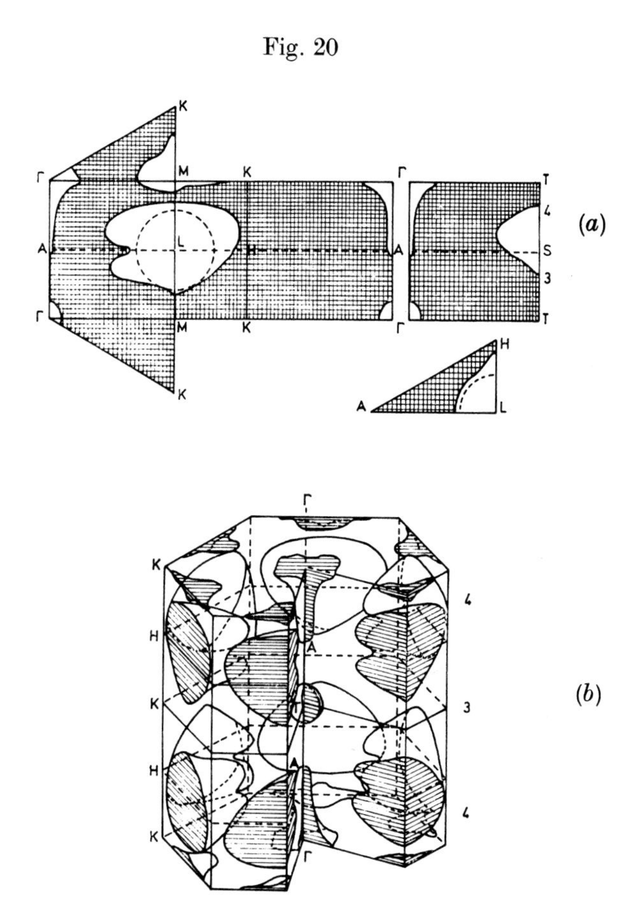

The Fermi surface of Zr in bands 3 and 4 according to Altmann and Bradley (1964) (in (*a*) regions of electrons are shaded and in (*b*) regions of holes are shaded).

Fig. 21

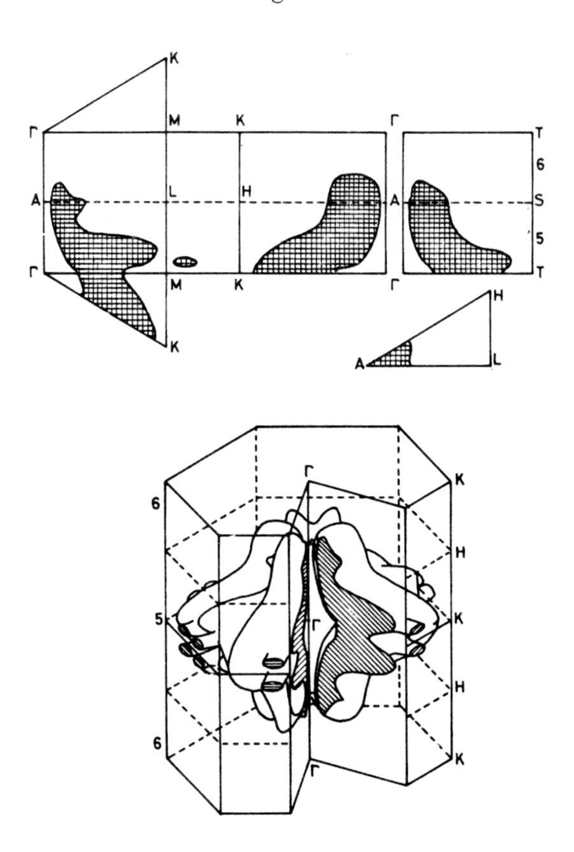

The Fermi surface of Zr in bands 5 and 6 according to Altmann and Bradley (1964) (regions of electrons are shaded).

no longer comes below the Fermi energy anywhere along ΓM so that the narrow waist of this pocket of electrons along ΓM vanishes. This Fermi surface bears very little resemblance to the free-electron Fermi surface shown in fig. 19 and was considerably more successful than the free-electron model in explaining the de Haas–van Alphen data of Thorsen and Joseph (1963). The Fermi surface of Altmann and Bradley was able to produce the correct values (up to a factor of 2) and the right angular dependence for four of the five measured periods, compared with one or, perhaps, two periods accounted for by the free-electron model with *ad hoc* modifications.

The A.P.W. band structure calculations performed by Loucks (1967) produced a Fermi surface that was substantially different from that calculated by Altmann and Bradley. The Fermi surface calculated by Loucks consisted, in the double-zone scheme, of a region of holes with axial

Fig. 22

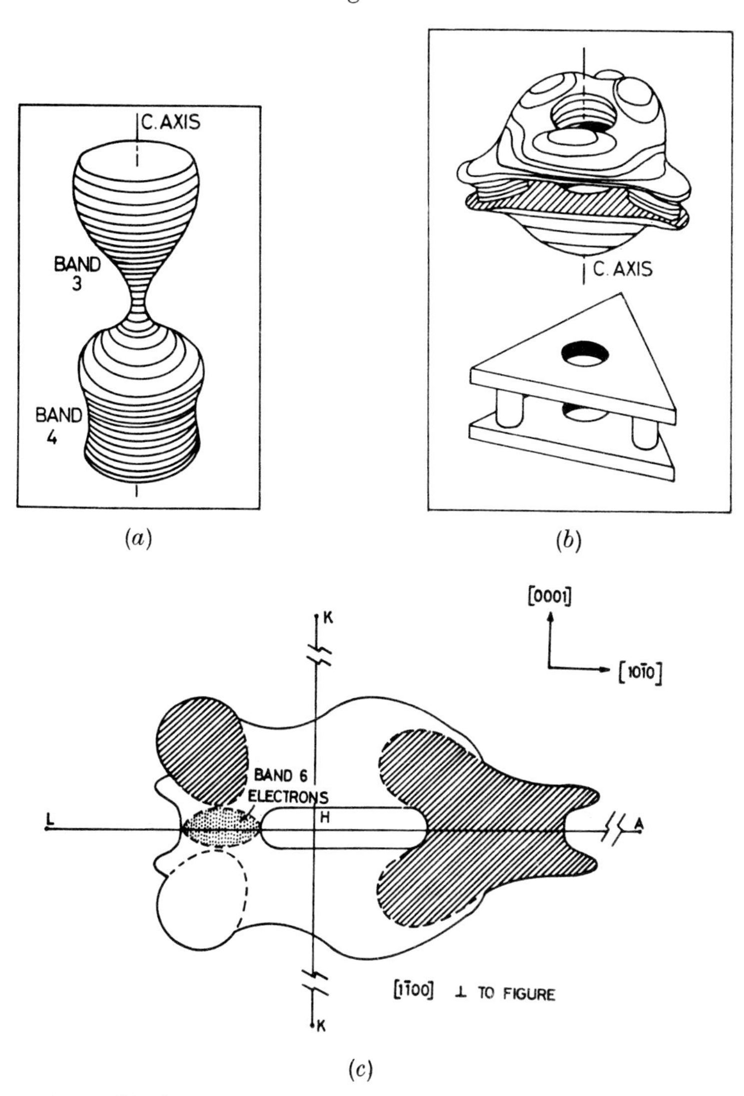

(*a*) (*b*)

(*c*)

(*a*) The regions of holes in bands 3 and 4 of Zr and (*b*) and (*c*) the regions of electrons in band 5 of Zr according to Loucks (1967).

rotational symmetry located along the [0001] axis in bands 3 and 4 (see fig. 22 (*a*)), an isolated region occupied by electrons around H in band 5 and small ellipsoidal pockets of electrons, also at H, in band 6 (see fig. 22 (*b*) and (*c*)). The band 6 pockets of electrons are contained within the ' pillars ' of the Fermi surface in band 5 in the single-zone scheme ; in the absence of spin–orbit coupling these two surfaces must touch in the plane AHL (see fig. 22 (*c*)). Further detailed cross sections of this calculated

Fermi surface are given by Loucks (1967). Loucks was also able, after a certain amount of modification of this calculated Fermi surface, to account semi-quantitatively for most of the features of the de Haas–van Alphen measurements of Thorsen and Joseph (1963). That two very different Fermi surfaces can both give semi-quantitative explanations of the same experimental data is an illustration of one or two of the weaknesses of the use of the de Haas–van Alphen effect for metals with complicated Fermi surfaces; first, the method only measures areas of cross section of the Fermi surface rather than caliper dimensions, and secondly only the orientation, but not the actual location of the cross section in the Brillouin zone, is directly available. Some further experimental measurements on the Fermi surface of Zr, preferably using one of the various size-effect methods, would be worth while.

The remaining work on Zr gives only indirect information concerning the Fermi surface of this metal. The calculated values of γ, the temperature coefficient of the electronic specific heat, of 1·8 mJ mole^{-1} deg^{-2} (Altmann and Bradley 1967) and 1·6 mJ mole^{-1} deg^{-2} (Loucks 1967) are compatible with the existence of a similar thermal effective mass, due to electron–phonon interactions, as pertains for neighbouring Y; the weighted mean experimental value of γ for Zr given by Gschneidner (1964) is $(2{\cdot}91 \pm 0{\cdot}12)$ mJ mole^{-1} deg^{-2}. Other existing work includes studies of the electrical and thermal conductivity (Kemp *et al.* 1956), the Hall effect (Berlincourt 1959) and its temperature dependence (Volkov *et al.* 1968), the magnetoresistance (Berlincourt 1959), optical properties (Graves and Lenham 1968 b, Eastman 1969 a), paramagnetic susceptibility (Kojime *et al.* 1961, Shimizu and Katsuki 1964, Shimizu *et al.* 1966), thermoelectric power (Vedernikov 1969) and a calculation by Gupta and Loucks (1968) of the expected angular dependence of positron annihilation radiation from Zr.

5.4. *Hafnium*

It appears that nothing at all has been done so far on the Fermi surface of Hf. Since the structure of Hf is the same as that of Ti and Zr we can make use of the fact that they are all in the same group of the periodic table to make qualitative predictions about the Fermi surface of Hf.

It will be recalled from § 4.2 that the Fermi surfaces of Sc and Y calculated using the A.P.W. method were very similar to each other, so that one might therefore expect the Fermi surface for Hf to be very similar to those of Ti and, more especially, Zr. This is, of course, not as helpful as it might be because, as we have seen in the previous section there are two rival proposed Fermi surfaces for Zr, each of which in some measure explains the somewhat scanty experimental measurements available for Zr.

Several of the usual macroscopic or indirect measurements have been performed on Hf (White and Woods 1957 d, Kojime *et al.* 1961, Kneip *et al.* 1963, Lenham 1967, Lenham and Treherne 1967, Graves and Lenham 1968 b, Eastman 1969 a, Vedernikov 1969). The experimental value of γ quoted by Gschneidner (1964) for Hf is $(2 \cdot 40 \pm 0 \cdot 24)$ mJ mole^{-1} deg^{-2}

§ 6. d-BLOCK : GROUP VA : VANADIUM, NIOBIUM AND TANTALUM

V	Vanadium	2.8.(8, 3) 2
Nb	Niobium	2.8.18.(8, 3) 2
Ta	Tantalum	2.8.18.32.(8, 3) 2

6.1. *Structures*

V, Nb and Ta all have the b.c.c. structure with lattice constants $a = 3 \cdot 024$ Å for V, $a = 3 \cdot 294$ Å for Nb and $a = 3 \cdot 30$ Å for Ta (Smithells 1967).

6.2. *The Rigid-band Model and the Transition Metals of Group VA*

A preliminary A.P.W. band structure calculation for V and several other b.c.c. transition metals along the line ΓΔH in the Brillouin zone performed by Mattheiss (1964) suggested that, for transition metals of the same structure, a rigid-band model gave a good approximation to the calculated band structures of these metals. As we shall see in § 7.2, the idea of a rigid-band model for transition metals had previously been used by Lomer (1962, 1964 a), particularly in connection with predicting the shapes of the Fermi surfaces of the group VIA metals Cr, Mo and W on the basis of the A.P.W. band structure calculations performed by Wood (1962) for Fe. The general features of these proposed Fermi surfaces were supported by the results of a direct A.P.W. calculation for W performed by Mattheiss (1965). Rigid-band-model arguments were then again applied by Mattheiss (1965) to the calculated band structure and Fermi surface

Fig. 23

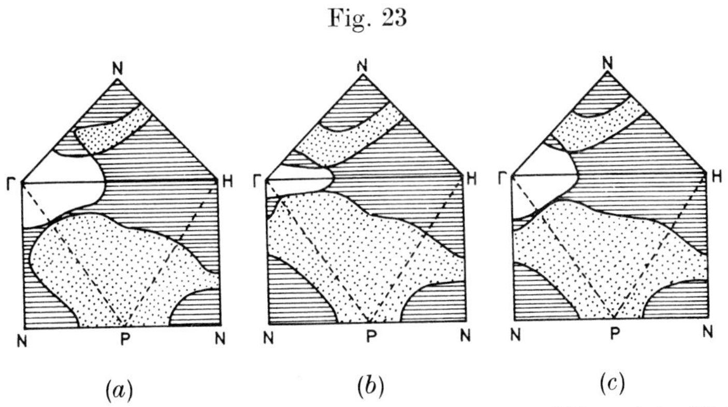

Cross sections of the Fermi surface predicted for group VA metals (V, Nb and Ta) on the rigid-band model by Mattheiss (1965) based on (*a*) A.P.W. energy bands for Fe and (*b*) and (*c*) two slightly different sets of A.P.W. energy bands for W.

for W to predict the general features of the Fermi surfaces of the metals V, Nb and Ta in group VA. Three slightly different Fermi surfaces for group VA metals were obtained by Mattheiss using Wood's results for Fe and two sets of bands obtained from A.P.W. calculations using two slightly different potentials in the band structure calculations for W, see fig. 23. These predictions are only semi-quantitative and are not able to distinguish between the three metals V, Nb and Ta. This proposed Fermi surface is illustrated in figs. 24 and 25 and consists of (i) an inner region of holes at Γ in band 2 which is illustrated in fig. 24, (ii) a multiply-connected set of hole tubes in band 3 along the ⟨100⟩ (or ΓΔH) directions, which has been described as a ' jungle gym ' (Mattheiss 1970) and which is illustrated in fig. 25, and (iii) ellipsoidal pockets of holes in band 3 at N which are also illustrated in fig. 25. The predictions based on the band structure of W suggest that the ellipsoids at N are isolated, see fig. 23 (*b*) and (*c*), whereas those based on the band structure of Fe suggest that these ellipsoids at N are joined to the multiply-connected tubular region by necks along ΓN, see fig. 23 (*a*). Further work is needed to distinguish between these two possibilities. It should be noted from fig. 23 that in the 100 and 110 planes the inner hole surface touches the multiply-connected tubular surface. The degeneracies that give rise to these contacts will be removed by the inclusion of spin–orbit coupling although they may be effectively restored, in V and Nb at least, by magnetic breakdown.

There is still in the Fermi surface proposed in figs. 24 and 25 for group VA metals a lingering resemblance to the free-electron model for a b.c.c. metal with five conduction electrons per atom. By extrapolation from the case of a b.c.c. metal of valence 4 shown in fig. 2 (*b*) of part I we should expect the free-electron Fermi surface for a b.c.c. metal of valence 5 to

Fig. 24

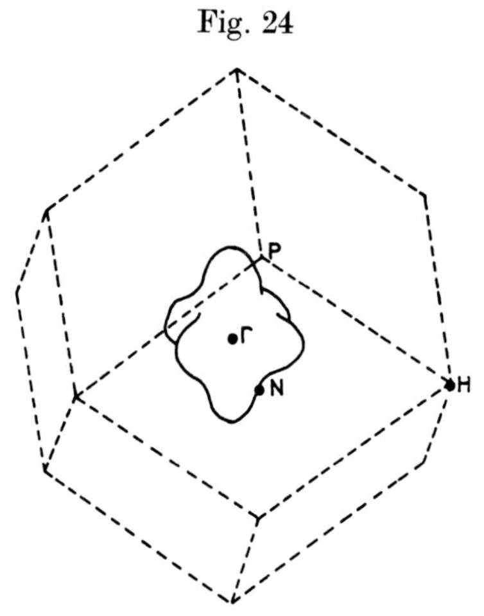

The region of holes around Γ with the Fermi surface proposed by Mattheiss (1965) for group VA metals (V, Nb and Ta).

Fig. 25

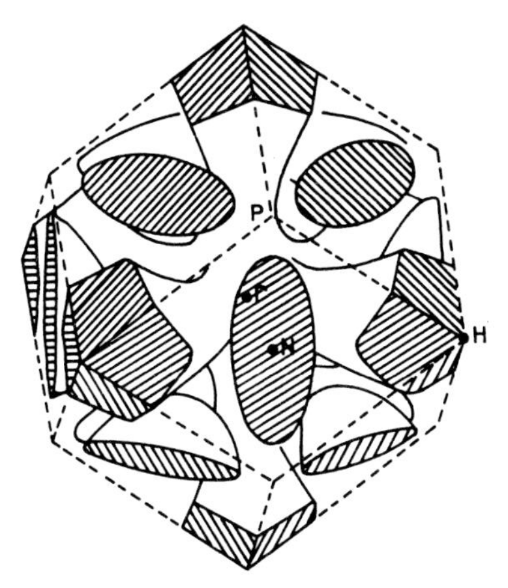

The ' jungle-gym ' hole tubes and the ellipsoidal pockets of holes at N with the Fermi surface proposed by Mattheiss (1965) for group VA metals (V, Nb and Ta).

have a region of holes in band 2 centred at Γ and with roughly the shape of an octahedron, quite fat arms of electrons in band 3 pointing from P towards Γ and with the pockets of electrons in band 4 somewhat enlarged. The region of holes in band 2 is not unlike the shape shown in fig. 24, while in fig. 25 the electrons are concentrated around the points P. Since band 2 is rather more full than in the free-electron model it is not too surprising that the pockets of electrons in band 4 in the free-electron model have vanished.

6.3. *Niobium and Tantalum*

Deegan and Twose (1967) calculated the band structure of Nb using the O.P.W. method and obtained a set of bands that were very similar to those used by Mattheiss (1965) in the calculation of the Fermi surfaces shown in figs. 23 (*b*) and (*c*) ; however, it was not possible to distinguish between these two possibilities. The work of Sharp (1969 a, b) on the observation of Kohn anomalies in Nb was in closer agreement with the Fermi surface shown in fig. 23 (*c*) than with the other possibilities in fig. 23. It therefore seems that in Nb the pocket of holes at N is closed so that fig. 23 (*a*) does not apply to Nb. The Fermi surfaces of Nb and Ta have been investigated experimentally principally by using the de Haas–van Alphen effect (Thorsen and Berlincourt 1961, Scott *et al.* 1968 a, b, Halloran *et al.* 1970, Scott and Springford 1970) and also by some galvano-magnetic measurements (Berlincourt 1959, Alekseevskiĭ *et al.* 1967, Fawcett *et al.* 1967, Reed and Soden 1968, Alekseevskiĭ *et al.* 1969). Open orbits were detected in Nb parallel to the four-fold axes (Alekseevskiĭ *et al.* 1967) as well as to the three-fold and two-fold axes (Alekseevskiĭ *et al.* 1969). Thorsen and Berlincourt (1961) observed two de Haas–van Alphen periods in each of the metals Nb and Ta. The two periods observed in Nb were both later ascribed to the ellipsoidal pockets of holes at N in band 3 by Scott *et al.* (1968 a, b) who also observed other periods which were ascribed to the multiply-connected tubular region of holes in band 3. No de Haas–van Alphen periods in Nb ascribable to the region of holes around Γ in band 2 were observed by Thorsen and Berlincourt (1961) or by Scott *et al.* (1968 a, b). The anisotropy of the galvanomagnetic properties of Nb was found to be very similar to that of Ta and for each of them the anisotropy of the magnetoresistance could be satisfactorily interpreted in terms of open orbits associated with the multiply-connected Fermi surface in band 3 proposed by Mattheiss. This was identified as being a multiply-connected region of holes rather than electrons from measurements of the Hall coefficient. New A.P.W. band structure calculations for Nb and Ta, including relativistic effects for Ta, have been performed by Mattheiss (1970). For Nb no significant differences from the Fermi surface illustrated in figs. 24 and 25 were obtained and, so long as spin–orbit coupling is neglected, the sheet of Fermi surface around Γ in band 2 and the multiply-connected ' jungle gym ' Fermi surface in band 3 are degenerate in the {110} and {100} symmetry planes. In the case of Ta spin–orbit coupling removes these degeneracies but the gaps which are introduced are quite small and can easily escape observation as a result of magnetic breakdown. Otherwise the Fermi surface of Ta calculated by Mattheiss (1970) was also very similar to that shown in figs. 24 and 25. Extensive new measurements of cross sections of the ellipsoids at N and the ' jungle gym ' for both Nb and Ta have been made by Halloran *et al.* (1970) using both the de Haas–van Alphen effect and magnetothermal oscillations. A detailed comparison between the calculated and measured extremal areas of cross section for both Nb and Ta are given in table V of Mattheiss (1970) and in most cases the agreement between theory and experiment is to within 5% and neither in Nb nor in Ta were oscillations observed experimentally that could be ascribed to the predicted region of holes around Γ in band 2. However, in further de Haas–van Alphen measurements on Nb by Scott and Springford (1970) some oscillations were observed which were ascribed to orbits involving the region of holes around Γ in band 2 ; it is now fairly certain that this piece of the Fermi surface does actually exist although Scott and Springford proposed a slight distortion of the octahedron from the shape given by Mattheiss. Important calculated and measured extremal areas of cross section are given in a table by Scott and Springford (1970).

Other results indirectly related to the electronic energy band structure of Nb and Ta include work on the paramagnetic susceptibility (Kojime *et al.* 1961 on Nb and Ta, Van Ostenburg *et al.* 1963 on Nb, Shimizu *et al.* 1966 on Nb), electrical and thermal conductivity (White and Woods 1957 d on Nb), magnetoresistance (Fawcett 1961 c on Nb and Ta), thermoelectric power (Vedernikov 1969 on Nb and Ta), positron annihilation (Dekhtyar *et al.* 1966 on Nb and Ta), x-ray and optical measurements (Claus and Ulmer 1963 on Nb and Ta, Merz and Ulmer 1966 on Nb and Ta, Lenham 1967 on Nb and Ta, Lenham and Treherne 1967 on Nb, Juenker *et al.* 1968 on Ta, Eastman 1969 a on Nb, Golovashkin *et al.* 1969 on Nb) and electronic specific heat ; measurements of γ for Nb range from 7·50 to 7·85 mJ mole^{-1} deg^{-2} (Boorse and Hirshfeld 1958, Chou *et al.* 1958, Hirshfeld *et al.* 1962, Morin and Maita 1963, Leupold and Boorse 1964, Van der Hoeven and Keesom 1964 a, Shen *et al.* 1965, Ferreira da Silva *et al.* 1966) and a mean value of (5·84 ± 0·31) mJ mole^{-1} deg^{-2} for Ta is quoted by Gschneidner (1964). Golovashkin *et al.* (1969) used their measurements of the optical constants n and κ of Nb over the range 0·4 to 10 microns to determine a pseudopotential for Nb.

We have now seen that the details of the Fermi surface of Nb are fairly well established and that it seems that the Fermi surface of Ta resembles that of Nb quite closely.

6.4. *Vanadium*

An A.P.W. band structure calculation for V was performed by Mattheiss (1964) along only one line in the Brillouin zone. Preliminary A.P.W. calculations by Anderson *et al.* (1969) have been reported using two different expressions for the exchange contribution to the potential ; tentative comparisons were made with preliminary de Haas–van Alphen results which had been ascribed to ellipsoidal pockets of holes at N. Little experimental evidence on V is available at present but it seems likely from the magnetoresistance data of Nelson *et al.* (1968) that the Fermi surface described in § 6.2 and illustrated in figs. 24 and 25 is fairly close to the true Fermi surface of V. The mean experimental value of γ, the temperature coefficient of the electronic specific heat, for V quoted by Gschneidner (1964) is (9·04 ± 0·22) mJ mole^{-1} deg^{-2}. Other results related to the electronic energy band structure of V include work on the paramagnetic susceptibility (Kojimé *et al.* 1961), electrical and thermal conductivity (White and Woods 1957 d), thermoelectric power (Vedernikov 1969), positron annihilation (Stewart 1957), and optical measurements (Lenham 1967, Lenham and Treherne 1967, Eastman 1969 a). Anomalies in the magnetic susceptibility, thermal expansion, electrical conductivity and mechanical properties of V have been observed in the range 220–240°K (Bolef *et al.* 1961, Burger and Taylor 1961, Taylor and Llewellyn Smith 1962, Smirnov and Finkel 1965, Suzuki and Miyahara 1965) ; the size of the anomaly in the electrical conductivity was found to be reduced by the application of pressure (Suzuki *et al.* 1966). Most investigators have tended to explain these anomalies in terms of antiferromagnetic ordering. Amitin *et al.* (1967) found no such anomaly in V in their measurements of the temperature dependence of the Hall effect and of the electrical resistivity.

§ 7. d-BLOCK : GROUP VIA : CHROMIUM, MOLYBDENUM AND TUNGSTEN

Cr	Chromium	2.8.(8, 5) 1
Mo	Molybdenum	2.8.18.(8, 5) 1
W	Tungsten	2.8.18.32.(8, 4) 2

7.1. *Structures*

Under ordinary conditions the three metals Cr, Mo and W have the b.c.c. structure, with lattice constants $a = 2{\cdot}89$ Å for Cr, $a = 3{\cdot}140$ Å for Mo and $a = 3{\cdot}16$ Å for W (Smithells 1967). At temperatures above 1840°C, Cr exhibits the f.c.c. structure with $a \simeq 3{\cdot}8$ Å ; this is called the β phase of Cr while the low temperature version is called the α phase. Other electrolytic or unstable forms of Cr and W exist but they do not concern us here. Cr undergoes a transition to an antiferromagnetically ordered state at a Néel temperature, T_N, of 311·5°K (for references see, for example, Falicov and Zuckermann (1967)). Below T_N the structure is not that of a simple two-sublattice antiferromagnet but one in which the spin polarization $\mathbf{P}(\mathbf{r})$ varies sinusoidally with a wave vector $\mathbf{q}$, that is, there is a spin-density wave (S.D.W.) (Overhauser 1962) which takes the form

$$\mathbf{P}(\mathbf{r}) = \mathbf{P}(\mathbf{q}) \exp(-i\mathbf{q} \cdot \mathbf{r}). \qquad (7.1.1)$$

References to the experimental evidence for this magnetic structure are cited by Koehler, Moon, Trego and Mackintosh (1966). We have already encountered spin-density waves in part I in connection with K, although it now seems likely that in K any special effects are due to variations in the electronic charge density (charge-density waves) rather than to any sinusoidal ordering of the spins of the electrons. Since Cr is antiferromagnetic there will be two sublattices, one consisting of the Cr atoms at the corners of the conventional b.c.c. unit cells and the other consisting of the atoms at the centres of these cells. The amplitude of the magnetic moment, or spin polarization, $\mathbf{P}(\mathbf{q})$ on one sublattice is exactly opposite to that on the other sublattice. The orientation of $\mathbf{P}(\mathbf{q})$ (that is, the polarization of the spin-density wave) may take any arbitrary value relative to $\mathbf{q}$; $\mathbf{q}$ in turn can take any arbitrary value so that the wavelength of the spin-density wave need not be commensurate with the crystal lattice of the metal. For Cr, if the temperature is lower than T_N but above a second transition temperature, T_S, of 115°K $\mathbf{P}(\mathbf{q})$ is perpendicular to $\mathbf{q}$, that is, the spin-density wave has transverse polarization ; below T_S the polarization becomes longitudinal, that is, $\mathbf{P}(\mathbf{q})$ is parallel to $\mathbf{q}$. T_S is called the 'spin-flip' temperature (see, for example, Matsumoto *et al.* (1969), and Meaden *et al.* (1969)). In Cr the wave vector $\mathbf{q}$ orients itself along the $\langle 100 \rangle$ type directions and its magnitude is given by

$$\mathbf{q} = \tfrac{1}{2}\mathbf{G}_0(1-\delta) \qquad (7.1.2)$$

where $\mathbf{G}_0$ is the basic reciprocal lattice vector in the [100] direction and δ varies with temperature from about 0·035 near T_N to about 0·05 at very low temperatures. The value of $\mathbf{q}$ is equal to the separation between two important pieces of the Fermi surface of Cr (see § 7.4 for details) and the theoretical justification for this is discussed, for example, by Herring (1966) and Roth *et al.* (1966). One would expect to find some magnetostrictive distortion along the direction of $\mathbf{q}$, however, when a single crystal of Cr is cooled below T_N it has several different choices of $\langle 100 \rangle$ type directions available so that antiferromagnetic domains will be formed in the crystal, where in different domains the magnetic ordering may occur with different directions, such as [100] and [010], for $\mathbf{q}$. This occurrence of antiferromagnetic domains is familiar in many antiferromagnetic non-metallic crystals (Cracknell 1969 b). Because of the domains any magnetostrictive distortions will tend to cancel out and any departure from cubic symmetry will be difficult to detect. However, if a single crystal of Cr is cooled below T_N in the presence of a magnetic field parallel to the [100] direction then it may be possible to produce a single-domain single crystal specimen of Cr with $\mathbf{q}$ parallel to [100], in which case the magnetostrictive distortion in the direction of $\mathbf{q}$ will reduce the symmetry of the crystal from cubic to tetragonal ; evidence of tetragonal symmetry in samples of field-cooled antiferromagnetic Cr has been found in several de Haas–van Alphen and magnetoresistance experiments (see, for example, Arko and Marcus (1964), Montalvo and Marcus (1964), Watts (1964 a, b), Graebner and Marcus (1966, 1968)) ; however, the actual magnitude of the distortion of the crystal from cubic symmetry was too small to be detected directly in x-ray diffraction measurements (Werner *et al.* 1966, Combley 1968) but the existence of the distortion was detected directly by Lee and Asgar (1969) using electrical resistance strain gauges.

7.2. *The Rigid-band Model and the Transition Metals of Group VIA*

The first successful model that was suggested for the Fermi surfaces of the metals Cr, Mo and W of group VIA was based on the use of the rigid-band model in conjunction with the A.P.W. band structure calculation performed for Fe by Wood (1962). Lomer (1962) observed that in previous band structure calculations for a number of f.c.c. d-block metals (Cu, Ag, Fe and Ni) the general appearance of the various band structures was quite similar although the actual widths of the bands might differ considerably. It therefore seemed that in the almost complete absence, at that time, of direct band structure calculations for Cr, Mo and W a good approximation to the band structures and Fermi surfaces of these metals could be obtained by using the rigid-band model. For the b.c.c. structure which is possessed by Cr, Mo and W there were very few band structure calculations available at that time for any d-block metals nearby in the periodic table and Lomer based his arguments almost exclusively on the band structure calculation by Wood (1962) for Fe but with some reference to the very early cellular calculation of Manning and Chodorow (1939) on

W and the tight-binding calculations of Asdente and Friedel (1961) on Cr. Later A.P.W. band structure calculations by Mattheiss (1964) along the line ΓΔH for a number of transition metals provided further support for Lomer's general idea of using a rigid-band model as a starting point in connection with the band structures of transition metals. An interpolation scheme was used to determine the value of the Fermi energy corresponding to six electrons per atom, see fig. 26. After some slight modification (Lomer 1964 a) the Fermi surface shown in fig. 27 was obtained; this Fermi surface consists entirely of closed surfaces and implies that the

Fig. 26

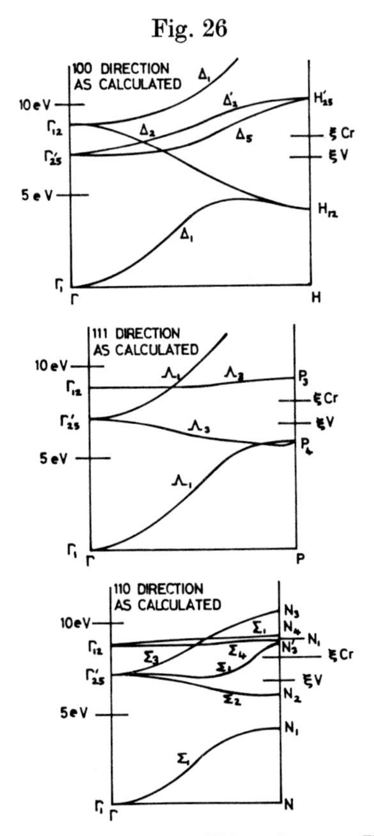

The position of the Fermi level for group VA and group VIA metals based on the rigid-band model and the A.P.W. calculations of Wood (1962) for Fe (Lomer 1962).

metal is compensated, with the number of holes in band 3 equal to the total number of electrons in bands 4 and 5. This Fermi surface encloses (i) in band 3 a large closed region of holes around H, (ii) in band 3 a smaller closed region of holes around N, (iii) in band 4 a large closed region of electrons around Γ, which resembles an octahedron with knobs on its corners and which is sometimes described as an electron ' jack ', and (iv) in band 5 a set of small electron pockets, or ' lenses ', located at a position

Fig. 27

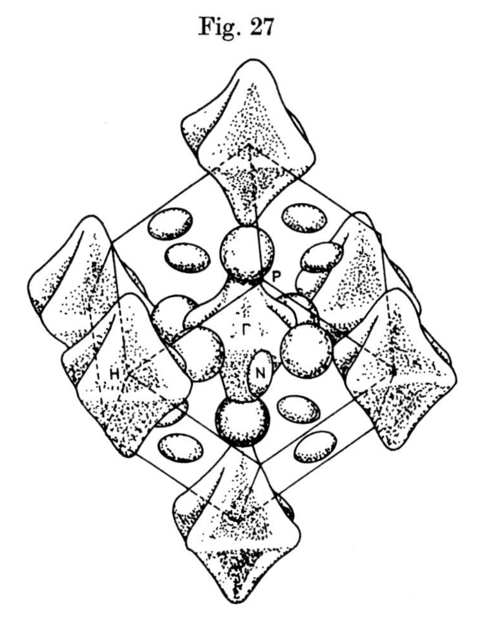

The Fermi surface of Mo proposed by Lomer (1962, 1964 a) (Leaver and Myers 1969).

on ΓH so as to be within the necks that connect the knobs to the body of the electron ' jack ', see the cross sections in fig. 28. While this Fermi surface was constructed for a metal of valence 6 and could therefore be expected to apply to all three metals Cr, Mo and W, it must be realized that for Cr it can only be expected to apply to the high-temperature non-magnetic phase and that for W it may be severely modified by spin–orbit coupling. The prediction of Fermi surfaces for group VIA metals consisting only of closed surfaces is consistent with the results of the magnetoresistance work on Mo and W (Fawcett 1961 c, 1962, Alekseevskiĭ *et al.* 1962). If the two sheets of Fermi surface enclosing the electron ' jack ' and the region of holes at H in Lomer's model do actually touch at a point along the line ΓH this might be expected to give rise to an infinitesimally small number of open orbits ; this number could become finite and produce an appreciable effect on the high-field magnetoresistance if magnetic breakdown occurs. Fawcett and Reed (1964 a) showed from their magnetoresistance measurements on Mo and W in fields up to 83 kG that the Fermi surfaces of these metals support less than 10^{-4} open orbits per atom for Mo and less than 10^{-7} open orbits per atom for W. The results of the size-effect measurements of Walsh and Grimes (1964) indicated that the electron ' jack ' at Γ and the hole ' octahedron ' at H do not touch but are separated by about 5% of the distance ΓH.

Since Cr has a complicated magnetically ordered structure below 311·5°K we shall discuss the Fermi surfaces of Mo and W first and then return to the more complicated case of Cr.

Fig. 28

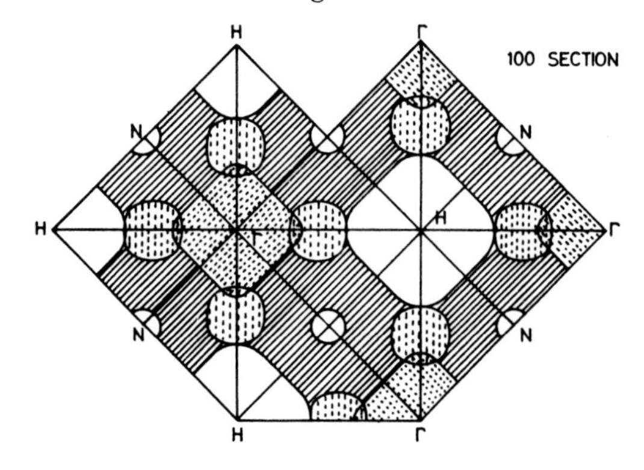

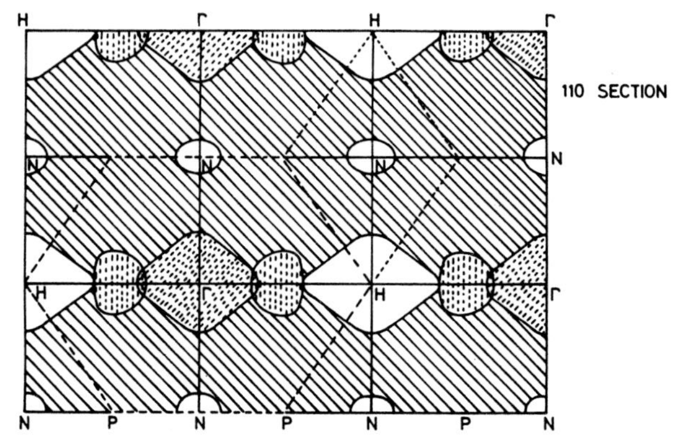

100 and 110 cross sections of the Fermi surface of group VIA metals proposed by Lomer (1964 a) ; the contact between the electron and hole surfaces along ΓH can be expected to be removed by spin–orbit coupling in W at least (Mattheiss and Watson 1964).

7.3. *Molybdenum and Tungsten*

A.P.W. band structure calculations were performed by Loucks (1965 a) for Mo and W and by Mattheiss (1965) for W ; Loucks found the calculated Fermi surfaces of Mo and W to be almost identical to one another and quite similar to that proposed by Lomer (1962, 1964 a). The Fermi surface for W calculated by Mattheiss (1965) was also very similar to that predicted by Lomer (1962, 1964 a).

On the experimental side a considerable amount of work has been done on the Fermi surfaces of both Mo and W. This includes work on the de Haas–van Alphen effect (Brandt and Rayne 1962, 1963, Sparlin and Marcus 1966, Girvan *et al.* 1968 on W only, Leaver and Myers 1969 on Mo only), radio-frequency size effect (Walsh and Grimes 1964, Boiko *et al.* 1967, 1969 on Mo only†) magnetoacoustic geometric resonances (Rayne and

† See note added in proof.

Sell 1962 on W only, Jones and Rayne 1964 a on Mo only, 1964 b, c, Rayne 1964 on W only, Bezuglyĭ *et al.* 1965†), anomalous skin effect (Fawcett and Griffiths 1962), magnetoresistance (Fawcett 1961 c, 1962, Alekseevskiĭ *et al.* 1962 on Mo only, Berthel 1964, Fawcett and Reed 1964 a†), and cyclotron resonance (Fawcett and Walsh 1962 on W only, Walsh 1964, Herrmann 1967 on Mo only, 1968). In the case of Mo the various experimental results are in good qualitative agreement with Lomer's model. de Haas–van Alphen oscillations have now been observed in Mo corresponding to all the four pieces of Fermi surface in the model proposed by Lomer (Sparlin and Marcus 1966, Leaver and Myers 1969). The detailed dimensions of the Fermi surface of Mo are given in table III of Sparlin and Marcus (1966) and also by Leaver and Myers (1969).

However, in spite of the good agreement between the various experimental results for Mo and the theoretical Fermi surface proposed by Lomer (1962, 1964 a) and supported by the non-relativistic A.P.W. calculations of Loucks (1965 a) and Mattheiss (1965), the agreement between these theoretical predictions and the experimental results for W was much less satisfactory, presumably as a result of spin–orbit coupling. Because of the probable importance of spin–orbit coupling in W (Mattheiss and Watson 1964, Walsh and Grimes 1964) an *ab initio* relativistic A.P.W. calculation for W was performed by Loucks (1965 b, 1966 a) and the calculated spin–orbit splitting of the Δ_5 band along ΓH was in good agreement with the experimental results of the r.f. size-effect measurements of Walsh and Grimes (1964). In the results of the relativistic A.P.W. calculation for W Loucks (1966 a) found that both the electron lenses along ΓH in band 5 and the holes at N in band 3 had disappeared and the size of the electron ' jack ' had been reduced. Later de Haas–van Alphen work indeed revealed no trace of the electron lens pockets in band 5 in W (Sparlin and Marcus 1966, Girvan *et al.* 1968). There is experimental evidence that although the pockets of holes at N in W do not actually disappear completely as predicted by the relativistic A.P.W. calculation they are, nevertheless, very much reduced from their sizes in Lomer's model and in the results of the non-relativistic A.P.W. band structure calculations of Loucks (1965 a) and Mattheiss (1965). For instance the de Haas–van Alphen measurements of Sparlin and Marcus (1966) showed that the linear dimensions of the pocket of holes at N in band 3 in W are about half the linear dimensions of the corresponding pocket in Mo. The reduction in size of the electron ' jack ' and of the hole ' octahedron ' at H is also supported by the de Haas–van Alphen results of Brandt and Rayne (1963), Sparlin and Marcus (1966) and Girvan *et al.* (1968). Fawcett and Griffiths (1962) showed from their anomalous skin effect results that the total area of the Fermi surface of W was smaller than that of Mo ; the actual measured areas were 1·74 atomic units for Mo and 1·66 atomic units for W. Loucks estimated the area of the calculated relativistic A.P.W.

† See note added in proof.

Fermi surface of W to be about 70% of that obtained from the non-relativistic calculations. The parameters of the Fermi surface of W are given in table III of Sparlin and Marcus (1966) and a very large number of de Haas–van Alphen frequencies, or areas of cross section, will be found in table I of Sparlin and Marcus (1966) and table I of Girvan *et al.* (1968). Thus we see that the inclusion of relativistic effects generally decreases the sizes of all the pieces of the calculated Fermi surface of W and leads to the complete disappearance of the small electron lens in band 5 along ΓH. Correspondingly the density of states at the Fermi surface will be expected to be lower in the relativistic calculation than in the non-relativistic one (Shimizu *et al.* 1962, Loucks 1965 b, 1966 a). Most of the experimental results for W are in better agreement with the calculated relativistic Fermi surface than with the non-relativistic one which, it will be recalled, was almost identical with that of Mo. Since relativistic effects should be almost negligible in Mo†, a comparison of experimental results on the Fermi surfaces of these two metals should indicate the extent of the relativistic effects.

Many of the usual experiments related to macroscopic properties that depend on the electronic structure of a metal have been performed on Mo and W. The paramagnetic susceptibility of Mo and W was calculated by Shimizu *et al.* (1962) and of Mo alone by Van Ostenburg *et al.* (1963) and Shimizu *et al.* (1966) while the temperature dependence of the susceptibility was measured by Kojime *et al.* (1961) for both metals. Other experimental work which has been performed on these metals includes measurements of the Knight shift (Aksenov 1958 on Mo), Hall effect (Volkenshteĭn *et al.* 1963 on W), size effects in the Hall effect and magnetoresistance (Soule and Abele 1969 on W), the thermal and electrical magnetoresistances (Van Witzenburg and Laubitz 1968 on W), positron annihilation (Lang *et al.* 1955 on W), Kohn anomalies (Walker and Egelstaff 1969 on Mo), and various optical and x-ray measurements (Claus and Ulmer 1963 on Mo and W, Merz and Ulmer 1966 on W, Juenker *et al.* 1968 on Mo and W, Eastman 1969 a on Mo). Calculated values of γ, the temperature coefficient of the electronic specific heat for Mo of 1·27 mJ mole^{-1} deg^{-2} and for W of 1·25 mJ mole^{-1} deg^{-2} are to be compared with mean experimental values for Mo of $(2{\cdot}10 \pm 0{\cdot}14)$ mJ mole^{-1} deg^{-2} and for W of $(1{\cdot}22 \pm 0{\cdot}15)$ mJ mole^{-1} deg^{-2} (Gschneidner 1964; for further references see Loucks (1965 a) and Rorer *et al.* (1965)). The thermoelectric power of Mo and W has been measured by a number of workers, see Vedernikov (1969).

7.4. *Chromium*

We have already discussed the structure of antiferromagnetic Cr in § 7.1. By analogy with the elements in the previous groups in the

† However, it has recently been shown that relativistic effects in Mo may be more important than was previously thought to be the case, see note added in proof.

periodic table we should expect the Fermi surfaces of the elements in the fourth and fifth periods, namely Cr and Mo, to be very similar to one another. Thus the Fermi surface for group VIA metals which was originally proposed by Lomer (1962, 1964 a) and supported by the non-relativistic A.P.W. calculations of Loucks (1965 a) and which was found to give a good description of the Fermi surface of Mo can be expected also to apply to non-magnetic Cr, that is, to Cr at temperatures above T_N. Slight differences from Lomer's model for the Fermi surface of non-magnetic Cr were obtained in the non-relativistic A.P.W. calculations of Loucks (1965 a) ; in Cr the pockets of holes at N in band 3 disappeared and the sizes of the knobs on the electron jack were also reduced. A somewhat different Fermi surface had been deduced by Asdente (1962) from the results of the tight-binding band structure calculation of Asdente and Friedel (1961) on Cr. However, direct experimental measurements of Fermi surface features at such high temperatures are almost impossible to obtain. Band structure calculations for antiferromagnetic Cr have been made by Switendick (1966) and by Asano and Yamashita (1967) (see below).

Fig. 29

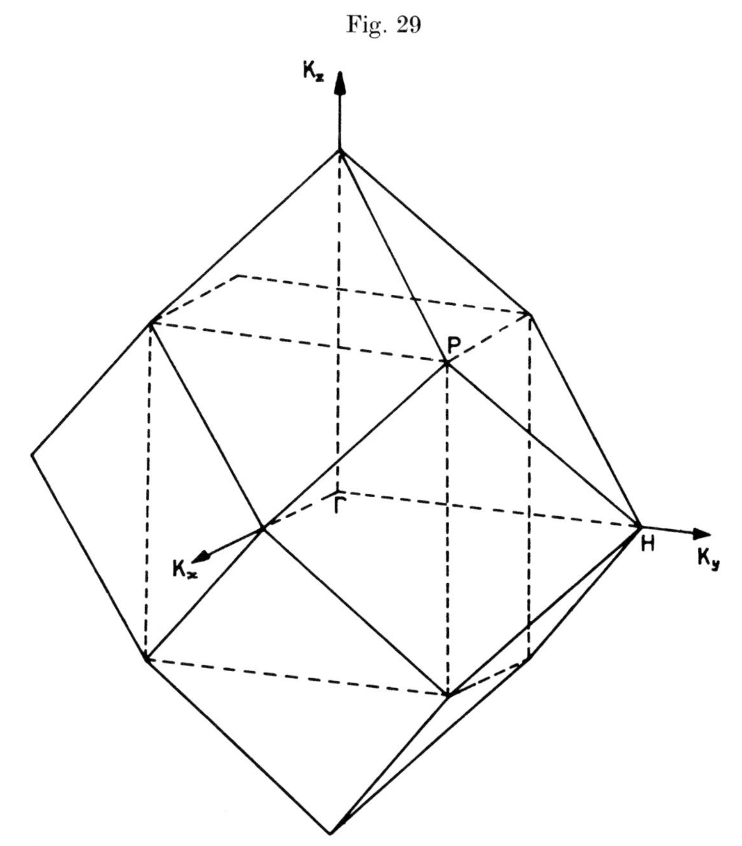

The Brillouin zones of non-magnetic Cr (continuous lines) and of antiferromagnetic Cr (broken lines).

Experimental work related to the determination of the Fermi surface topology of Cr has, of necessity, been carried out almost exclusively below T_N and the results show that there is considerable departure from the Lomer model, presumably as a result of the magnetic ordering (see, for example, Brandt and Rayne (1962, 1963)). It is therefore necessary to consider what effect the presence of an oriented spin-density wave will have on the Fermi surface. The presence of an oriented axial vector $\mathbf{P}(\mathbf{r})$, the spin polarization, will necessarily reduce the symmetry of the Cr crystal and will therefore reduce quite substantially the degeneracies in the electronic band structure (Falicov and Ruvalds 1968, Cracknell 1969 a, 1970). Any magnetostrictive distortions will generally be found to be compatible with this very much reduced symmetry of the crystal in its magnetically ordered phase. It was noticed by Lomer (1962) that the magnitude of $\mathbf{q}$ was just of the size that corresponds to the separation of opposite surfaces of the 'jack' at Γ and the region of holes around H in the Fermi surface of the non-magnetic form of Cr. Similar relationships between the wave vector $\mathbf{q}$ of a periodic magnetic structure and the 'nesting' of portions of the Fermi surface occur in a number of the rare-earth metals (see § 10.2) and the theoretical reasons for this are discussed

Fig. 30

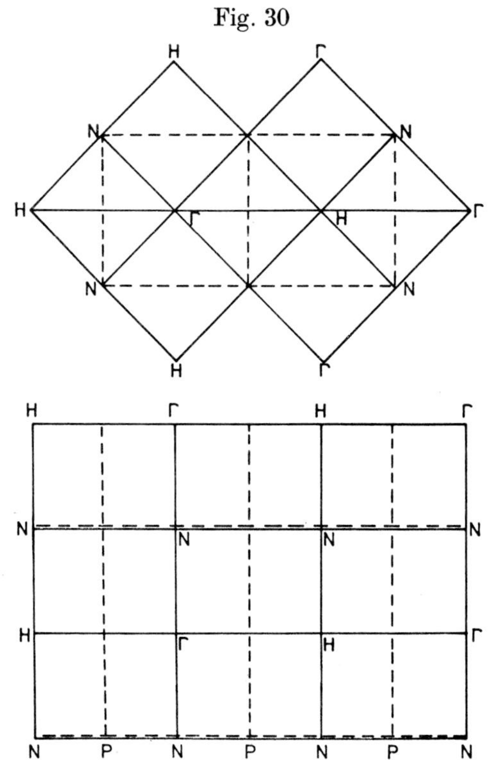

The Brillouin zone boundaries for the antiferromagnetic phase of Cr indicated (by broken lines) on the cross sections shown in fig. 28.

by Herring (1966) and Roth *et al.* (1966). The term ' nesting ' is employed to describe the situation in which two approximately parallel sheets of the Fermi surface can be made to coincide by displacing either of them through an appropriate distance in $\mathbf{k}$ space (Andersen and Loucks 1968). In their calculation of the band structure of antiferromagnetic Cr Asano and Yamashita (1967) considered Cr to have a simple two-sublattice antiferromagnetic structure where $\mathbf{P}(\mathbf{q})$ was of opposite sign on the two sublattices. In this case the size of the unit cell will be doubled, as happens in quite a large number of non-metallic antiferromagnetic crystals (see, for example, Cracknell (1969 b)). The volume of the Brillouin zone is therefore halved, see fig. 29, and in particular the points that were H points on the corners of the Brillouin zone now become Γ points at the centre of the new Brillouin zone for the antiferromagnetic phase. The relationship of the new Brillouin zone boundaries to some of the principal cross sections of the Fermi surface of non-magnetic Cr is shown in fig. 30.

A very extensive set of de Haas–van Alphen measurements on antiferromagnetic Cr was performed by Graebner and Marcus (1968) who assumed that the electron ' jack ' at Γ in band 4 and the hole ' octahedron ' at H in band 3 both disappeared completely in antiferromagnetic Cr. Graebner and Marcus were then able to explain all their results in terms of the holes at N and the electron lens along ΓΔH together with the existence of the spin-density wave with wave vector $\mathbf{q}$. At first sight this disappearance of the ' jack ' and ' octahedron ' seems rather strange in view of the fact that for Mo, and therefore presumably also for non-magnetic Cr, they were very much larger than the other two pieces of Fermi surface and also in view of the fact that in Loucks' non-relativistic band structure calculation for non-magnetic Cr it was the holes at N that vanished and not the holes at H. However, it will be recalled that the size of the Brillouin zone in the antiferromagnetic state is reduced by a factor of two because the volume of the unit cell is doubled. Because the antiferromagnetic unit cell contains two Cr atoms there are 12 conduction electrons per unit cell so that in the Brillouin zone the equivalent of six bands must be filled. The points H and Γ now coincide so that the hole ' octahedron ' at H becomes transferred to Γ and becomes nearly filled up using the electrons from the octahedral body of the ' jack ' and the electron lens in band 5 leaving the approximately spherical pockets of electrons that were the knobs on the corners of the jack. This leads to an unoccupied region corresponding to a thin shell of which the inner surface is that of the body of the original electron jack and with its outer surface formed by the surface of the original octahedron. Indeed one such surface was found in the band structure calculations for antiferromagnetic Cr performed by Asano and Yamashita (1967) using the Greens function method, see fig. 31 (*a*). Quite a small change in the choice of the Fermi energy could cause most of this thin shell to vanish except for the small regions near X, see fig. 31 (*b*). It can therefore be seen qualitatively how, from the electron jack at Γ, the hole octahedron at H and the electron lens in band 5

Fig. 31

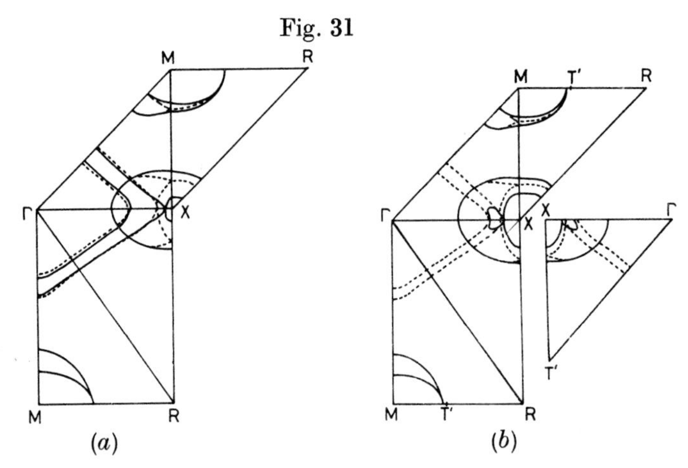

Fermi surface cross sections for Cr according to Asano and Yamashita (1967) for two slightly different choices of E_F (solid lines are for the antiferromagnetic phase and broken lines for the paramagnetic phase).

along ΓH in the paramagnetic case all that is likely to remain in the magnetic phase is two pockets of electrons around X. The holes at N simply become folded back to give two pockets of holes in bands 5 and 6 at the same point, which happens, however, to be labelled M in the magnetic Brillouin zone. Therefore we can see that the assumption of the disappearance of the electron jack and the hole octahedron that was made by Graebner and Marcus (1968) on an apparently *ad hoc* basis to explain the de Haas–van Alphen data is a natural consequence of the antiferromagnetic ordering and the consequent changes in the sizes of the unit cell and Brillouin zone. Because of the folding back of the outer regions of the Brillouin zone of the non-magnetic phase into the smaller magnetic Brillouin zone there would be pockets of electrons at X in both bands 7 and 8 (formerly band 4 in the Brillouin zone of non-magnetic Cr) and there would be pockets of holes at M (formerly called N) in both bands 5 and 6 (formerly band 3). Because de Haas–van Alphen type oscillations only give extremal areas of cross section rather than caliper dimensions of the Fermi surface and also do not locate the positions of these cross sections in the Brillouin zone, there is often a danger that more than one proposed Fermi surface may lead to ‘ good qualitative agreement ’ with the experimental data. To eliminate this possibility it would therefore be useful to have some direct measurements of caliper dimensions of the Fermi surface of antiferromagnetic Cr to test that the various pockets of electrons and holes, of which the shapes have been determined by Graebner and Marcus (1968), are actually at X for the electrons and at M (formerly called N in the non-magnetic phase) for the holes.

The detailed description of all the reductions in the degeneracy of the electronic band structure of Cr due to the presence of an axial vector $\mathbf{P}(\mathbf{r})$ associated with the magnetic ordering does not concern us here. The only degeneracy in the band structure of non-magnetic Cr which is of significance for the topology of the Fermi surface in the model due to Lomer and to Loucks is the two-fold degenerate Δ_5 band which leads to the contact along $\Gamma\Delta$H between the hole octahedron around H in band 3 and the electron jack around Γ in band 4. In the magnetically ordered phase the original Δ points are not now all equivalent. In the temperature range $T_S \leqslant T \leqslant T_N$, where $\mathbf{P}(\mathbf{q})$ is perpendicular to $\mathbf{q}$, this degeneracy is lifted because of the loss of the four-fold rotation axis of symmetry. For $T \leqslant T_S$ when $\mathbf{P}(\mathbf{q})$ is parallel to $\mathbf{q}$ the four-fold rotation axis of symmetry returns for the Δ points that are on ΓH lines parallel to $\mathbf{q}$ but is not present for the other Δ points on ΓH lines normal to $\mathbf{q}$. However, unless the splitting of the Δ_5 band due to the magnetic ordering is quite large one can expect that in practice magnetic breakdown will restore the original connectivity.

Although in principle a considerable number of degeneracies in the band structure of non-magnetic Cr can be expected to be lifted as a result of the reduction in the symmetry due to magnetic ordering, nevertheless in practice the magnetostrictive distortions are so small that the changes in the potential $V(\mathbf{r})$ can be expected to be small. The changes in the energies $E(\mathbf{k})$ as a result of changes in $V(\mathbf{r})$ induced by the magnetostrictive distortion of the crystal can therefore be expected to be quite small. However, for a few wave vectors $\mathbf{k}$ in the Brillouin zone that have special relationships to the wave vector $\mathbf{q}$ of the spin-density wave the effect of the presence of the spin-density wave becomes quite important. The presence of the spin-density wave means that in the crystal there is now a second periodic potential, due to the spin-density wave, superposed on the original periodic potential of the crystal lattice ; these two periodic potentials are generally incommensurate. It is probably easier to visualize what happens by using a one-dimensional lattice ; the electronic

Fig. 32

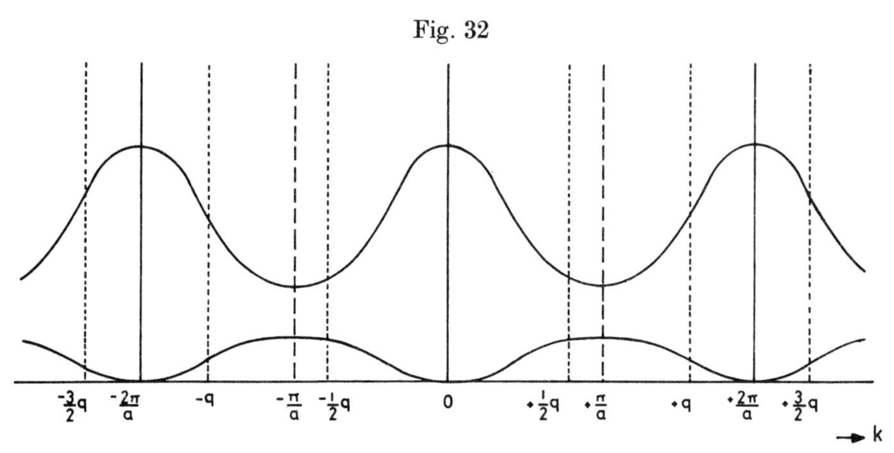

The band structure of a hypothetical one-dimensional metal with magnetic ordering vector $\mathbf{q}$.

energy bands of a hypothetical one-dimensional crystal are shown in fig. 32 in the repeated zone scheme. An electron is free to 'jump' by adding integer multiples of the reciprocal lattice vector $\mathbf{G}$ to its wave vector since

$$E(\mathbf{k}) = E(\mathbf{k} + m\mathbf{G}) \quad . \quad . \quad . \quad . \quad . \quad (7.4.1)$$

for all $\mathbf{k}$, where m is an integer. The reciprocal lattice vector associated with the potential due to the spin-density wave is $\mathbf{q}$ so that a second set of repeated zones with width $|\mathbf{q}|$ appear, see fig. 32. The potential due to the spin-density wave is not strong enough to impose on all the wave vectors $\mathbf{k}$ its own periodicity condition

$$E(\mathbf{k}) = E(\mathbf{k} + n\mathbf{q}), \quad . \quad . \quad . \quad . \quad . \quad . \quad (7.4.2)$$

where n is an integer. However, for those particular wave vectors for which this condition is accidentally satisfied by the original energy bands the interaction of the spin-density wave with the translational motion of the electron is sufficiently strong to cause the electron to jump by some integer multiple of $\mathbf{q}$ since no appreciable energy change is involved (Lomer 1964 b). A particularly simple case is of course when $\mathbf{k} = -\frac{1}{2}\mathbf{q}$ when the condition $E(\mathbf{k}) = E(\mathbf{k} + \mathbf{q})$ will be satisfied because $E(\mathbf{k}) = E(-\mathbf{k})$. If one evaluates the perturbation in the energy bands due to the spin-density wave then gaps in the energy bands will occur at all the positions $\mathbf{k} = \pm\frac{1}{2}\mathbf{G}$; $\pm\frac{1}{2}\mathbf{q}$; $\pm\mathbf{G}$; $\pm\mathbf{q}$; $\pm\frac{1}{2}(\mathbf{G}+\mathbf{q})$; $\pm\frac{1}{2}(\mathbf{G}-\mathbf{q})$ or in general at $\mathbf{k} = \frac{1}{2}(m\mathbf{G} + n\mathbf{q})$ in the repeated zone scheme. Values of these gaps were calculated for small values of m and n by Falicov and Zuckermann (1967). When all the energy bands are 'folded back' into the original Brillouin zone there is then a large number of energy gaps introduced throughout this zone ; one can only hope that for large values of m and n the gaps are sufficiently small as to be insignificant, as was indeed indicated by the calculations of Falicov and Zuckermann (1967). If one is concerned with de Haas–van Alphen type experiments the gaps produced by the perturbation in the potential caused by the spin-density wave are liable to be small enough for magnetic breakdown to occur across them. The correct pictorial description of the electron orbits can therefore be obtained by considering the Fermi surface cross section of the non-magnetic metal and allowing electrons to 'jump' by wave vectors $\mathbf{q}$ as well as by the usual $\mathbf{G}$. Thus for the Fermi surface cross section shown in fig. 33 the presence of $\mathbf{q}$ effectively changes the connectivity to that shown in fig. 34. In addition to the original orbit of area A shown in fig. 33 there is then the possibility of an extended orbit along $P_1Q_1P_2Q_2P_3Q_3$ . . . and of a very small closed orbit just involving $A_1Q_1B_1S_1$. An orbit of exactly twice the area A can be followed around the path $A_2Q_2P_3Q_3P_4B_4S_4R_4S_3A_3Q_3B_3S_3R_3S_2$. Many other complicated orbits are obviously possible. It is therefore not surprising that a large number of harmonics (up to the sixth harmonic) as well as various combination frequencies were observed in the work of Wallace and Bohm (1968) on magnetoacoustic quantum oscillations in magnetically ordered Cr. This extended construction was also used by

Fig. 33

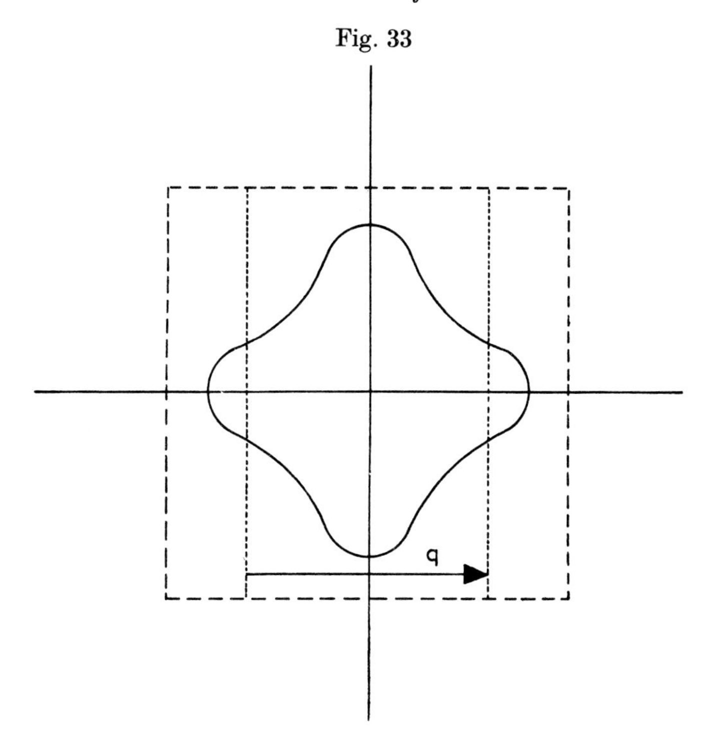

Fig. 34

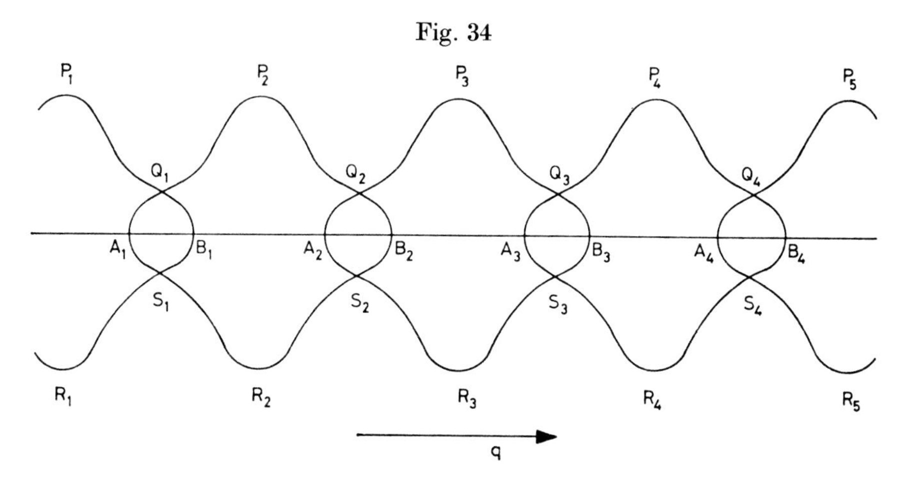

Graebner and Marcus (1968) in the interpretation of their de Haas–van Alphen data for antiferromagnetic Cr. This extended construction is also consistent with the observation of open orbits parallel to **q**, but in no other direction, in the high-field magnetoresistance measurements of Arko *et al.* (1966, 1968) on antiferromagnetic Cr. de Haas–van Alphen type oscillations (or quantum oscillations) in the magnetoresistance of anti-ferromagnetic Cr were investigated by Arko *et al.* (1969) and there were found to be considerable differences from the results of the ordinary de Haas–van Alphen data ; these differences remain largely unexplained although there are several possible causes.

In addition to the various pieces of work already mentioned there are several others relevant to the electronic structure of Cr. These include studies of the relative stabilities of various kinds of spin-density waves (Shibatani *et al.* 1969) and measurements of quantum oscillations in the ultrasonic attenuation (Wallace *et al.* 1965), the temperature dependence of the ultrasonic attenuation through the Néel point (O'Brien and Franklin 1966), the thermoelectric power (Vedernikov 1969), the total Fermi surface area using the anomalous skin effect (Fawcett and Griffiths 1962), the dependence of the magnetic susceptibility on field and temperature (Kostina *et al.* 1963), positron annihilation (Stewart 1957), optical measurements (Hughes and Lawson 1967, Graves and Lenham 1968 b, Lapeyre and Kress 1968, Eastman 1969 a, b), x-ray measurements (Borisov and Nemoshkalenko 1961, Ershov and Brytov 1967, Krivitskiĭ and Nikiforov 1967), the magnetoresistance (Borovik and Volotskaya 1959, Arajs and Dunmyre 1965 a†), the Hall effect (Amitin and Kovalevskaya 1968†), the thermal conductivity (Harper *et al.* 1957), and the electrical conductivity (Harper *et al.* 1957, Sambongi and Mitsui 1968). The value of γ, the temperature coefficient of the electronic specific heat, of 1·26 mJ mole^{-1} deg^{-2} calculated by Loucks is to be compared with experimental values ranging from 1·51 to 1·59 mJ mole^{-1} deg^{-2} (see references given by Loucks (1965 a) ; also Estermann *et al.* (1952) and Friedberg *et al.* (1952)). Kohn anomalies in Cr were detected by Costello and Weymouth (1969) in x-ray scattering measurements on both paramagnetic and antiferromagnetic Cr.

§ 8. d-BLOCK : GROUP VIIA : MANGANESE, TECHNETIUM AND RHENIUM

Mn	Manganese	2.8.(8, 5) 2
Tc	Technetium	2.8.18.(8, 6) 1
Re	Rhenium	2.8.18.32.(8, 5) 2

8.1. *Structures*

Mn exists in a variety of forms most of which have quite complicated structures. α-Mn, which is stable at temperatures up to 742°C, has a complicated cubic structure which is described by the space group I$\bar{4}$3m($T_d{}^3$) and has 58 atoms in the conventional unit cell or 29 atoms in the fundamental unit cell. For α-Mn the lattice constant $a = 8{\cdot}90$ Å (Smithells 1967). At low temperatures α-Mn is antiferromagnetic with a Néel temperature of 100°K (Shull and Wilkinson 1953, Kasper and

† See note added in proof.

Roberts 1956, Arrott and Coles 1961). The other forms of Mn, β-Mn, γ-Mn and δ-Mn, are stable in various temperature ranges between 742°C and the melting point (1244°C), although one or more of these forms may be metastable below 742°C. We shall not discuss these high-temperature forms of Mn any further, except to note that Fletcher (1969) has calculated the band structure of the high-temperature f.c.c. form of γ-Mn which at low temperatures becomes tetragonal. The remaining two metals in this group both have the h.c.p. structure with lattice constants $a = 2{\cdot}74$ Å, $c = 4{\cdot}39$ Å for Tc and $a = 2{\cdot}76$ Å, $c = 4{\cdot}45$ Å for Re (Smithells 1967). Tc is, as is well known, the element of lowest atomic number for which there are no stable isotopes.

8.2. *Manganese*

Since the structure of Mn is so complicated there are obvious difficulties in the way of discussing the band structure and Fermi surface of this metal. The number of atoms per unit cell is an order of magnitude larger than we have encountered in any other metal so far ; even the fundamental unit cell for Ga, which was considered in § 5.3 of part I, only contains four atoms. The volume of the unit cell in Mn is 352·5 Å^3 whereas the volume of the fundamental unit cell of Cr, which is next to Mn in the periodic table and has the ordinary b.c.c. structure is only 12·1 Å^3. The volume of the Brillouin zone for Mn is therefore very small, namely only (12·1/352·5) or 3·4% of the volume of the Brillouin zone of Cr. Correspondingly the cross-sectional areas and linear dimensions are also only 10·5% and 32·5% of the corresponding cross-sectional areas and linear dimensions, respectively, of the Brillouin zone of Cr. This means that any piece of the Fermi surface of Mn, unless it is multiply-connected, must be very small and, therefore, may prove difficult to measure experimentally, particularly if one uses techniques such as the de Haas–van Alphen effect which involve cross-sectional areas rather than linear dimensions. From the point of view of band structure calculations for Mn if one assumes, as we have done in connection with the neighbouring metals V and Cr in the periodic table, that the 4s electrons and all the 3d electrons are in the conduction band then with 29 atoms in the fundamental unit cell, this means that there must be the equivalent of $29 \times \frac{7}{2} = 101\frac{1}{2}$ full bands. To make a realistic band structure calculation involving so many bands to obtain a meaningful Fermi surface is clearly a mammoth undertaking, while the existence of antiferromagnetic ordering below 100°K is a further complication. One might try to simplify the problem by ascribing some or all of the five 3d electrons in Mn to the ion cores and thereby reduce the number of conduction electrons to be considered ; however, experience with neighbouring metals in the periodic table suggests that this is not likely to be very realistic.

In view of the above difficulties that hamper both experimental and theoretical determinations of the Fermi surface of Mn it is not very surprising that little has been attempted. Indeed, in searching the literature the author has failed to find an account of any piece of work that is directly related to the determination of the Fermi surface of Mn, although measurements of some properties indirectly related to the electronic band structure do exist (White and Woods 1957 b, Gschneidner 1964, Vedernikov 1969).

8.3. *Technetium*

Since Tc has no stable isotopes it is not entirely surprising that its Fermi surface has not been studied experimentally. There is no obvious reason why band structure calculations for Tc should not be performed but, again, nothing appears to have been done so far in this connection. Since the structure of Tc is the same as that of Re, which we shall discuss in the next section, there is every reason to suppose that the Fermi surface of Tc is very similar to that of Re, which has been determined quite accurately, except that one would expect spin–orbit coupling effects in Tc to be less marked than in Re.

The Knight shift in Tc has been measured by Jones and Milford (1962) and an estimated value of γ of 4·06 mJ mole^{-1} deg^{-2} is given by Gschneidner (1964).

8.4. *Rhenium*

Re is the only one of the three metals in group VIIA for which any significant work related to its Fermi surface has been performed. Magnetoresistance measurements (Alekseevskiĭ *et al.* 1962, Alekseevskiĭ *et al.* 1963, Fawcett and Reed 1964 b) showed that open orbits are possible in Re parallel to the [0001] direction, so that the Fermi surface is multiply-connected in the z direction. The results also indicated that open orbits can occur in the plane normal to the [0001] direction ; however, it was thought possible that these open orbits only arose as a result of magnetic breakdown. Several sets of de Haas–van Alphen measurements have been made on Re (Thorsen and Berlincourt 1961, Joseph and Thorsen 1963 a, 1964 c). With the magnetic field parallel to [10$\bar{1}$0] Thorsen and Berlincourt (1961) observed four periods corresponding to extremal areas of cross section of 5·79, 1·52, 1·34 and 0·077 × 10^{15} cm^{-2} and considerable evidence of magnetic breakdown was obtained by Joseph and Thorsen (1963 a). The periods observed by Joseph and Thorsen (1964 c) were interpreted as arising primarily from two pieces of Fermi surface ; these consisted of a set of small ellipsoids and a set of dumb-bell like surfaces each of which was similar to two intersecting distorted spheres. The angular dependences of the observed periods required that both these pieces of Fermi surface are situated on lines parallel to $\langle 10\bar{1}0 \rangle$ directions. Certain of their observed periods were interpreted as breakdown periods between these two surfaces and, since two-fold degeneracies in an h.c.p. metal with strong spin–orbit coupling only survive along the line AL itself, it was concluded that these two pieces of Fermi surface are situated on the line AL. Magnetoacoustic geometric resonance measurements by Jones and Rayne

Fig. 35

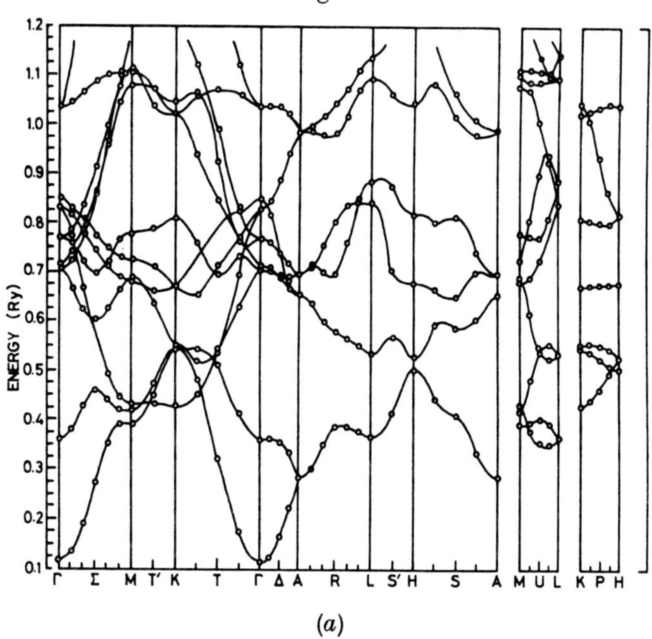

(*a*)

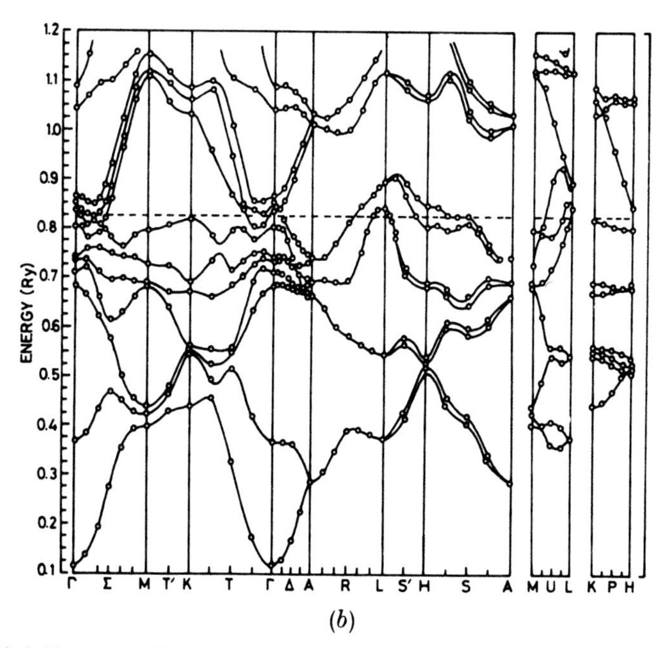

(*b*)

The relativistic energy bands for Re (*a*) neglecting spin–orbit coupling and (*b*) including spin–orbit coupling (Mattheiss 1966).

(1965 a, b) provided caliper dimensions that were in reasonable agreement with this model.

A relativistic A.P.W. band structure calculation for Re was performed by Mattheiss (1966), see fig. 35, and this produced a Fermi surface that consisted of five sheets. Closed regions occupied by holes were found in bands 5, 6 and 7 and a closed region occupied by electrons in band 9. In band 8 Mattheiss obtained a Fermi surface that enclosed a region occupied by electrons and corresponding roughly to a cylinder with its axis directed parallel to [0001]. Cross sections of these various pieces of Fermi surface are shown in fig. 36 (*a*); the sensitivity of the various features of this Fermi surface to small changes in the Fermi energy is indicated in fig. 36 (*b*) where a change of $+0{\cdot}005$ Ry has been made in the Fermi energy. The total number of electrons and holes should be equal, making Re a compensated metal and rough calculations of the volumes of the various pieces of Fermi surface showed that this condition was fairly well satisfied.

Fig. 36

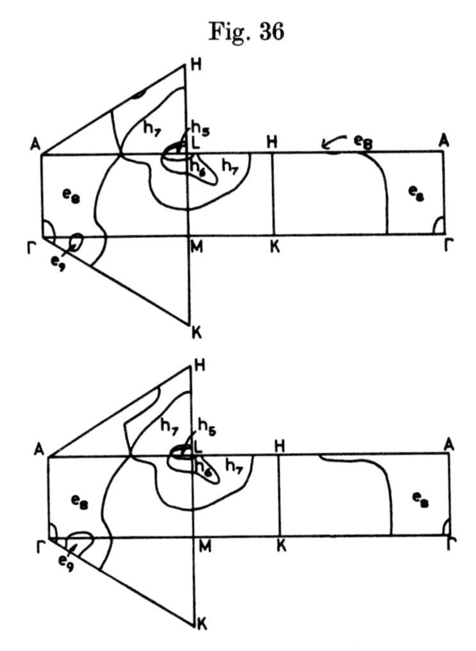

Cross sections of the Fermi surface of Re for two slightly different choices of E_F. e_n denotes a region of electrons in band n and h_n denotes a region of holes in band n (Mattheiss 1966).

The closed hole surfaces obtained by Mattheiss (1966) in bands 5 and 6 are illustrated in figs. 37 (*a*) and (*b*) and are quite similar to the two surfaces which were deduced from the de Haas–van Alphen data by Joseph and Thorsen and which are illustrated in fig. 37 (*c*). A detailed comparison between the calculated and measured extremal areas of cross section of these two surfaces is given in table II of Mattheiss (1966); the agreement

Fig. 37

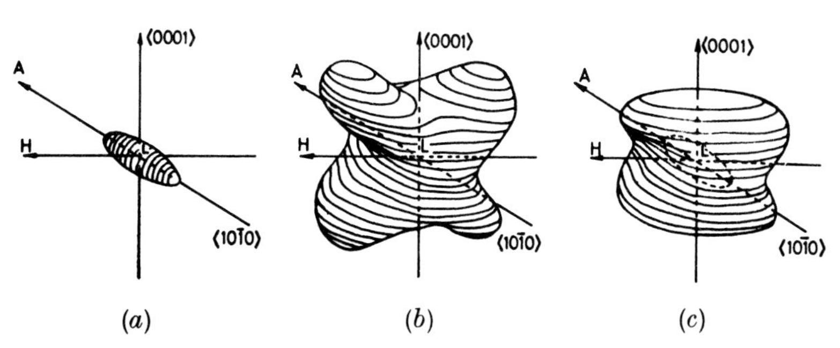

The Fermi surface of Re : (*a*) holes in band 5, (*b*) holes in band 6 and (*c*) the ellipsoid and dumb-bell model of Joseph and Thorsen (Mattheiss 1966).

between the calculated and measured dimensions was particularly good in the case of the ellipsoid in band 5. The larger pieces of Fermi surface are illustrated in fig. 38. The closed pockets of electrons in band 7 are located at L and the multiply-connected vertical cylinder of electrons passes through Γ and A ; these two regions touch along the line AL. A small roughly ellipsoidal region around Γ has been scooped out of the cylinder of electrons and is unoccupied. Spin–orbit coupling effects are rather important in Re, as they also are in W as we saw in § 7.3, and this was illustrated by Mattheiss (1966) by constructing a Fermi surface based on a non-relativistic A.P.W. calculation for Re and this was found to differ in several important details from the Fermi surface shown in fig. 36 (*a*).

In further de Haas–van Alphen measurements on Re Thorsen *et al.* (1966) observed a period that was readily assigned to a closed extended orbit in the plane ΓALM involving the band 7 hole surface at L and the

Fig. 38

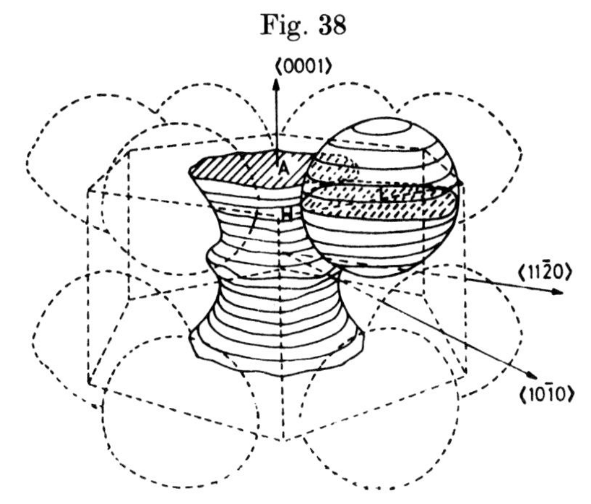

The Fermi surface of Re, holes in band 7 and electrons in band 8 (Mattheiss 1966).

band 8 electron surface, the theoretical and experimental areas being 2·80 Å^{-2} and 2·86 Å^{-2} respectively. Most of the remaining de Haas–van Alphen periods that have been observed can be interpreted in terms of the holes in band 7 although a few corresponding to the magnetic field close to [0001] appear to be associated with the region of electrons in band 8. According to the calculation by Mattheiss (1966) the Fermi surface in band 9 was thought to be toroidal, but there was some doubt as to whether this piece of Fermi surface was actually multiply-connected or not since it is very sensitive to small changes in the Fermi energy. Thorsen *et al.* (1966) observed two de Haas–van Alphen periods which they assigned to this piece of Fermi surface in band 9, but they were unable to detect several other extremal orbits that were predicted by Mattheiss (1966) on the Fermi surface in band 9. Thorsen *et al.* (1966) suggested some slight modifications to the shape of this torus to explain their results ; they proposed that the torus has the approximate shape of six spherical balls connected together to form a ring and that these balls are deformed so as to be concave towards the inside of the ring. In addition to the open orbits in the [0001] direction which we have already mentioned in connection with the early magnetoresistance work on Re, further open orbits were observed in the magnetoresistance work of Reed *et al.* (1965). These included a set of open orbits, resulting from magnetic breakdown, when the magnetic field is in the $(10\bar{1}0)$ plane, but not in the $[1\bar{2}10]$ direction. These orbits were explained by Mattheiss as arising from magnetic breakdown between the band 7 holes at L and the cylinder of electrons in band 8. For the special case when the magnetic field is parallel to $[1\bar{2}10]$ these orbits close to give the extended orbits (of area 2·86 Å^{-2}) which were observed in the de Haas–van Alphen results of Thorsen *et al.* (1966) which we have already mentioned. Two-dimensional aperiodic open orbits were also observed centred on the $\langle 11\bar{2}0\rangle$ axes ; on the Fermi surface calculated by Mattheiss (1966) these arise as combinations of the other two kinds of open orbits which both involve the cylindrical region of electrons in band 8. Reed *et al.* (1965) had suggested that their magnetoresistance results indicated the presence of two separate sheets of Fermi surface that were multiply-connected parallel to [0001], however, only one such surface appears in the results of the calculations of Mattheiss (1966).

A further magnetoacoustic study of the Fermi surface of Re using both geometric resonances and the quantum oscillations of the attenuation has been performed by Testardi and Soden (1967). The results were able to provide considerable further experimental details of the caliper dimensions of the hole region at L in band 7 and the electron cylinders in band 8, but were not able to provide much extra information, beyond that already available, relating to the smaller pieces of Fermi surface. No conclusive evidence related to the topology of the region of electrons in band 9, if it exists at all, was obtained from this work. Detailed experimental caliper dimensions for a number of principal cross sections of the Fermi surfaces in bands 5, 6, 7 and 8 are given in figs. 5–9 of Testardi and Soden (1967).

A number of caliper dimensions were assigned to the small ellipsoidal pocket of holes at Γ within the electron cylinder in band 8.

The experimental value of γ, the temperature coefficient of the electronic specific heat, of 2·47 mJ mole^{-1} deg^{-2} for Re obtained by Morin and Maita (1963) was compatible with the band structure value of γ calculated by Mattheiss (1966) and a thermal effective mass of 1·46 due to electron–phonon interactions (McMillan 1968). A slightly lower mean experimental value of γ of $(2{\cdot}40 \pm 0{\cdot}14)$ mJ mole^{-1} deg^{-2} is quoted by Gschneidner (1964). The electrical and thermal resistivities of Re at low temperatures have been measured by Schriempf (1967). The temperature dependence of the magnetic susceptibility of Re has been measured by Kojime *et al.* (1961), Volkov *et al.* (1968), who also measured the temperature dependence of the Hall effect in Re, Wunsch *et al.* (1968), and Volkenshteĭn *et al.* (1969). Measurements of the thermoelectric power of Re are described by Vedernikov (1969) and some optical measurements have been made by Juenker *et al.* (1968).

Therefore, in conclusion, we can see that the general features of the Fermi surface of Re calculated by Mattheiss (1966) have been verified experimentally, except that there is still some doubt about the exact shape of the region of electrons in band 9.

§ 9. d-block : Group VIII : the Iron Group Metals

Fe Iron 2.8.(8, 6) 2	Co Cobalt 2.8.(8, 7) 2	Ni Nickel 2.8.(8, 8) 2
Ru Ruthenium 2.8.18.(8, 7) 1	Rh Rhodium 2.8.18.(8, 8) 1	Pd Palladium 2.8.18.(8, 10)
Os Osmium 2.8.18.32.(8, 6) 2	Ir Iridium 2.8.18.32.(8, 9)	Pt Platinum 2.8.18.32.(8, 9) 1

There are various possible ways to divide these elements into smaller sets and both straightforward horizontal and vertical division have been used quite commonly. A particularly attractive scheme is to consider together the three elements in the first row, which are all ferromagnetic, and then to divide the remaining six elements vertically into three sets of two metals each (Sidgwick 1950) ; we shall follow this scheme.

9.1. *Structures*

Fe has the b.c.c. structure (α-Fe ; $a = 2.86$ Å) below 900°C and above 1400°C, while the f.c.c. form (γ-Fe ; $a = 3{\cdot}56$ Å) is stable between 900°C and 1400°C. The remaining elements in this group either have the h.c.p. structure, namely Co ($a = 2{\cdot}5059$ Å, $c = 4{\cdot}0659$ Å), Ru ($a = 2{\cdot}70$ Å, $c = 4{\cdot}27$ Å) and Os ($a = 2{\cdot}73$ Å, $c = 4{\cdot}31$ Å), or the f.c.c. structure, namely Ni ($a = 3{\cdot}52$ Å), Rh ($a = 3{\cdot}80$ Å), Pd ($a = 3{\cdot}88$ Å), Ir ($a = 3{\cdot}83$ Å) and Pt ($a = 3{\cdot}92$ Å). The h.c.p. structure of Co (α-Co) is only stable below about 390°C and above this temperature this metal exhibits the f.c.c. structure (β-Co ; $a = 3{\cdot}54$ Å). Some of the other elements also exhibit different structures under unusual conditions.

9.2. *Iron, Cobalt and Nickel*

The electronic properties of Fe, Co and Ni have been studied very extensively over the years. Fe was one of the metals for which band structure calculations were performed in the very early days (Greene and Manning 1943, Manning 1943) and a large number of calculations have been performed on these three metals since then. Of course, these metals are of particular interest because they exhibit ferromagnetism. However, many of the early band structure calculations for these metals completely ignored the fact of the ferromagnetic ordering ; that is, the Hamiltonian, $\mathscr{H}$, used in calculating the band structure contained no terms attributable to the large internal magnetic field $\mathbf{B}_{\text{int}}$ which must exist inside each domain of a sample of the metal below its Curie temperature. It is then assumed that the only effect of the internal field $\mathbf{B}_{\text{int}}$ is to cause a uniform separation of $g\beta|\mathbf{B}_{\text{int}}|$ to appear between the spin-up and spin-down bands at each wave vector $\mathbf{k}$. This may give quite a good first approximation to the true band structure of one of these metals. However, even if the band structure has been calculated sufficiently accurately it is also necessary to know the value of the energy difference, $g\beta|\mathbf{B}_{\text{int}}|$, between the spin-up and spin-down electrons with the same $\mathbf{k}$ and in the same band before one can determine the Fermi energy and hence the shape of the Fermi surface (Slater 1936). Because $g\beta|\mathbf{B}_{\text{int}}|$ is quite large the topological features of the Fermi surface for spin-up electrons can be expected to be completely different from those of the Fermi surface for spin-down electrons. Although the actual magnitudes of any changes in the band structure due to the inclusion of magnetic terms in the Hamiltonian, $\mathscr{H}$, may be quite small, nevertheless the inclusion of these extra terms may cause quite drastic reductions in the essential degeneracies of a ferromagnetic metal, as we have already mentioned in § 1. These reductions in the degeneracies of the band structure may have important repercussions on the possible connectivities of the Fermi surface. On the experimental side, doubts used to be expressed as to whether the methods commonly used in measuring the shapes of Fermi surfaces could be used for ferromagnetic metals (Pippard 1960, Shoenberg 1960 b), however, with the magnetoresistance measurements of Fawcett and Reed (1962) on Ni it became apparent that direct information about the shape of the Fermi surface of a ferromagnetic metal could be determined in the usual way. Shortly after this Anderson and Gold (1963) unequivocably observed conventional de Haas–van Alphen oscillations in Fe although, as might be expected, the oscillations were found to be periodic in $|\mathbf{B}|^{-1} = |\mathbf{H} + 4\pi\mathbf{M}_s|^{-1}$ rather than in just $|\mathbf{H}|^{-1}$, where $\mathbf{M}_s$ is the saturation magnetization (see also Kittel (1963)). Since then a number of other direct measurements related to the Fermi surfaces of Fe, Co and Ni have been made using the standard methods. There are substantial difficulties in the way of preparing suitable single-domain single-crystal specimens of Fe and Co because of the phase transitions which occur between ordinary temperatures and the melting points and the specimens are commonly in the form of whiskers (see, for example, Gold (1964)). It is easier to prepare pure single-crystal specimens of Ni than it is for Fe and Co because of the absence of inconvenient phase transitions in Ni ; consequently more experimental work has been done in connection with the Fermi surface of Ni than has been done for Fe and Co. A review of experimental work related to the determination of the Fermi surfaces of Fe and Ni has been given by Gold (1968), with particular reference to galvanomagnetic and de Haas–van Alphen measurements.

Several measurements of the x-ray and neutron scattering factors for Fe have been made (Weiss and De Marco 1958, Nathans *et al.* 1959, Batterman *et al.* 1961, Borisov and Nemoshkalenko 1961, Sirota and Olekhnovich 1961, Shull 1963) to study the spatial distribution of the charge density and spin density of the conduction electrons ; this gives some indication of the numbers of s-like and d-like electrons that are present. Initially attempts were made to determine the relative numbers of s-like and d-like electrons in Fe ; Weiss and De Marco (1958) deduced that the number of d electrons in the conduction band of Fe was very much smaller (only $2{\cdot}3 \pm 0{\cdot}3$) than the corresponding number in Cu ($9{\cdot}8 \pm 0{\cdot}3$) (see also Weiss and Freeman (1959)). However, Batterman *et al.* (1961) re-measured the x-ray scattering factors for Fe and Cu and deduced that there is in fact no such large discrepancy between the numbers of d electrons in Fe and Cu. Of course, the use of atomic wave functions in calculating the scattering factors is unrealistic because, as is well known, no conduction electron energy band in a metal is ever pure s-like or pure d-like but always consists of some admixture of these and other atomic states (Stern 1961, Wakoh and Yamashita 1966, De Cicco and Kitz 1967). The differences between the spin-up and spin-down band structures and Fermi surfaces in a ferromagnet can be investigated by positron annihilation experiments using polarized beams of positrons (Hanna and Preston 1958, Berko and Zuckerman 1964, Mijnarends and Hambro 1964). Walmsley (1962) suggested another method for investigating the polarization of the electrons at the Fermi surface in a ferromagnetic metal. If there is an appreciable net polarization there should be a change in the Fermi level with applied magnetic field ; this would appear as a change in the contact potential and could be measured relative to a non-magnetic metal. Walmsley (1962) made this measurement on a capacitor, with one plate of Fe and the other of Cu, in a magnetic field and concluded that the magnetic moments of the electrons at the Fermi surface in Fe are completely aligned in the direction of the magnetization. Belson (1966) repeated and extended this experiment and concluded, within the accuracy of the experiment, that no such large shift of the Fermi level occurs and that the ratio of the numbers of spin-up and spin-down electrons at the Fermi surface is less than 1·18 for Fe.

Band structure calculations on Fe have been performed by a number of workers (Greene and Manning 1943, Manning 1943, Callaway 1955, 1959, 1960, Suffczyński 1957, Stern 1959, Wood 1960, 1962, Gersdorf 1962, Abate and Asdente 1965, Wakoh and Yamashita 1966, Asdente and Delitala 1967, Kobayasi and Matumoto 1967, Cornwell *et al.* 1968, Hubbard and Dalton 1968, Shirokovskiĭ and Kulakova 1968, Hum and Wong 1969) using a variety of methods. An interpolation scheme based on the method of Slater and Koster (1954) and using the results of the band structure calculation of Wood (1960) has been used to calculate a density

Fig. 39

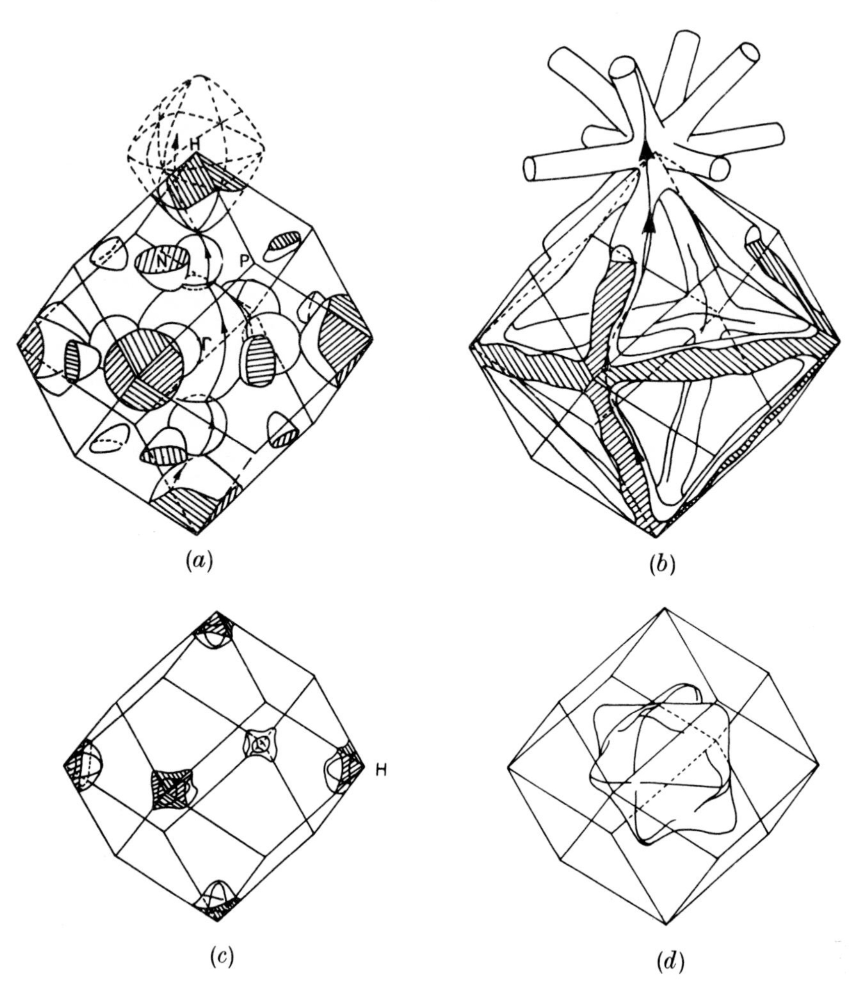

The Fermi surface of Fe calculated by Wakoh and Yamashita (1966), (*a*) spin-up (electrons), (*b*) and (*c*) spin-down (holes), and (*d*) spin-down (electrons).

of states curve for Fe (Wohlfarth and Cornwell 1961, Cornwell and Wohlfarth 1962). The A.P.W. band structure calculation by Wood (1962) for Fe is probably more famous for the use that was made of it by Lomer (1962, 1964 a) in connection with the group VIa metals, Cr, Mo and W, using the rigid band model, rather than for any light that it sheds on the nature of the Fermi surface of Fe itself, see § 7.2. In most of the band structure calculations just mentioned no attempt was made to deduce the shape of the Fermi surface. Asdente and Delitala (1967) made some predictions of the shape of the Fermi surface of Fe in both the paramagnetic and ferromagnetic phases. However, these calculations were based on the tight-binding method assuming complete separation into s-bands and d-bands and neglecting any hybridization either between the s and d-bands or with other states ; this hybridization is actually sufficiently important to make very large changes to their predicted Fermi surfaces. In the Green's function calculation by Wakoh and Yamashita (1966) proper account was taken of the different crystal potentials $V(\mathbf{r})$ that apply to spin-up and spin-down electrons ; separate band structures and densities of states were then calculated for the spin-up and spin-down electrons. As a result of their calculations Wakoh and Yamashita (1966) obtained 2·9 electrons per atom in the spin-up bands and 5·1 electrons per atom in the spin-down bands and the corresponding calculated Fermi surface is shown in fig. 39. The Fermi surface for the spin-up bands shown in fig. 39 (*a*) is very similar to the Fermi surfaces of the group VIa metals ; this is not entirely surprising because in the group VIa metals there are three electrons per atom in the spin-up or spin-down bands and this is very close to the number of electrons per atom (2·9) in the spin-up bands of Fe obtained by Wakoh and Yamashita. The Fermi surface shown in fig. 39 (*a*) for the spin-up electrons in Fe exhibits all the general features of the Fermi surface of (non-magnetic) Cr that we described in § 7.2, namely the electron 'jack' at Γ, the electron lenses along ΓH, the octahedral region of holes around H and the smaller regions of holes around N. The Fermi surface for the spin-down bands shown in figs. 39 (*b*), (*c*) and (*d*) is quite different from this and consists of two pockets of holes at H (fig. 39 (*c*)), a multiply-connected tubular region of holes (fig. 39 (*b*)), and an isolated region of electrons at Γ (fig. 39 (*d*)). The Fermi surface shown in fig. 39 is capable of supporting open orbits in certain directions and these have in fact been detected experimentally in magnetoresistance measurements (Lüthi 1960, Reed and Fawcett 1964 a, c, Bazan 1965, Isin and Coleman 1965, 1966, Coleman and Isin 1966, Vautier and Colombani 1966). Difficulties in the way of experimental measurements on the Fermi surface of Fe include not only complications caused by the presence of domains but also difficulties in obtaining high-purity single-crystal specimens of Fe. Reed and Fawcett observed open orbits along ⟨110⟩ and ⟨001⟩ axes although the absence of complete saturation of the magnetoresistance, when the magnetic field was at any one of the minima, suggested that either the number of open orbits is very small or that they result from magnetic breakdown. The Fermi surface calculated by Wakoh and Yamashita (1966) explains the existence of these open orbits by ascribing those in $\langle 001 \rangle$ directions to the spin-up Fermi surface and those in $\langle 110 \rangle$ directions to the multiply-connected spin-down Fermi surface ; examples of these open orbits are illustrated, respectively, in figs. 39 (*a*) and 39 (*b*). Because of the experimental difficulties associated with the preparation of good samples of pure single-crystal Fe only relatively few de Haas–van Alphen measurements exist for Fe. Anderson and Gold (1963) observed de Haas–van Alphen periods corresponding to areas of cross section of $4{\cdot}41 \times 10^{-2}$ Å^{-2} for $\mathbf{B}\|[100]$ and $3{\cdot}52 \times 10^{-2}$ Å^{-2} for $\mathbf{B}\|[110]$. Assuming that these periods arise from the lens along ΓH of the spin-up Fermi surface the corresponding calculated areas of cross section are $2{\cdot}1 \times 10^{-2}$ Å^{-2} and $1{\cdot}8 \times 10^{-2}$ Å^{-2} respectively (the complete set of calculated areas of cross section is given in table V of Wakoh and Yamashita (1966)). Although these two theoretical areas are only about half the experimental values, the directional property that the value for $\mathbf{B}\|[100]$ is larger than that for $\mathbf{B}\|[110]$ is common to both the experimental and theoretical results. Only quite a small change (about 0·002 Ry) in the calculated Fermi energy would be necessary to improve the agreement between theory and experiment. Wakoh and Yamashita (1966) also made calculations of the x-ray and neutron scattering factors based on their band structure calculations rather than on atomic wave functions.

There are several other miscellaneous pieces of work related to the electronic band structure of Fe. Calculated values of γ, the temperature coefficient of the electronic specific heat, include 7·5 mJ mole^{-1} deg^{-2} (Callaway 1955) and 4·2 mJ mole^{-1} deg^{-2} (Wakoh and Yamashita 1966) ; experimental values include :

$(4{\cdot}98 \pm 0{\cdot}06)$ mJ mole^{-1} deg^{-2} Gschneidner (1964)
$(4{\cdot}741 \pm 0{\cdot}014)$ mJ mole^{-1} deg^{-2} Dixon *et al.* (1965).

The experimental determinations are impeded by the presence of the spin-wave contribution to the specific heat which follows a $T^{3/2}$ law ; this spin-wave contribution was clearly exhibited and separated from the γT contribution by Dixon *et al.* (1965). The existence of the spin waves in ferromagnetic Fe can be verified directly by inelastic neutron scattering (Schweiss *et al.* 1967). From their x-ray emission measurements on Fe Tomboulian and Bedo (1961) deduced a density of states curve which did not exhibit the double peak that is familiar in the results of many of the band structure calculations on Fe (see also Nikiforov (1961), Nikiforov and Blokhin (1964)), but the double peak was present in the photoemission density of states (Spicer 1966) and in the x-ray measurements of the density of states (Nemnonov *et al.* 1966). Peaks were obtained at 0·35, 2·4 and 5·5 ev below the Fermi level from the photoemission measurements of Blodgett and Spicer (1967). Other work which has been done on Fe includes studies of the thermal and electrical conductivities (Kemp *et al.* 1956, Colquitt and Goodings 1964, Chari 1967), the thermoelectric power

(Schröder and Giannuzzi 1969, Vedernikov 1969), positron annihilation (Dekhtyar *et al.* 1966), the temperature dependence of the electrical resistance at low temperatures (Semenenko *et al.* 1962), and further galvanomagnetic measurements (Christopher *et al.* 1967, Dheer 1967). O.P.W. calculations of the effect of a pressure of 10^8 atmospheres (10^5 kbar) on the band structure and density of states of Fe were performed by Henry (1962) ; the relevance to geophysics should be apparent. The results for the density of states suggested a reduction of magnetization with increasing pressure, which has been observed experimentally.

Relatively little work appears to have been performed in connection with the determination of the band structure and Fermi surface of Co. With an atomic structure of . . . $3d^74s^2$ and with two atoms per unit cell in the h.c.p. form which is stable at ordinary temperatures, there will be the equivalent of 18 bands to be considered and for a band structure calculation this represents no mean undertaking ; however, an A.P.W. calculation has been performed by Hodges and Ehrenreich (1968) for both the f.c.c. and h.c.p. phases of (paramagnetic) Co†. On the experimental side there is the difficulty mentioned before of preparing pure single-domain single-crystal specimens of Co because of the phase transition at about 390°C. No measurements of properties that give direct information about the shape of the Fermi surface of Co appear to have been made so far. It was deduced by Weiss and De Marco (1958) from measurements of the X-ray scattering factor that there are $(8{\cdot}4 \pm 0{\cdot}03)$ d electrons per atom in metallic Co. The density of states, $n(E)$, in Co has been investigated by optical methods (Yu and Spicer 1966, 1968 a, Yu *et al.* 1968) and X-ray methods (Borisov and Nemoshkalenko 1961, Tomboulian and Bedo 1961) and a theoretical density of states curve was produced by Wong *et al.* (1969) (see also Wong (1970)). Measurements of the electronic specific heat of Co have yielded γ values which include :

$4{\cdot}74$ mJ mole^{-1} deg^{-2}	Heer and Erickson (1957, 1958)
$(5{\cdot}6 \pm 0{\cdot}3)$ mJ mole^{-1} deg^{-2}	Arp *et al.* (1959)
$(4{\cdot}44 \pm 0{\cdot}05)$ mJ mole^{-1} deg^{-2}	Walling and Bunn (1959)
$(4{\cdot}72 \pm 0{\cdot}09)$ mJ mole^{-1} deg^{-2}	Cheng *et al.* (1960)
$(4{\cdot}38 \pm 0{\cdot}01)$ mJ mole^{-1} deg^{-2}	Dixon *et al.* (1965).

Although Dixon *et al.* (1965) successfully isolated a spin-wave $T^{3/2}$ contribution to the specific heat in both Fe and Ni this contribution was masked in Co by the nuclear term. However, the existence of spin waves in ferromagnetic Co has been verified directly by inelastic neutron scattering measurements (Pickart *et al.* 1967). The temperature dependence of the electrical resistance of Co at very low temperatures was measured by

† The results of a new band structure calculation for Co have recently been published, see note added in proof.

Semenenko *et al.* (1963). Measurements of the galvanomagnetic properties of Co are described by Masumoto *et al.* (1966) and Cheremushkina and Vasil'eva (1966) and of the thermoelectric power by Schröder and Gianuzzi (1969) and Vedernikov (1969) and on positron annihilation by Čížek and Adam (1969). Belson (1966) measured the change in the contact potential of Co in a magnetic field and concluded that the ratio of spin-up to spin-down electrons at the Fermi surface in Co is less than 1·18. The use of the x-ray and neutron scattering factors to determine the number of d-like electrons and their orientations has been applied to Ni; Weiss and De Marco (1958) deduced that Ni possesses ($9{\cdot}7 \pm 0{\cdot}3$) d electrons per atom, while Weiss and Freeman (1959) deduced that there were 5·0 3d electrons with spin up and 4·4 3d electrons with spin down. Hodges, Lang, Ehrenreich and Freeman (1966) used a pseudopotential to calculate the neutron scattering factor for ferromagnetic Ni and were able to reproduce the charge and spin distributions obtained from the experimental results of Mook and Shull (1966). There have been a number of other band structure calculations performed for Ni using a variety of methods but nearly all of them have been based on the assumption that the band structure of ferromagnetic Ni can be obtained by calculating the band structure for the paramagnetic phase and then simply moving the origin of the spin-up bands relative to the spin-down bands in order to obtain the band structure for the ferromagnetic state (Fletcher 1952, Gersdorf 1962, Phillips and Mattheiss 1963, Yamashita *et al.* 1963, Friedel *et al.* 1964, Hubbard 1964, Mattheiss 1964, Phillips 1964, Wakoh and Yamashita 1964, Hodges and Ehrenreich 1965, Wakoh 1965, Snow *et al.* 1966, Connolly 1967, Lenglart 1967, 1968, Allan *et al.* 1968, Zornberg 1968, Callaway and Zhang 1970). In view of the difficulties associated with the performance of band structure calculations that include magnetic contributions *ab initio* there have been a number of studies of interband transitions aimed at investigating the ordering of the levels at L (Ehrenreich *et al.* 1963, Krinchik and Gan'shina 1966, Hanus *et al.* 1967, Eastman and Krolikowski 1968, Krinchik *et al.* 1968). Magneto-optic measurements which are capable of distinguishing between optical transitions involving spin-up and spin-down electrons have also been used to investigate the ordering of the levels at L (Krinchik and Nuralieva 1959. Phillips and Mattheiss 1963, Cooper *et al.* 1964, Krinchik 1964 a, b, Phillips 1964, Cooper and Ehrenreich 1964, Martin *et al.* 1964, Cooper 1965, Ehrenreich 1965, Afanas'eva *et al.* 1966, Krinchik and Nurmukhamedov 1965, Martin *et al.* 1965, Hanus *et al.* 1967, 1968, Krinchik and Artem'ev 1967, 1968, Krinchik *et al.* 1968). Optical measurements on Ni have been made by several workers (Tomboulian and Bedo 1961, Blodgett and Spicer 1965, 1966, Spicer 1966, Graves and Lenham 1968 b, Stoll 1969). Investigations of the spin distributions in Ni using polarized positrons have been performed by Sedov (1965) and by Mihalisin and Parks (1966, 1967, 1969). The measurements of the contact potential in a magnetic field which we have already mentioned in connection with Fe and Co have also been performed for Ni by Belson (1966) with again the conclusion that the ratio of spin-up to spin-down electrons at the Fermi surface in Ni is less than 1·18.

Direct information concerning the shape of the Fermi surface in ferromagnetic Ni has come primarily from measurements of the magnetoresistance and of the de Haas–van Alphen effect. The existence of open orbits in Ni was demonstrated in the magnetoresistance work of Lüthi (1959), while in a series of magnetoresistance measurements on ferromagnetic Ni Fawcett and Reed observed open orbits in the ⟨100⟩, ⟨110⟩ and ⟨111⟩ directions (Fawcett and Reed 1962, 1963, Reed and Fawcett 1964 b). They showed that these results could be explained by assuming that one sheet of the Fermi surface of ferromagnetic Ni is similar to the Fermi surface of Cu and consists of a sphere pulled out to touch the 111 faces of the Brillouin zone, although in Ni the contact areas are considerably smaller than in Cu. This similarity with the Fermi surface of Cu suggests that the number of electrons per atom in the bands corresponding to one of the spin polarizations in Ni is close to the number of electrons with each polarization in Cu, that is 5·5. The value of the Hall constant obtained by Reed and Fawcett (1964 b) corresponds to about one electron per atom ; this suggests that Ni is uncompensated which, since there are eight electrons per atom, is a direct consequence of the magnetic ordering and the separation of the spin-up and spin-down band structures. Magnetoresistance and Hall-effect measurements on Ni have also been made by a number of workers (Goureaux *et al.* 1958, Colombani and Goureaux 1959, Goureaux and Colombani 1959, 1960, Marcus and Langenberg 1963, Marsocci 1965, Ehrlich *et al.* 1967, Ehrlich and Rivier 1968). The angular variation of the de Haas–van Alphen oscillations observed by Joseph and Thorsen (1963 b, 1964 d) and Gold (1964) in Ni were compatible with the suggestion of

Fig. 40

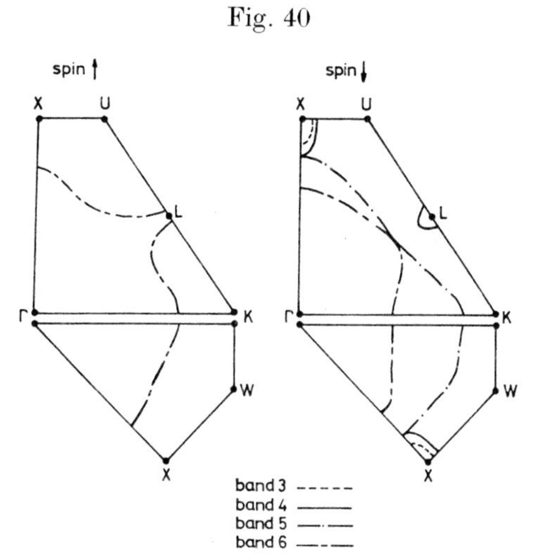

Cross sections of the Fermi surface of Ni (Tsui 1967).

Fig. 41

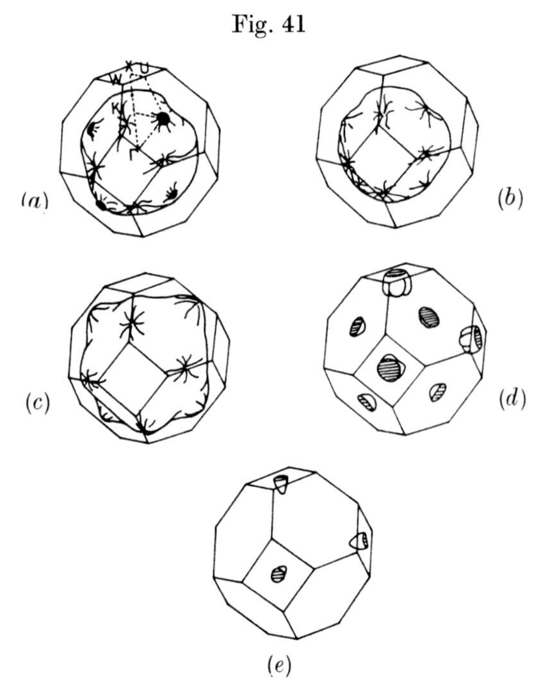

The Fermi surface of ferromagnetic Ni calculated by Wakoh and Yamashita (1964), (*a*) spin-up (electrons), (*b*) and (*c*) spin-down (electrons at Γ), (*d*) and (*e*) spin-down (holes at L and X).

Fawcett and Reed that one sheet of the Fermi surface of Ni is topologically similar to the Fermi surfaces of the noble metals, where the periods actually observed were assigned to the ⟨111⟩ necks. Tsui and Stark (1966 b) used the band structure of paramagnetic Ni calculated by Hanus (1962), together with the known value of the saturation magnetization and the fact that the Fermi surface of ferromagnetic Ni is not compensated but encloses an excess of approximately one electron per atom (Reed and Fawcett 1964 b), to construct a schematic band structure for ferromagnetic Ni, see fig. 2. In this schematic band structure the multiply-connected sheet of Fermi surface which we have already described is assumed to be for the spin-up electrons and is then the only sheet of Fermi surface for the spin-up bands. There are then several sheets of Fermi surface for the spin-down bands, namely two large closed electron pockets centred at Γ in bands 5 and 6, one small hole pocket at L in band 4 and two small hole pockets at X in bands 3 and 4 (see also Tsui (1967)). All these pieces of Fermi surface, except the small pocket of holes at L, were also obtained in the A.P.W. band-structure calculation of Connolly (see figs. 6 and 7 of Connolly (1967)). Cross sections of all these pieces of Fermi surface are shown in fig. 40 ; this is quite similar to the Fermi surfaces deduced earlier by Ehrenreich *et al.* (1963) based on the same A.P.W. results of Hanus (1962) by Phillips (1964) and by Wakoh and Yamashita

(1964). The Fermi surface calculated by Wakoh and Yamashita is illustrated in fig. 41. Further de Haas–van Alphen periods were also observed and were assigned to an ellipsoidal pocket of holes which was assumed to be one of the two pockets of holes at X (Tsui and Stark 1966 b, Stark and Tsui 1968). This surface was parametrized by the expression

$$k^2 = \frac{k_\phi^2 k_z^2}{k_z^2 + (k_\phi^2 - k_z^2)\cos^2\theta} + k_2 \cos^2\theta - k_4 \cos^4\theta \quad . \quad . \quad (9.2.1)$$

(Tsui 1967), where $k_\phi = k_0 + k_4 \cos 4\phi + k_8 \cos 8\phi$, k is the length of the wave vector from X to a point on the Fermi surface specified by the usual polar angles θ and ϕ referred to XW, [100], and XΓ, [001], as x and z axes ; values of the parameters are given in table III of Tsui (1967). These de Haas–van Alphen periods were also observed by Hodges *et al.* (1967). Previously an interpolation scheme, sometimes called the '*combined interpolation scheme*', had been developed by Hodges, Ehrenreich and Lang (1966) and applied to the determination of the band structure of Ni using the results of existing A.P.W. calculations as the basis of the interpolation scheme. This scheme is based on an initial separation between d electrons and s electrons ; the d electrons were then described by tight-binding wave functions while the s electrons were described by plane waves ; hybridization between d and s electrons, correlation between the electrons, spin–orbit coupling and magnetic effects were also included. When the resulting band structure was compared with the shape of the roughly ellipsoidal Fermi surface at the point X determined experimentally it was concluded by Hodges *et al.* (1967) that the importance of spin–orbit coupling was quite considerable ; this is a little surprising in a metal of such low atomic number as that of Ni. The angular dependence of the amplitude of the de Haas–van Alphen oscillations associated with this pocket of holes at X, as observed by Tsui (1967) was also rather strange, see fig. 42. However,

Fig. 42

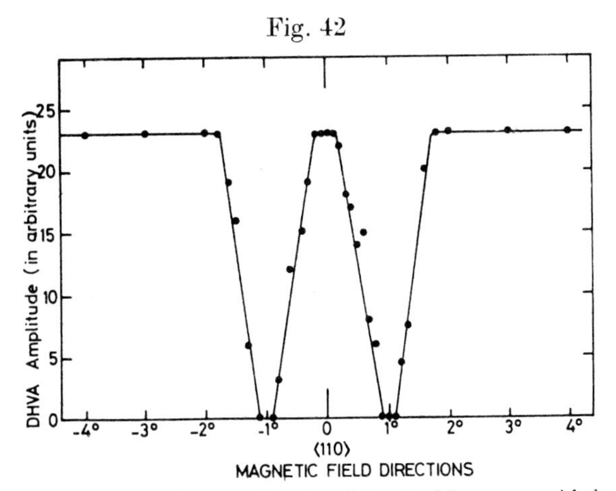

The angular dependence of the amplitude of the de Haas–van Alphen oscillations associated with the pocket of holes at X in Ni (Tsui 1967).

rather than assuming a large spin–orbit interaction in a metal of such low atomic number it is possible to show that both the shape of this pocket of holes at X as well as the curious angular dependence of the amplitude of the de Haas–van Alphen oscillations can be explained in terms of the lifting of many of the band structure degeneracies by the internal magnetic field (Falicov and Ruvalds 1968, Ruvalds and Falicov 1968, Zornberg 1968, Cracknell 1969 a, 1970) together with the existence of magnetic breakdown (for details of the explanation see Ruvalds and Falicov (1968)). No de Haas–van Alphen oscillations corresponding to a second pocket of holes at X or to a pocket of holes at L were observed experimentally (Tsui and Stark 1966 b, Hodges *et al.* 1967, Tsui 1967). A model Hamiltonian for ferromagnetic Ni was constructed by Zornberg (1968, 1970) and used as the basis of an interpolation scheme for Ni ; important dimensions of the pocket of holes at X calculated on this scheme are compared with the experimental values of the dimensions in table 1 of Zornberg (1968).

Pawel and Stansbury (1965) quote values of γ, the temperature coefficient of the electronic specific heat, for Ni ranging from 5·23 mJ mole^{-1} deg^{-2} (Busey and Giauque 1952) to 7·28 mJ mole^{-1} deg^{-2} (Keesom and Clark 1939, Rayne and Kemp 1956, Walling and Bunn 1959, Gupta *et al.* 1964). Dixon *et al.* (1965), who obtained some evidence of a $T^{3/2}$ term in the specific heat of Ni due to the spin waves, obtained a value of $(7{\cdot}028 \pm 0{\cdot}007)$ mJ mole^{-1} deg^{-2} for γ. The experimental values are to be compared with the band structure value of 3·76 mJ mole^{-1} deg^{-2} (Ehrenreich *et al.* 1963) which shows a substantial value for the enhancement of γ. Calculations of the electron–phonon interaction in Ni were made by Hasegawa *et al.* (1965) using their own calculated band structure and the known form of the phonon dispersion relations. In addition to the direct magnon contribution to the specific heat, of a term proportional to $T^{3/2}$, in a ferromagnetic metal we should also expect the interactions between the magnons and the translational motions of the electrons to lead to an enhancement of γ similar to, and in addition to, the usual enhancement of γ due to electron–phonon interactions (Bennemann 1967). The presence of spin waves in ferromagnetic Ni has, of course, been verified directly by inelastic neutron scattering measurements (Pickart *et al.* 1967). Other work indirectly related to the determination of the electronic band structure of Ni includes discussions of the thermal and electrical conductivities (Kemp *et al.* 1956, Colquitt and Goodings 1964, White and Tainsh 1967), the Knight shift (Dutta Roy and Subrahmanyam 1969), and the thermoelectric power (Schröder and Giannuzzi 1969, Vedernikov 1969), infra-red and ultra-violet measurements (Krinchik and Gorbacher 1961, Biondi and Guobadia 1968, Yarovaya and Timchenko 1968), ion-neutralization spectroscopy measurements of the density of states (Hagstrum and Becker 1967), further x-ray photo-emission measurements (Nikiforov and Blokhin 1964) and positron annihilation measurements (Čížek and Adam 1969). The effect of stress (up to 700 kbar) on the magnetoresistance of Ni was investigated experimentally by Bagchi and Cullity

(1967). Almost all our discussion of the electronic band structure of Ni has been in connection with the ferromagnetic phase because nearly all of the available experimental techniques require the use of temperatures very much lower than the Curie temperature. However, the density of states at the Fermi surface, $n(E_F)$, can be related to the diffusion constant, D, for protons in the solid ; Nikulin (1966 b) was able to show that $n(E_F)$ in the paramagnetic region is 7% greater than in the ferromagnetic region.

In connection with all three metals Fe, Co and Ni we have mentioned the existence of spin-wave excitations in the ferromagnetic phases in connection with their contribution to the specific heat of a ferromagnetic metal. In addition to this and to the direct observation of spin waves in inelastic neutron scattering measurements (for a collection of references see, for example, Shirane *et al.* (1968)) the spin waves can be expected to contribute to many of the other macroscopic properties of a magnetic metal such as the electrical resistivity, thermal conductivity, and various other transport properties of the metal.

9.3. *Ruthenium and Osmium*

Ru and Os have the same structure (h.c.p.) as Re and do not possess any spontaneous magnetization. Since no band structure calculations for Ru or Os appear to exist it is tempting to use the rigid-band model and the relativistic A.P.W. band structure calculated by Mattheiss (1966) for Re. This can be expected to give a good description of the Fermi surface of Os, which only differs from Re by one in its atomic number so that spin–orbit coupling effects will be quite similar in the two metals. This approach will be slightly less appropriate to Ru because of the difference between the importance of spin–orbit coupling in Ru and Re. From the band structure of Re shown in fig. 35 it can be seen that it is only at Γ, L and H that there are any bands just above the Fermi energy for Re. If one electron per atom is added to the band structure shown in fig. 35 we can expect the regions h_5, h_6, and h_7 in fig. 36 to contract and the regions e_8 and e_9 to expand. de Haas–van Alphen oscillations in Ru were observed by Coleridge (1966 a) and interpreted on the basis of the band structure shown in fig. 35 ; Coleridge ascribed one set of oscillations to a small pocket of holes at Γ and further oscillations to two fairly large closed regions of electrons around Γ, the smaller of which is probably ‘ waisted ’. This corresponds to raising the Fermi energy in fig. 35 by slightly less than 0·02 Ry. Such a change in E_F brings it very close to the next band at L so that whether the two pockets of holes h_5 and h_6 at L in fig. 36 vanish completely depends very critically on the exact form of the band structure around L and the value of the Fermi energy. However, there is a drawback to this interpretation because the $n(E)$ curve given in fig. 5 of Mattheiss (1966) suggests that the increase in the number of conduction electrons per atom from 7 to 8 would raise E_F by about 0·1 Ry rather than 0·02 Ry. This would lead to two large closed regions of electrons around

Γ in bands 9 and 10 and one large region of holes around M and K in band 8. It is clear that it would be desirable to have some more detailed experimental data for Ru† and Os.

Other existing experimental work related to the band structure and Fermi surface of Ru and Os includes measurements of γ, the temperature coefficient of the electronic specific heat, of 3·3 mJ mole^{-1} deg^{-2} for Ru (Gschneidner 1964) and 2·35 mJ mole^{-1} deg^{-2} for Os (Gschneidner 1964, Andres and Jensen 1968), the low temperature electrical and thermal resistivities of Ru and Os (White and Woods 1958), the thermoelectric power (Vedernikov 1969), and the temperature dependence of the paramagnetic susceptibility (Kojime *et al.* 1961 on Ru, Weiss and Kohlhaas 1967 on Ru and Os).

Fig. 43

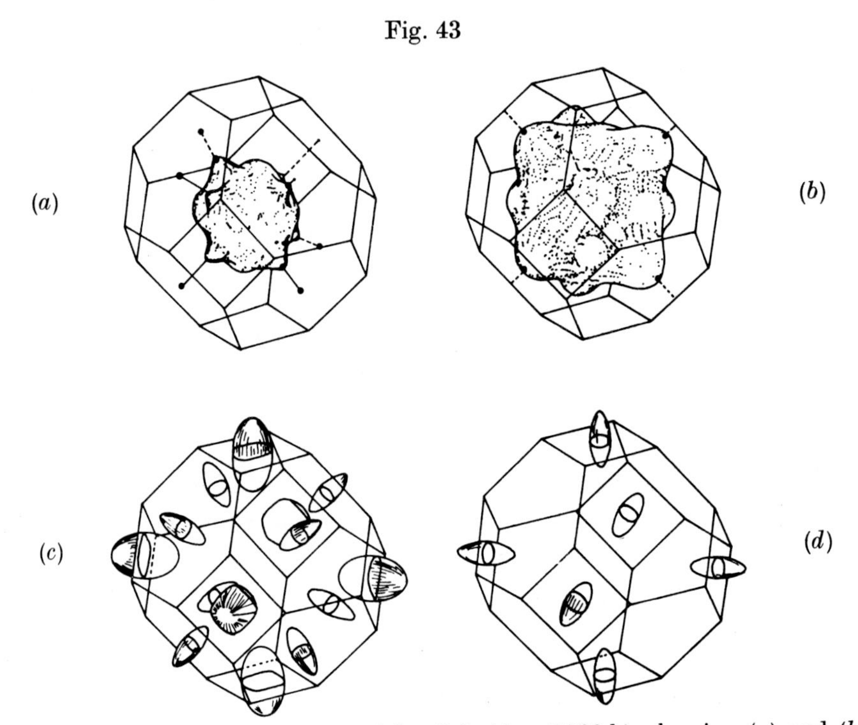

The Fermi surface of Rh proposed by Coleridge (1966 b), showing (*a*) and (*b*) regions of electrons and (*c*) and (*d*) pockets of holes.

9.4. *Rhodium and Iridium*

Extensive de Haas–van Alphen measurements were performed on Rh by Coleridge (1965, 1966 b) and detailed measurements of the dimensions of a number of sheets of the Fermi surface were obtained. Since there were, at that time, no band structure calculations available for Rh

† Further results for Ru obtained by Coleridge have now been published, see note added in proof.

Coleridge employed the rigid-band model and used the band structure of Ni (Hanus 1962, Wakoh and Yamashita 1964), which was the nearest available f.c.c. metal in the periodic table for which the band structure had been calculated (see § 9.2). On the basis of this band structure, but with one less electron per atom, and with one or two quite small *ad hoc* modifications Coleridge deduced that there were in all five closed sheets of Fermi surface in Rh ; Coleridge (1966 b) showed that detailed dimensions of four of these five sheets agreed rather nicely with his experimental results. The Fermi surface of Rh proposed by Coleridge is illustrated in fig. 43.

Fig. 44

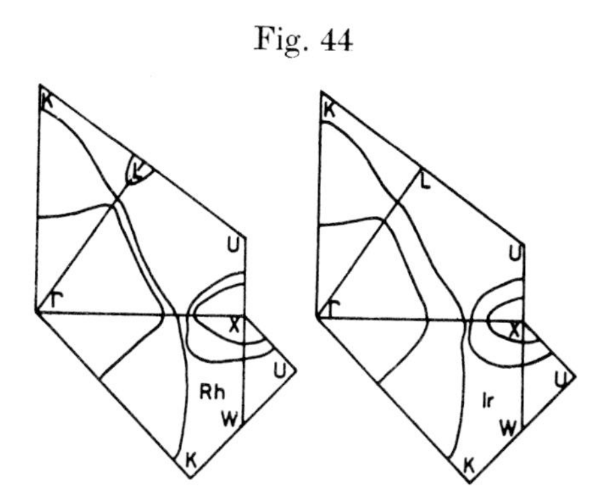

Cross sections of the Fermi surfaces of Rh and Ir calculated by Andersen and Mackintosh (1968).

Ab initio relativistic A.P.W. band structure calculations for the metals Rh and Ir, which both have the f.c.c. structure, were performed by Andersen and Mackintosh (1968). As one would expect, relativistic effects were found to be more important in the heavier metal, Ir. Sections of their calculated Fermi surfaces are shown in fig. 44. For Rh the calculated Fermi surface possessed the same general features that had been suggested previously by Coleridge, namely two large closed electron surfaces at Γ, two closed hole surfaces at X and one small closed hole surface at L. For Ir the calculated Fermi surface is very similar to that of Rh, except that the small pocket of holes at L has vanished. Extremal areas of cross section of these calculated Fermi surfaces for Rh and Ir are given in table 3 of Andersen and Mackintosh (1968). For Rh the experimental cross-sectional areas obtained by Coleridge (1965, 1966 b) and Ketterson *et al.* (1968) are also given in that table and the agreement between experiment and theory is quite close. For Ir the experimental cross sections of the two pockets of holes at X obtained using the de Haas–van Alphen effect by Grodski and Dixon (1969) and Hörnfeldt (1970) are given in table 1 of Hörnfeldt (1970).

There are several other pieces of work related to the band structure and Fermi surface of Rh and Ir ; these include optical measurements

(Bolotin and Chukina 1967 on Rh), X-ray measurements of the density of states (Claus and Ulmer 1965 on Rh and Ir), and measurements of the magnetoresistance (Fawcett 1961 c on Rh), thermoelectric power (Vedernikov 1969 on Rh and Ir), positron annihilation (Chen *et al.* 1967 on Rh), and of the temperature dependences of the paramagnetic susceptibility (Kojime *et al.* 1961 on Rh and Ir, Shimizu and Katsuki 1964 on Rh, Weiss and Kohlhaas 1967 on Rh and Ir), Knight shift (Seitchik *et al.* 1965 on Rh), and thermal and electrical conductivities (White 1956 on Rh and Ir, White and Woods 1957 a on Rh and Ir). Experimental values of γ, the temperature coefficient of the electronic specific heat, for Rh and Ir showed similar values of electron–phonon enhancement to those found in neighbouring metals in the periodic table. Calculated values of γ of 3·2 mJ mole^{-1} deg^{-2} for Rh and 2·4 mJ mole^{-1} deg^{-2} for Ir obtained by Andersen and Mackintosh (1968) are to be compared with the experimental values of $(4{\cdot}6 \pm 0{\cdot}4)$ mJ mole^{-1} deg^{-2} for Rh and $(3{\cdot}15 \pm 0{\cdot}05)$ mJ mole^{-1} deg^{-2} for Ir (Gschneidner 1964).

9.5. *Palladium and Platinum*

A maximum in the magnetic susceptibility of metallic Pd was observed by Hoare and Matthews (1952) between 80 and 100°K and, following the work of Lidiard (1954), there was some speculation as to whether this maximum might be associated with an antiferromagnetic phase transition. However, no confirmation of the existence of an antiferromagnetic structure was obtained from neutron-diffraction experiments. Crangle and Smith (1962) measured the specific heat of Pd from 65 to 105°K and found no evidence of any specific heat anomaly which could be ascribed to a magnetic transition. It is now generally accepted that both Pd and Pt are non-magnetic.

Magnetoresistance measurements by Alekseevskiĭ and Gaĭdukov (1960) and Fawcett (1961 c) revealed the existence of open orbits in $\langle 100 \rangle$ directions in Pt and similar results were also obtained later for Pd (Alekseevskiĭ *et al.* 1962, Alekseevskiĭ *et al.* 1964). The sign of the Hall coefficient indicates that the multiply-connected Fermi surface encloses a region of holes (Alekseevskiĭ *et al.* 1964). de Haas–van Alphen measurements of the Fermi surface of Pd have been made by Vuillemin and Priestley (1965) and Vuillemin (1966). The results were interpreted on the basis of the rigid-band model using the band structures calculated by Segall (1962) for Cu and by Mattheiss (1964) for non-magnetic Ni; this approach suggested that the Fermi surface of Pd consists of (i) a large closed region of electrons centred at Γ, (ii) small closed ellipsoidal regions of holes at X and (iii) a multiply-connected region of holes passing through the points X and W. Of these three surfaces, the dimensions of the region of electrons at Γ and the small pockets of holes at X were measured in considerable detail by Vuillemin and Priestley. The region of electrons at Γ is roughly spherical in shape with bumps along [111] and [100]; the detailed dimensions of this sheet of the Fermi surface of Pd are given in table II of Vuillemin (1966). Although no de Haas–van Alphen periods directly attributable to the multiply-connected hole region were observed, the general topological features of this surface are consistent with the results of the magnetoresistance measurements that we mentioned previously. Pd is a compensated metal, since there is an even number of conduction electrons per atom, and the number of electrons, or holes, was found to be about 0·36 per atom. Mueller and Priestley (1966) applied the inversion scheme devised by Mueller (1966), which uses cubic harmonics, to the inversion of the de Haas–van Alphen data of Vuillemin and Priestley for the electron Fermi surface of Pd centred at Γ ; this region was found to contain (0·364 ± 0·001) electrons per atom and the values of the coefficients used in the expansion are given by Mueller and Priestley (1966).

In measurements of the de Haas–van Alphen effect in Pt Stafleu and de Vroomen (1965) demonstrated the existence of small ellipsoidal pockets which were interpreted as being regions of holes at X because of their close similarity to those previously found in Pd. Ketterson *et al.* (1966) found two further sets of de Haas–van Alphen periods in Pt. One of these sets of periods could be associated with a Fermi surface that was very similar to the region of electrons around Γ in Pd which was described above. The cross-sectional areas in Pt were found to be somewhat larger than in Pd so that this region encloses about 0·40 electrons per atom. The other oscillations observed by Ketterson *et al.* (1966) in Pt were tentatively assigned to a multiply-connected region of holes similar to that just described for Pd and the existence of which in Pt as well was indicated by galvanomagnetic measurements (Borovik and Volotskaya 1958, Lüthi 1959, Alekseevskiĭ and Gaĭdukov 1960, Fawcett 1961 c). More recent de Haas–van Alphen measurements (Ketterson and Windmiller 1968, Windmiller and Ketterson 1968 a, Windmiller *et al.* 1969) showed that all the three sheets of Fermi surface that occur in Pd also occur in Pt. The pockets of holes at X were found to be ellipsoids of revolution prolate along ΓX and further oscillations were observed which could be ascribed to the multiply-connected region of holes enclosing Γ and X ; the form of this multiply-connected region of holes calculated by Mueller (1967) is illustrated in fig. 1 of Ketterson and Windmiller (1968). Magnetoacoustic quantum oscillations were observed in Pt by Fletcher *et al.* (1967) and two periods were obtained. The periods were in reasonable agreement with the de Haas–van Alphen periods obtained by Stafleu and de Vroomen (1965) and Ketterson *et al.* (1966) ; one period was ascribed to the pockets of holes at X and the other to the multiply-connected region of electrons passing through X and W. The conduction-electron g factor for electrons on the sheet of Fermi surface centred at Γ was determined by Windmiller and Ketterson (1968 b) from the spin-splitting of the de Haas–van Alphen oscillations. Measurements of the effect of pressure on the cross-sectional areas of the Fermi surfaces of Pd and Pt were made by Vuillemin and Bryant (1969) using the de Haas–van Alphen effect.

A number of band structure calculations for Pd and Pt have been performed using a variety of methods (Friedel *et al.* 1964 on Pd and Pt, Freeman, Furdyna and Dimmock 1966 on Pd, Lenglart *et al.* 1966 on Pd, Allan *et al.* 1968 on Pd and Pt, Andersen and Mackintosh 1968 on Pd and Pt, Lenglart 1968 on Pd and Pt†). In the non-relativistic A.P.W. calculation for Pd performed by Freeman, Furdyna and Dimmock (1966) two regions of holes and one compensating region of electrons were obtained which agreed quite well with the experimental results of the de Haas–van Alphen and magnetoresistance measurements. The computed density of states led to a value of γ equal to about half the experimental value indicating an electron–phonon interaction which is of a magnitude that is typical for metals in this part of the periodic table. Experimental measurements of the specific heat give γ values which include for Pd $(10{\cdot}0 \pm 0{\cdot}7)$ mJ mole^{-1} deg^{-2} (Gschneidner 1964) and $(9{\cdot}57 \pm 0{\cdot}07)$ mJ mole^{-1} deg^{-2} (Boerstoel *et al.* 1964) and for Pt $(6{\cdot}41 \pm 0{\cdot}026)$ mJ mole^{-1} deg^{-2} (Budworth *et al.* 1960), 6·54 mJ mole^{-1} deg^{-2} (Andres and Jensen

Fig. 45

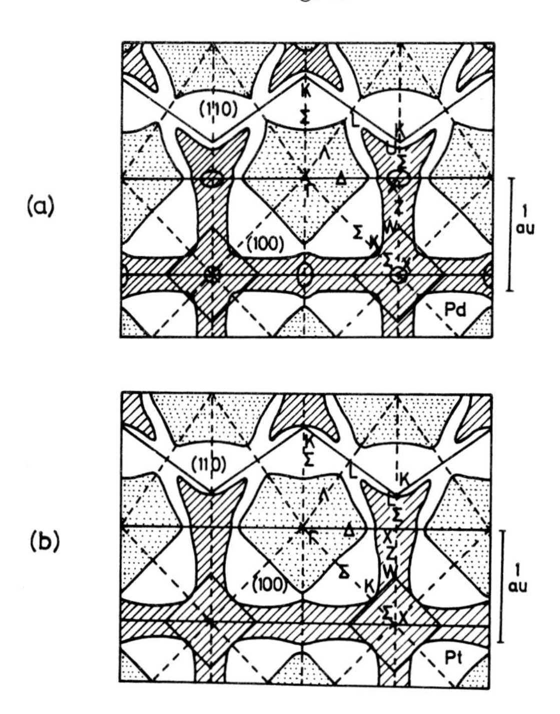

Cross sections of the Fermi surfaces of (*a*) Pd and (*b*) Pt calculated by Andersen and Mackintosh (1968).

---

† See note added in proof.

1968), $(6{\cdot}507 \pm 0{\cdot}006)$ mJ mole^{-1} deg^{-2} (Dixon *et al.* 1965), $(6{\cdot}56 \pm 0{\cdot}03)$ mJ mole^{-1} deg^{-2} (Shoemake and Rayne 1968) and $(6{\cdot}59 \pm 0{\cdot}03)$ mJ mole^{-1} deg^{-2} (Berg 1969). Shimizu *et al.* (1963) made use of experimental specific heat data on Pd and a number of alloys to deduce the density of states curve on the basis of the rigid-band model. The relativistic A.P.W. band structure calculation of Andersen and Mackintosh (1968) for both Pd and Pt again reproduced Fermi surfaces for both metals with these same general features ; cross sections of the calculated Fermi surfaces are shown in fig. 45. Detailed comparisons between the various calculated and experimental values of the cyclotron masses and of the extremal areas of cross section of the Fermi surfaces of Pd and Pt are given in table 2 of Andersen and Mackintosh (1968). The calculated and measured extremal areas of cross section of the Fermi surface generally agreed to within a few per cent while the enhancement of the cyclotron mass varied between about 1·2 and 1·7. It therefore seems that the shapes of the Fermi surfaces of Pd and Pt are now well established.

In addition to the experimental work which has given direct information about the shapes of the Fermi surfaces of Pd and Pt there have been a considerable number of other measurements of properties that depend, in one way or another, on the electronic band structures of these metals. Since the electrical resistance of Pt is widely used in thermometry and in the definition of the scale of temperature it would be too tedious to enumerate all the experimental work related to the determination of power-law variations of the resistivity of Pt with temperature. Other work includes investigations of the density of states by optical measurements (Bolotin *et al.* 1967 on Pd, Yu and Spicer 1966, 1968 b on Pd, Vehse *et al.* 1970 on Pd) and x-ray measurements (Claus and Ulmer 1964 on Pt, 1965 on Pd and Pt, Eggs and Ulmer 1968 on Pd), the magnetic susceptibility (Kojime *et al.* 1961 on Pd and Pt, Lenglart 1967 on Pd, Nemnonov and Sorokina 1967 on Pd, Weiss and Kohlhaas 1967 on Pd and Pt, Shimizu *et al.* 1969 on Pd), electrical conductivity (White 1956 on Pt, Schriempf 1968 on Pd), thermal conductivity (Kemp *et al.* 1955 on Pd, White and Woods 1957 c on Pt, Schriempf 1968 on Pd), thermoelectric power (Vedernikov 1969 on Pd and Pt), Hall effect (Wilding 1967 on Pd†), positron annihilation (Lang *et al.* 1955 on Pd and Pt, Lang and De Benedetti 1957 on Pd), and the Knight shift (Clogston *et al.* 1964 on Pt, Seitchik *et al.* 1964 on Pd, Matzkanin and Scott 1966 on Pt, Kushida and Rimai 1966 on Pt). Measurements of the magnitude, sign and temperature dependence of the Knight shift and susceptibility in Pt were used by Clogston *et al.* (1964) to determine the contributions to each arising from the spin paramagnetism of the different bands, the orbital paramagnetism and the core diamagnetism and, incidentally, to demonstrate that on the B.C.S. theory of superconductivity (Bardeen *et al.* 1957) Pt should not be superconducting at any temperature.

---

† See note added in proof.

§ 10. f-block : Lanthanides (Rare-earth Metals) and Actinides

La	Lanthanum	2.8.18.(2, 6, 10)(2, 6, 1) 2
Ce	Cerium	
⋮	⋮	2.8.18.(2, 6, 10, n)(2, 6, 1) 2
Lu	Lutetium	

where n = (atomic number—57)

Ac	Actinium	2.8.18.32.(2, 6, 10)(2, 6, 1) 2
Th	Thorium	2.8.18.32.(2, 6, 10, 1)(2, 6, 1) 2
Pa	Protactinium	2.8.18.32.(2, 6, 10, 2)(2, 6, 1) 2
U	Uranium	2.8.18.32.(2, 6, 10, 3)(2, 6, 1) 2
etc.		2.8.18.32.(2, 6, 10, n)(2, 6, 1) 2

where n = (atomic number—89).

10.1. *Structures*

From table 1 of part I it can be seen that there are two series of f-block elements in the periodic table. The first series of f-block elements (from La to Lu), which are sometimes called the lanthanides, exhibit a variety of structures, while the second series (from Ac onwards), which are sometimes called the actinides, contains a small number of naturally occurring elements together with a large number of artificial elements. The metals of the series of rare-earth elements exhibit a variety of structures which are listed below (McHargue and Yakel 1960, Major and Leinhardt 1967, Smithells 1967, Samsonov 1968, Bucher *et al.* 1969) :

La (α)	double h.c.p.	$a=3{\cdot}770$ Å, $c=12{\cdot}159$ Å at room temperature
(β)	f.c.c.	$a=5{\cdot}31$ Å (above 340°C)
Ce (β)	h.c.p.	$a=3{\cdot}65$ Å, $c=5{\cdot}96$ Å (95°K to 263°K)
(γ)	f.c.c.	$a=5{\cdot}1612$ Å (above 263°K)
(α)	f.c.c.	$a=4{\cdot}84$ Å (at 15 kbar, or below 95°K under normal pressure)
Pr	f.c.c.	$a=5{\cdot}15$ Å
	double h.c.p.	$a=3{\cdot}672$ Å, $c=11{\cdot}835$ Å
Nd	f.c.c.	$a=?$
	double h.c.p.	$a=3{\cdot}66$ Å, $c=11{\cdot}80$ Å
Pm	?	?
Sm (α)	D_{3d}^5	$a=8{\cdot}996$ Å, $\alpha=23°\ 13'$ (below 1190°K)
(β)	b.c.c.	$a=4{\cdot}07$ Å (1190°K to 1345°K)
Eu	b.c.c.	$a=4{\cdot}5820$ Å
Gd	h.c.p.	$a=3{\cdot}63$ Å, $c=5{\cdot}79$ Å
Tb	h.c.p.	$a=3{\cdot}59$ Å, $c=5{\cdot}66$ Å
Dy	h.c.p.	$a=3{\cdot}58$ Å, $c=5.65$ Å
Ho	h.c.p.	$a=3{\cdot}56$ Å, $c=5{\cdot}62$ Å

Er	h.c.p.	$a=3{\cdot}55$ Å, $c=5{\cdot}58$ Å
Tm	h.c.p.	$a=3{\cdot}52$ Å, $c=5{\cdot}56$ Å
Yb	f.c.c.†	$a=5{\cdot}47$ Å
Lu	h.c.p.	$a=3{\cdot}51$ Å, $c=5{\cdot}56$ Å.

The naturally occurring elements in the second series of f-block elements have rather complicated structures in some cases :

Ac	f.c.c.	$a=5{\cdot}311$ Å
Th	f.c.c.	$a=5{\cdot}08$ Å
(ξ)	orthorhombic	$a=9{\cdot}820$ Å, $b=8{\cdot}164$ Å, $c=6{\cdot}681$ Å
Pa	$D_{4h}{}^{17}$I4/mmm	$a=3{\cdot}925$ Å, $c=3{\cdot}238$ Å
U (α)	$D_{2h}{}^{17}$ Cmcm	$a=2{\cdot}854$ Å, $b=5{\cdot}869$ Å, $c=4{\cdot}955$ Å.

The four U atoms in the unit cell of α-U are at $[000\,;\ \frac{1}{2}\frac{1}{2}0]+0y\frac{1}{4}\,;\ 0\bar{y}\frac{3}{4}$ where $y=0{\cdot}105$. There are several other U structures.

La, Pr and Nd have the so-called ' double ' hexagonal close-packed structure with four atoms in the unit cell which are located at (0, 0, 0), $(\frac{1}{3}, \frac{2}{3}, \frac{1}{4})$, $(0, 0, \frac{1}{2})$ and $(\frac{2}{3}, \frac{1}{3}, \frac{3}{4})$ where the coordinates are referred to $\mathbf{t}_1=a\mathbf{i}$, $\mathbf{t}_2=\frac{1}{2}a\mathbf{i}+\frac{1}{2}\sqrt{3}a\mathbf{j}$, and $\mathbf{t}_3=c\mathbf{k}$. Fleming *et al.* (1968) related the fact of the existence of this double h.c.p. structure to one of the features of their calculated Fermi surfaces of these metals (see § 10.2). The transition between the two phases γ and α of Ce, which both have the same structure (f.c.c.) but different values of the lattice constant (or between the h.c.p. β phase of Ce and the α phase) is thought to correspond to the emptying of the 4f shell by the raising of its one 4f electron into the conduction band (Lawson and Tang 1949, Itskevich 1962).

Many of the rare-earth metals between La and Lu exhibit various magnetically ordered structures, some of which are ferromagnetic and some of which are antiferromagnetic (see, for example, Yosida (1964), Martin (1967)). The existence of these magnetic phases was first suspected from the existence of anomalies in various macroscopic properties such as the specific heat (Parkinson *et al.* 1951, Griffel *et al.* 1956, Jennings *et al.* 1957, Parkinson and Roberts 1957, Kurti and Safrata 1958, Gerstein *et al.* 1967, Lounasmaa and Sundström 1966, 1967), thermal expansion (Birss 1960), magnetic susceptibility (Trombe 1945, 1951, 1953, Elliott *et al.* 1954, Behrendt *et al.* 1958, Belov *et al.* 1961, Belov and Ped'ko 1962, Hegland *et al.* 1963), electrical resistivity (James *et al.* 1952, Elliott 1954, Colvin *et al.* 1960, Curry *et al.* 1960, Hall *et al.* 1960, Hegland *et al.* 1963, Elliott and Miner 1967) and magnetoresistance (Belov *et al.* 1964), but has now been established from neutron diffraction measurements (for example, Cable *et al.* 1961 on Er, Wilkinson, Child, McHargue, Koehler and Wollan 1961 on Ce, Wilkinson, Koehler, Wollan and Cable 1961 on Dy, Koehler *et al.* 1962 a on Tm, 1962 b on Tb, Ho and Tm, Olsen *et al.* 1962 on Eu, Koehler *et al.* 1963 on Tb, Cable *et al.* 1964 on Pr, Moon *et al.* 1964 on Nd,

---

† See note added in proof.

Table 6. Magnetic structures of rare-earth metals

Element	Atomic number	Structure
Ce	58	Antiferromagnetic $T_N \doteqdot 4{\cdot}2$°K
Pr	59	Double h.c.p. Sinusoidal antiferromagnetic $T_N \doteqdot 24$°K, sublattice magnetization normal to *c* axis f.c.c. Ferromagnetic $T_C \doteqdot 8{\cdot}7$°K
Nd	60	As Pr but with $T_N \doteqdot 19$°K, for double h.c.p. and $T_C = 29$°K for f.c.c.
Eu	63	Helical antiferromagnetic $T_N \doteqdot 91$°K axis parallel to cube edge
Gd	64	Ferromagnetic $T_C = 294$°K (see text)
Tb	65	Ferromagnetic in basal plane $T_C = 218$°K Helical antiferromagnetic $T_N = 230$°K
Dy	66	Ferromagnetic in basal plane $T_C = 86$°K Helical antiferromagnetic $T_N = 179$°K
Ho	67	Conical ferromagnetic with moment along *c* axis $T_C = 19$°K Helical antiferromagnetic $T_N = 133$°K
Er	68	Conical ferromagnetic with moment along *c* axis $T_C = 20$°K Sinusoidally modulated antiferromagnetic $T_N = 80$°K
Tm	69	4,3,4,3 ferrimagnetic at 4°K changing to sinusoidal antiferromagnetic with $T_N = 56$°K

Nereson *et al.* 1964 on Eu, Koehler, Cable, Wilkinson and Wollan 1966 on Ho, Cable and Wollan 1968 on Gd, Umebayashi *et al.* 1968 on Tb and Ho under pressure, and Will 1968 on Tb and Er). The magnetic structures of many of these metals are summarized in table 6. Some of the structures are simple ferromagnetic structures with all the individual magnetic moments parallel to the *c* axis (Gd) or in the plane normal to the *c* axis (Tb and Dy). It has also recently been reported that the f.c.c. phases of Pr and Nd become ferromagnetic at low temperatures (8·7°K and 29°K respectively) (Bucher *et al.* 1969). Actually the magnetization in Gd is only parallel to the *c* axis between 294°K and 232°K ; below 232°K Gd remains ferromagnetic but the direction of the magnetization moves away from the *c* axis to a maximum deviation of about 65° near 180°K and then moves back to within about 32° of the *c* axis at low temperatures (Cable and Wollan 1968). The anomalous behaviour of the electrical resistivity of Gd in the region of 210°K was correlated with the rapid change of the magnetization direction with temperature (Nellis and Legvold 1969). More complicated conical ferromagnetic structures are found in Ho and Er in which the magnetic moment of each individual ion has the same *c* component while the components normal to the *c* axis are arranged helically, see fig. 46 (*a*). In each plane normal to the

Fig. 46

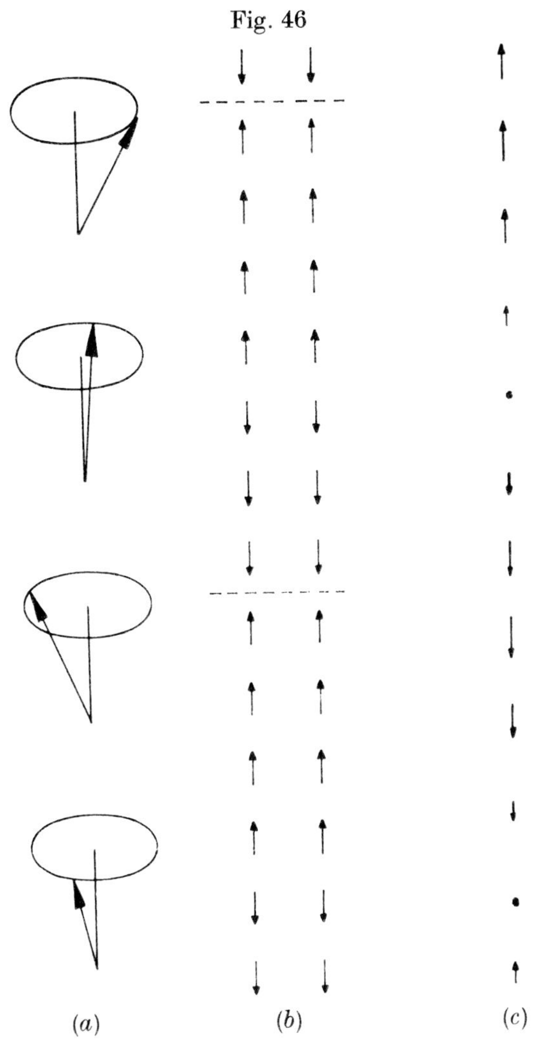

Some types of magnetic ordering (*a*) conical ferromagnetism, (*b*) 4, 3, 4, 3 ferrimagnetism, and (*c*) sinusoidal antiferromagnetism.

c axis the magnetic moments are all parallel, that is, each individual sheet is ferromagnetic. If the components of the spins in fig. 46 (*a*) parallel to the *c* axis vanish the structure will have no net magnetic moment and it is described as a ' helical antiferromagnetic ' structure (Eu, Tb, Dy and Ho). At 4°K Tm has a 4, 3, 4, 3 ferrimagnetic structure, see fig. 46 (*b*), which changes progressively, as the temperature is raised, into a sinusoidal antiferromagnet with no resultant magnetic moment, see fig. 46 (*c*). Sinusoidal antiferromagnetism is also exhibited by Pr and Nd but for these two metals the direction of the spins is normal, instead of parallel, to the *c* axis. If an external magnetic field is applied to an antiferromagnetic metal there may be a critical field above which the ordering becomes ferromagnetic; this has been observed, for example, in Dy, Ho and Er (Behrendt *et al.* 1958, Flippen 1964). The mechanism governing the transition from the ferromagnetic phase to the helical antiferromagnetic phase in both Tb and Dy has been discussed by Cooper (1967) and in Tb alone by Jackson and Doniach (1969). As we shall see in § 10.2 there is an intimate relationship between the type of magnetic ordering which occurs in any given rare-earth metal and the details of the geometry of the Fermi surface of that metal. Given the shape of the Fermi surface of a metal and considering the exchange interaction between the 4f electrons and the conduction electrons it is possible to predict the value of **q**, the wave vector associated with the magnetic ordering, and the temperature dependence of **q** (Elliott and Wedgwood 1964, Freeman, Dimmock and Watson 1966 a, b, Specht 1967, Evenson and Liu 1968, 1969, Keeton and Loucks 1968, Harris 1969, Jackson and Doniach 1969).

10.2. *The Rare-earth Metals* (*Lanthanum to Lutetium*) (*the Lanthanides*)

Relatively little work has so far been performed on direct measurements of the dimensions of the Fermi surfaces of the rare-earth metals but quite a large number of indirect measurements have been made (see § 10.3).

As we have already mentioned in § 1 the electronic structure of a rare-earth atom, outside a Xe core, is $4f^n5d^16s^2$, where $n=$ (atomic number—57) and goes from 0 for La to 14 for Lu. The configuration of the outermost electrons for each of these metals is therefore the same; the atoms are normally trivalent and there are correspondingly assumed to be three electrons per atom in the conduction band of the metallic solid element. By analogy with the transition metals the 4f bands can be assumed to be very narrow so that the 4f electrons are highly localized, while the three outer electrons form a typical broad metallic conduction band that can be regarded as a 5d–6s hybrid band. The rare-earth metals can therefore be regarded as being composed of trivalent positive ions in a sea of three conduction electrons per atom. Useful confirmation of this comes from measurements of the magnetic form factor in neutron diffraction experiments. For example, for Tm measurements of the magnetic form factor were in good agreement with the $4f^{12}$ form factor which is in accord with the general expression just given with $n=12$ for Tm (Brun and Lander 1969), leaving the three $5d^16s^2$ electrons to go into the conduction band. The close similarities among the electronic structures of the heavy rare-earth metals are also illustrated by x-ray isochromat measurements (Bergwall 1965). The position of the narrow 4f band relative to the conduction band varies considerably through the series of rare-earth metals (Gupta and Loucks 1969). The 4f band is thought to be above the Fermi energy in La and below the Fermi energy in metals from Pr onwards and only in Ce is the 4f band close enough to the Fermi energy to be of importance†. The

† See note added in proof.

α phase of Ce is thought to be associated with the increase of the number of conduction electrons per atom from three to four at the expense of the 4f shell which thereby becomes empty (Itskevich 1962, Likhter and Venttsel' 1962). For the free atoms of the rare-earth elements there are one or two exceptions to the electronic structure with valence 3 which we have just described ; Ce, Pr, Tb and probably Nd can be tetravalent while Sm, Eu, Yb and perhaps Tm can be divalent. The reasons for the departures from valence three in these particular elements are discussed by Sidgwick (1950). Yb becomes divalent by using one of its outermost electrons to complete the 4f shell while Eu becomes divalent by using one of its outermost electrons to form an exactly half full 4f shell which is also very stable. Yb is therefore rather different from most of the other rare-earth metals (Lounasmaa 1963). Yb is soft, its density (7·0 g cm^{-3}) is lower than the densities of the neighbouring elements Tm (9·3 g cm^{-3}) and Lu (9·8 g cm^{-3}) and its crystal structure is f.c.c. rather than h.c.p. That Yb has fewer conduction electrons per atom is indicated by the very low value of its electronic specific heat ; γ for Yb is only $(2{\cdot}92 \pm 0{\cdot}01)$ mJ $mole^{-1}$ deg^{-2} (Lounasmaa 1963) compared with values of about 10 mJ $mole^{-1}$ deg^{-2} for most other rare-earth metals. Angular correlation curves for the γ-rays produced by positron annihilation in Ce, Gd and Yb were obtained by Gustafson and Mackintosh (1964) ; the curve for Yb was narrower than the curves for Ce and Gd by an amount corresponding to a difference of about one conduction electron per atom. These curves correlate well with the results of Rodda and Stewart (1963) who found that the positron lifetimes in Eu and Yb are significantly longer than in the other rare-earth metals.

In view of the above discussion we would therefore expect that the Fermi surfaces of the quite large number of rare-earth metals which have the h.c.p. structure and a valence of three will be very similar to the Fermi surfaces of Sc and Y which were described in § 4.2, while the Fermi surface of Eu, which is divalent and b.c.c., should be very similar to that of Ba and the Fermi surface of Yb, which is divalent and f.c.c.† should be very similar to those of Ca and Sr. Recent calculations, mostly using the A.P.W. method, do indeed show that the Fermi surfaces of the heavy rare-earth metals and of Y (Loucks 1966 b) are very much alike. Non-relativistic A.P.W. band structure calculations have been performed for La and Lu (Freeman, Dimmock and Watson 1966 a), Ce (Mukhopadhyay and Majumdar 1969), Gd (Dimmock and Freeman 1964, Freeman, Dimmock and Watson 1966 a) and Tm (Freeman, Dimmock and Watson 1966 a, b) while relativistic A.P.W. band structure calculations have been performed for a large number of the rare-earth metals (Keeton and Loucks 1966 a on Lu, Fleming *et al.* 1968 on La, Pr and Nd, Keeton and Loucks 1968 on Gd, Dy, Er and Lu, Mackintosh 1968 on Tb, Jackson 1969 on Tb and Mukhopadhyay and Majumdar 1969 on Ce). As one might expect, by

† See note added in proof.

analogy with the transition metals, the energy bands for the 5d and 6s electrons, which are the conduction electrons, depart very considerably from the free-electron energy bands. With three conduction electrons per atom and two atoms per unit cell in the h.c.p. structure the equivalent of three bands must be full. The Fermi surfaces in bands 3 and 4 in the double-zone scheme for Gd and Tm, calculated non-relativistically, were almost identical and are shown in fig. 47. Although there are some detailed differences from the Fermi surface of Y described in § 4.2, the general features are quite similar to those of the Fermi surface calculated for Y ; that is, the Fermi surface is entirely in bands 3 and 4 and, in the double-zone scheme, is multiply-connected in a rather complicated fashion.

Fig. 47

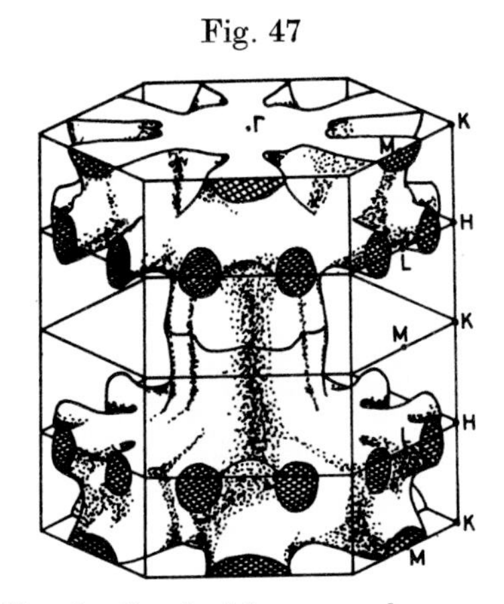

The Fermi surface of Tm, in the double-zone scheme, calculated by Freeman, Dimmock and Watson (1966 b) (holes shaded).

Since the atomic numbers of the rare-earth metals are quite large it is natural to expect that spin–orbit coupling effects will be quite significant ; in particular, of course, many degeneracies in the electronic band structure at the special points of symmetry in the Brillouin zone will be lifted (see, for example, Keeton and Loucks (1966 a) on Lu). In particular, the lifting of these degeneracies means that the double-zone scheme will no longer be appropriate for the description of the Fermi surfaces of these metals. In their relativistic A.P.W. calculations Keeton and Loucks (1968) found that the inclusion of relativistic effects did not alter the calculated Fermi surfaces very significantly from the results of the non-relativistic calculations, whereas for the heavier rare-earth metals the inclusion of relativistic effects does produce significant effects. Therefore, while fig. 47 probably describes fairly accurately the Fermi surface of any of the lighter rare-earth metals that happen to have the h.c.p. structure, significant changes arise in this Fermi surface, as a result of

Fig. 48

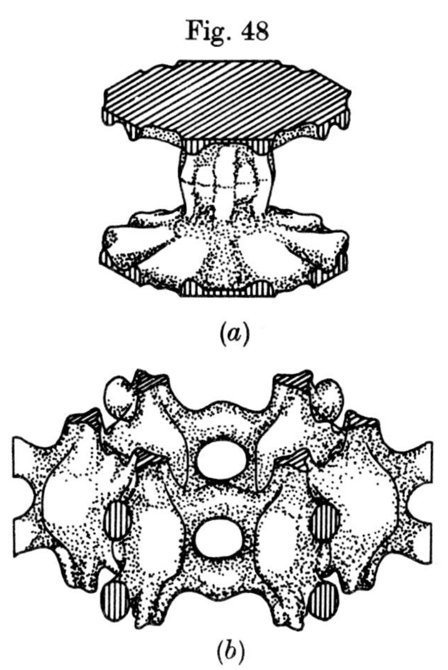

The Fermi surface of Tb shown in the single-zone scheme, (*a*) holes in band 3 and (*b*) electrons in band 4 (Jackson 1969).

relativistic effects, when it comes to be applied to the heavier h.c.p. rare-earth metals. The particular changes which occur are, according to Keeton and Loucks (1968) that the arm at M disappears and a ' webbing ' appears between the two arms at L ; this Fermi surface is then even more similar to that calculated by Loucks (1966 b) for Y, see fig. 15. We should thus expect, for h.c.p. rare-earth metals, a steady change from the Gd type Fermi surface, shown in fig. 47, to the Y type Fermi surface shown in fig. 15, as one moves through the rare-earth metals in the direction of increasing atomic number. For intermediate rare-earth metals such as Tb (Jackson 1969) and Dy (Keeton and Loucks 1968) both the arms at M and the webbing at L will be present ; the Fermi surface of Tb calculated by Jackson (1969) is shown, in the single-zone scheme, in fig. 48. Cross sections of the Fermi surfaces in bands 3 and 4 of Gd, Dy, Er and Lu, calculated by the relativistic A.P.W. method, are given in fig. 4 of Keeton and Loucks (1968). Therefore, it now seems that for those rare-earth metals that possess the h.c.p. structure the general features of the Fermi surface are fairly clear.

It is now appropriate to consider the relationship between the electronic band structure and the various complicated forms of magnetically ordered structure which may exist. In the sections on the iron-group metals we have already discussed the effects of ferromagnetic ordering on the band structure and Fermi surface of a metal. There is no change in the size of the Brillouin zone, although the size of the basic domain (Bradley and Cracknell 1970) of the Brillouin zone may be increased, but many of the essential degeneracies that existed in the paramagnetic phase will be removed by the introduction of ferromagnetic ordering. However, if relativistic effects are included in the band structure calculations for the paramagnetic phase there are very few essential degeneracies left anyway even before the magnetic ordering is introduced (see, for example, Keeton and Loucks (1966 a)). The effect of ferromagnetic ordering on the actual shapes of the bands can be expected to be quite small. However, as we saw in connection with the iron-group metals, the introduction of ferromagnetic ordering gives rise to a large internal field $\mathbf{B}_{int}$ which causes a large separation by $g\beta|\mathbf{B}_{int}|$ between the spin-up and spin-down bands and this in turn causes drastic changes in the shape of the Fermi surface. Completely different Fermi surfaces will then apply to the spin-up and spin-down bands. Little work appears to have been done for the rare-earth metals on the estimation of the magnitude of this separation for simple ferromagnetic ordering and the study of the consequent shapes of the Fermi surfaces for the spin-up and spin-down bands. For the other more complicated forms of magnetic ordering which can occur in the rare-earth metals there are two rather different possibilities, depending on whether the period of the magnetic order is commensurate or incommensurate with the crystal lattice of the metal.

If the magnetic structure is commensurate with the crystal lattice the size of the unit cell of the magnetically-ordered metal will, almost inevitably, be several times larger than the unit cell of the paramagnetic phase of the metal. By increasing the size of the unit cell of the metal the size of the Brillouin zone will be correspondingly reduced. For example, between 40°K and 50°K Tm has a sinusoidal antiferromagnetic structure in which the magnitudes of the magnetic moments, which in each horizontal layer are ferromagnetically aligned in the z direction, vary sinusoidally with a period of $7c$. The size of the unit cell is therefore increased by a

Fig. 49

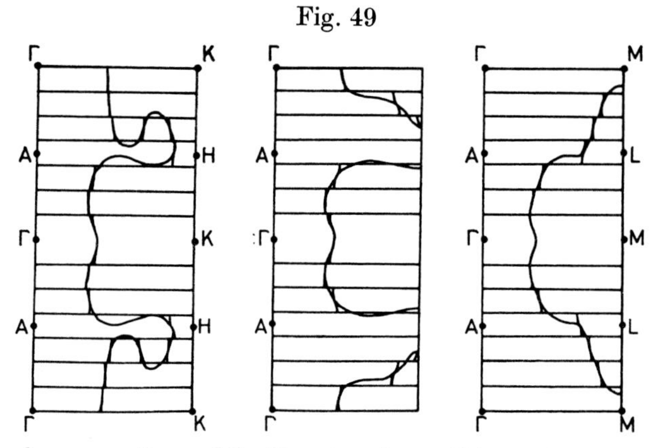

Some vertical cross sections of the Fermi surface of Tm with magnetic ordering; the horizontal lines denote magnetic superzone boundaries at $k_z = \pm n(2\pi/7)$ (Freeman, Dimmock and Watson 1966 b).

factor of seven and new *superzone boundaries* appear in the reciprocal lattice at $k_z = \pm n(2\pi c/7)$. These new superzone boundaries are shown in fig. 49 ; this figure now represents an extended zone scheme for the antiferromagnetic structure and the true Brillouin zone only extends from $k_z = -(2\pi c/7)$ to $k_z = +(2\pi c/7)$ and the entire band structure and Fermi surface can be folded back into this region. A particularly suitable metal for studying the effect of magnetic superzone boundaries is Dy because between 86°K and 178°K it exhibits a simple helical structure, with no magnetic moment parallel to the c axis, and if a sufficiently large magnetic field is applied normal to c it will effect a change to simple ferromagnetic ordering so that the superzone boundaries will disappear ; it is therefore possible to study directly the effects of the controlled appearance and disappearance of the superzone boundaries (Wilding and Lee 1965).

If the period of the magnetic ordering is incommensurate with the crystal lattice of the metal the situation is very similar to that described in § 7.4 in connection with antiferromagnetic Cr and we may expect to find some relationship between the wave vector $\mathbf{q}$ associated with the magnetic ordering and the separation between some important points on the Fermi surface ; in the case of Cr the wave vector $\mathbf{q}$ corresponds to the separation along the $\langle 100 \rangle$ axes between the octahedron of holes around H and the body of the electron jack around Γ. There is a qualitative difference between the cases of Cr and of the rare-earth metals in that the magnetic moments in Cr are associated with the conduction electrons themselves, whereas the magnetic moments in the rare-earth metals are generally thought not to be associated with the conduction electrons but with the inner incomplete shell of 4f electrons. As we have seen in § 7.4 the relationship between the spin-density waves in antiferromagnetic Cr and the Fermi surface of that metal was first studied by Lomer (1962, 1964 a). These ideas were extended to the magnetic rare-earth metals by Mackintosh (1962) and Miwa (1963) and discussed in considerable detail by Elliott and Wedgwood (1963, 1964) and Herring (1966). Much of the initial interest was concerned with attempting to explain the temperature dependence of the electrical resistivity of rare-earth metals and, in particular, the anomalies which occur at the transition temperatures. Since the changes in periodicity introduced by the magnetic ordering are in the c axis direction it is not surprising that these anomalies occur principally for the component of the resistivity along this direction. The helical or other complicated magnetic structure found in one of these rare-earth metals gives rise to an exchange field with a lower symmetry than that of the crystal lattice. The effect of this exchange field on the conduction electrons is to introduce new Brillouin zone boundaries, see fig. 49, and to distort the Fermi surface. Elliott and Wedgwood (1963) calculated this distortion and the scattering of the conduction electrons by the spin disorder and included the effect of the variation of the pitch of the helix with temperature. They obtained qualitative agreement with the experimental measurements for Dy, Ho, Er and Tm. Better agreement than this could hardly have been expected at that stage because there was little information available about the real Fermi surfaces of these metals and the calculations had to be based on the free-electron model. This discussion has been extended to the case of the electrical resistivity, thermal conductivity and Seebeck effect in Tm using the calculated Fermi surface shown in fig. 47 (Freeman, Dimmock and Watson 1966 b, Edwards and Legvold 1968). As we hinted in § 10.1, the detailed shapes of the Fermi surfaces of the rare-earth metals are very closely linked with the origins of the magnetic ordering which may occur. Keeton and Loucks (1968) noticed that there appears to be a close correlation between the wave vector **q** associated with the magnetic ordering and the thickness, in the c axis direction, of the ' webbing ' which appears between the arms near L in the hole Fermi surface of h.c.p. rare-earth metals heavier than Gd.

Table 7. Comparison of the magnetic wave vector (in units of π/c) as determined from experiment and from the minimum separation of the webbing

	Y	Lu	Er	Dy	Gd
Experiment†	0·54	0·53	0·57	0·49	0
Theory‡	0·49	0·45	0·54	0·46	0

† Nigh *et al.* (1964), Koehler (1965).
‡ Keeton and Loucks (1968).

The values of **q** determined experimentally from neutron diffraction work for a number of metals are compared in table 7 with the thickness of the webbing calculated by Keeton and Loucks (1968) and illustrated in fig. 50. The webbing is absent from the Fermi surface of Gd and, in practice, Gd is observed not to have an antiferromagnetic phase. Similar work on Tb (Harris 1969, Jackson 1969, Jackson and Doniach 1969) led to calculated values of **q** that were in semi-quantitative agreement with the experimentally determined value of $0{\cdot}224\pi/c$ (Yosida and Watabe 1962). Assuming exchange interactions of the Ruderman–Kittel–Kasuya–Yosida (R.K.K.Y.) type between the electrons in the 4f shells of different ion cores, via the conduction electrons (Ruderman and Kittel 1954, Kasuya 1956, Yosida 1957) it is possible to calculate the total energy of the system. The value of the total energy will be dependent on the value of **q** so that the actual value of **q** can be found by minimization. This was done by Elliott and Wedgwood (1964) for Dy, Ho, Er and Tm using the free-electron Fermi surface, by Watson *et al.* (1968) for a number of the heavy rare-earth metals, and by Evenson and Liu (1968, 1969) for Gd, Dy, Er and Lu using the Fermi surfaces calculated by Keeton and Loucks (1968). Evenson and Liu obtained values of **q** for Dy, Er and Lu

Fig. 50

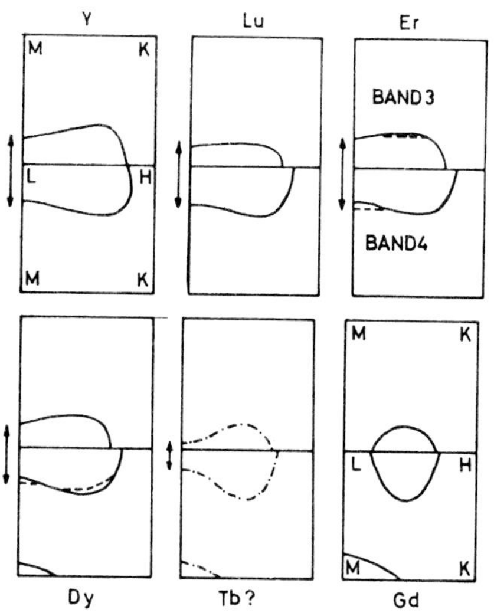

Cross sections of various Fermi surfaces showing the webbing feature (Keeton and Loucks 1968).

that were in quite good agreement with the observed **q** values for the antiferromagnetic ordering in these metals.

Having discussed in considerable detail the Fermi surfaces of those rare-earth metals that possess the h.c.p. structure we now turn to the remaining rare-earth metals which possess other structures. Fleming *et al.* (1968) used the relativistic A.P.W. method to calculate the band structures and Fermi surfaces of the metals La, Pr and Nd which have the double h.c.p. structure described in § 10.1. Since there are four atoms per unit cell there will have to be the equivalent of six full bands. Cross sections of the calculated Fermi surface for La are shown in fig. 51 and the Fermi surface consists of (i) a nearly circular cylinder of holes centred along ΓA in band 5 and another cylinder of holes centred along ΓA in band 6 which tapers from a nearly hexagonal cross section in the ΓKM plane to a circular cross section in the AHL plane, (ii) a multiply-connected region of electrons in band 7 which encloses H, K and M and has a nearly circular cross section near H but spreads out to produce a shallow shelf centred at M in the ΓKM plane, and (iii) small ellipsoidal pockets of electrons along AH in band 7 and along HK in band 8. The Fermi surfaces of Pr and Nd can be expected to be similar to that of La. La, of course, is non-magnetic but Pr and Nd do exhibit magnetic ordering. For Pr and Nd the wave vector **q** associated with the magnetic ordering is normal to the *c* axis (Cable *et al.* 1964, Moon *et al.* 1964) and was assumed by Fleming *et al.* (1968) to be associated with the shelf-like part of the multiply-connected region of electrons in band 7. We have already seen that in the heavy rare-earth metals (beyond Gd) which have the ordinary h.c.p. structure there are very flat

Fig. 51

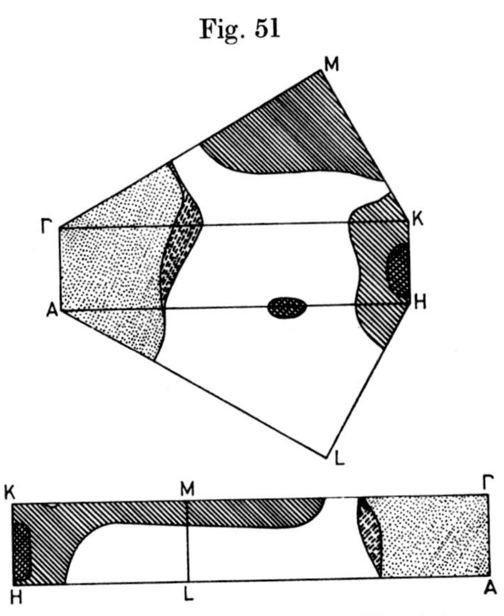

Cross sections of the Fermi surface of La (Fleming *et al.* 1968).

pieces of Fermi surface normal to the c axis and approximately halfway between the central plane and the top surface of the double zone, see figs. 47 and 48. These flat pieces can be eliminated and the total energy can therefore be reduced by the introduction of a periodic potential with period $2c$ and consequently new superzone boundaries appear at $\pm \frac{1}{2}\pi/c$. Therefore Fleming *et al.* (1968) suggested that the reason for the existence of the double h.c.p. structure was because it produces just such a potential with a periodicity of $2c$ in the z direction and thereby lowers the energy of the system. The relationship between the nesting of various pieces of this calculated Fermi surface and the observed wave vector $\mathbf{q}$ of the magnetic ordering in Pr and Nd was discussed by Fleming *et al.* (1969). Relativistic band structure calculations for the high-temperature f.c.c. phases of La and Pr have been performed by Myron and Liu (1970). Their predicted Fermi surface was entirely contained in band 2 and was multiply-connected in a rather complicated way ; it bore no obvious resemblance to the Fermi surface of Al, a simple metal with the same structure and valence ; this is a little surprising considering the close resemblance which appears to exist between the Fermi surfaces of Yb† and Ca (or Sr) or between the Fermi surfaces of Eu and Ba (see below). The only remaining rare-earth metals not covered by the above discussion of the h.c.p. and double h.c.p. structures are Pm, Sm, Eu, Yb and the f.c.c. phases of La, Ce and Pr. Nothing is known about the Fermi surfaces of Pm, Sm and the f.c.c. phases of La, Ce and Pr.

† It now appears that the important form of Yb is the h.c.p. phase rather than the f.c.c. phase which is discussed in this paragraph, see note added in proof.

We have already mentioned that Yb commonly has a valence of two rather than three because it prefers to form a complete 4f shell. Yb metal is then an f.c.c. metal with two electrons per atom and one can therefore expect that its band structure and Fermi surface will resemble quite closely the band structures and Fermi surfaces of Ca, Sr and the hypothetical f.c.c. phase of Ba (see § 4.4 of part I)†. This is consistent with the results of the magnetoresistance measurements of Datars and Tanuma (1968) which showed that no open orbits are possible on the Fermi surface of Yb. From the relativistic A.P.W. band-structure calculation for Yb by Johansen and Mackintosh (1970) the energy bands of Yb were more closely similar to those of the hypothetical f.c.c. phase of Ba (Johansen 1969) than of Ca and Sr (Vasvari *et al.* 1967, Johansen 1970, private communication). The Fermi surface of Yb should therefore consist of small pockets of electrons around L in band 2 and small regions of holes around W and K(U) in band 1 and Yb can be expected, like Ca and Sr, to become semiconducting at high pressures. Experimental evidence that at ordinary pressures the Fermi surface of Yb consists of small pockets of carriers was obtained in the de Haas–van Alphen work of Tanuma *et al.* (1967) and experimental evidence of a semiconducting transition in Yb at high pressure was obtained by Souers and Jura (1963), Jerome and Rieux (1969) and McWhan *et al.* (1969). The measured de Haas–van Alphen periods in Yb were about one order of magnitude larger than the periods in Ca, which indicates that the pockets of carriers in Yb are smaller than those in Ca.

Eu is similar to Yb in having only two conduction electrons per atom rather than three ; in this case it is because one extra electron is used to make the 4f shell exactly half full. Since Eu has the b.c.c. structure it is therefore not unreasonable to suppose that just as the band structure and Fermi surface of Yb are similar to those of Ca and Sr, so also the band structure and Fermi surface of Eu should be very similar to that of Ba which has been discussed in § 4.4 and the *note added in proof* in part I. This Fermi surface is unlike those of Ca, Sr and Yb in that the pockets of carriers are not particularly small, so that Eu is not a semi-metal and we should not expect a transition to a semiconducting state under high pressure. Experimental evidence for the similarity between the band structures of Ba and Eu was found in the very close similarities between the optical reflectance and transmission spectra of these two metals (Müller 1965, 1966, 1967, Schüler 1965). A comparison of the fusion curves and compressibility data also suggests this relationship (Jayaraman 1964). This similarity between the Fermi surfaces of Eu and Ba was confirmed by the direct relativistic A.P.W. calculation of Andersen and Loucks (1968) on Eu in which the Fermi surface of Eu was found to be almost identical to that of Ba ; that is, it consists of a pocket of holes at P,

---

† It now appears that the important form of Yb is the h.c.p. phase rather than the f.c.c. phase which is discussed in this paragraph, see note added in proof.

Fig. 52

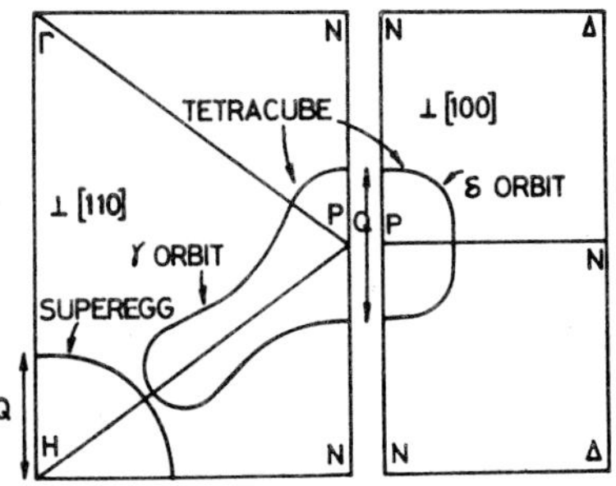

Cross sections of the Fermi surface of Eu (Andersen and Loucks 1968).

Fig. 53

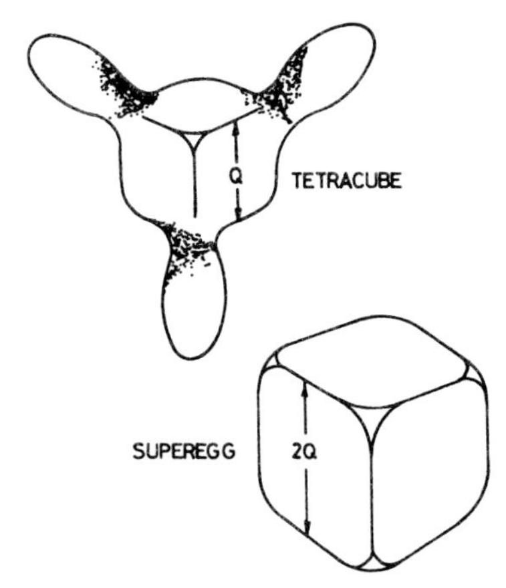

The Fermi surface of Eu (Andersen and Loucks 1968).

described as a ' tetracube ', and a pocket of electrons at H, described as a ' super-egg '. Cross sections of these two pieces of Fermi surface are shown in fig. 52 and the surfaces are illustrated in fig. 53. The region of holes at P is shaped like a rounded cube and has ellipsoidal projections arranged tetrahedrally on four of the eight corners, hence the name ' tetracube '. The region of electrons at H is shaped like a cube with rounded corners. The reason for the introduction of the name ' super-egg ' to describe this piece of Fermi surface is explained by Andersen and Loucks (1968) as follows. In many design applications Piet Hien, the Danish writer and inventor, has used simple shapes which mediate between the round and the rectangular ; in two dimensions the curves are called ' super-ellipses ' and in three dimensions he chose to call them ' super-eggs '. One family of such surfaces can be described by the equation

$$x^n + y^n + z^n = \text{const.}, \qquad (10.2.1)$$

where for $n = 2$ the surface is a sphere and for $n \to \infty$ the surface becomes a cube. Intermediate values of n will describe a family of surfaces which are super-eggs and, in particular, Andersen and Loucks (1968) found that with $n = 13/4$ eqn. (10.2.1) described quite closely the shape of their calculated region of electrons at H. Some extremal areas of cross section, and hence predicted de Haas–van Alphen frequencies, for both the super-egg and tetracube pieces of Fermi surface in Eu are given in table I of Andersen and Loucks (1968) but so far no de Haas–van Alphen type measurements appear to have been made on Eu. Andersen and Loucks (1968) associated the wave vector $\mathbf{q}$ of the helical magnetic ordering with the nesting of the tetracube with itself, see fig. 53, which gives a value of $\mathbf{q}$ quite close to the experimental value. It is an interesting coincidence that the opposite faces of the super-egg are separated by a distance that is almost exactly equal to $2\mathbf{q}$.

10.3. *Miscellaneous Work on the Electronic Properties of the Lanthanides*

The large amount of work related to the electronic properties of the rare-earth metals includes work on the electrical resistivity (James *et al.* 1952 on La, Ce, Pr and Nd, Colvin *et al.* 1960 on Gd, Tb, Dy, Ho, Er, Tm and Lu, Hall *et al.* 1960 on Dy, Green *et al.* 1961 on Er, Itskevich 1962 on Ce at pressures up to 10 kbar, Volkenshteĭn and Fedorov 1963 on Dy and Er, Arajs and Colvin 1964 b on Tb, Volkenshteĭn and Startsev 1964 on Gd and Yb, Arajs and Dunmyre 1965 b on Er, Volkenshteĭn, Dyakina, Novoselov and Startsev 1966 on Dy, Major and Leinhardt 1967 on Ce, Nagasawa and Sugawara 1967 on Pr, Edwards and Legvold 1968 on Tm, Lodge and Taylor 1969 on Dy, Nikol'skiĭ *et al.* 1969 on Ho, Watabe and Kasuya 1969 on La, Pr and Nd), magnetic susceptibility (Trombe 1945, 1951, 1953 on Dy, Elliott *et al.* 1954 on Dy, Lock 1957 on Pr, Behrendt *et al.* 1958 on Dy, Arajs 1960 a on Sm, 1960 b on Tm, Arajs and Miller 1960 on Nd, Ho and Er, Colvin *et al.* 1961 on Ce and Eu, Nigh *et al.* 1963 on Gd, Telesnin *et al.* 1962 on Gd, Hudgins and Pavlovic 1965 on Dy, Nagasawa and Sugawara 1967 on Pr, and Féron and Pauthenet 1969 on Gd), magnetoresistance (Belov and Nikitin 1962 on Tb, Dy and Ho, Babushkina 1964 on Gd, 1965 a on Ho, Wilding and Lee 1965 on Dy, Nikol'skiĭ and Eremenko 1966 on Er, Volkenshteĭn and Dyakina 1966 on Gd, Nagasawa and Sugawara 1967 on Pr and Nd), Hall effect (Kevane *et al.* 1953 on La, Ce, Pr, Nd, Gd, Dy and Er, Andersen *et al.* 1958 on Sm, Tm, Yb and Lu, Likhter and Venttsel' 1962 on Ce at high pressures, Volkenshteĭn and Fedorov 1963 on Dy and Er, Babushkina 1965 b on Gd and Tb, Volkenshteĭn and Fedorov 1965 on Nd and Sm, Volkenshteĭn and Dyakina 1966 on Gd, Volkenshteĭn and Fedorov 1966 on Ho, Lee and Legvold 1967 on Gd and Lu, Abel'skiĭ and Irkhin 1968, Rhyne 1968 on Dy), Knight shift (Blumberg *et al.* 1960 on La, Torgeson and Barnes 1964 on La, Zamir and Schreiber 1964 on La, Volkenshteĭn, Grigorova and Fedorov 1966 on Dy), thermal conductivity (Rosenberg 1955 on La and Ce, Arajs and Colvin 1964 a on Gd, 1964 b on Tb, Colvin and Arajs 1964 on Dy, Aliev and Volkenshteĭn 1965 a on Ho and Er, 1965 b on Tm, Yb and Lu, Arajs and Dunmyre 1965 b on Er, Gallo 1965 on Gd, Tb and Dy, Mamiya *et al.* 1965 on La, Powell and Jolliffe 1965 on Ce, Nd, Sm, Gd, Tb, Dy, Ho and Er, Golubkov *et al.* 1966 on La, Nikol'skiĭ and Eremenko 1966 on Er, Powell 1966 on Gd, Tb and Dy, Rao 1967 on Dy, Edwards and Legvold 1968 on Tm, Karagezyan and Rao 1968 on Tb), positron annihilation (Mackintosh *et al.* 1965 on Ho, Williams *et al.* 1966 on Ho and Er, Hohenemser *et al.* 1968 on Gd, Williams and Mackintosh 1968 on Gd, Tn, Dy, Ho and Er), and various optical and x-ray measurements (Bergwall 1965 on Gd to Lu, Blodgett *et al.* 1965 on Gd, Eastman 1969 a on Gd, Hodgson and Cleyet 1969 on Gd). The variation of the electrical resistance of most of the rare-earth metals as a function of pressure has been discussed extensively by Drickamer (1965). The conduction electron densities at the nuclei in Sm and Dy were determined by Henning *et al.* (1968). Miwa (1963) pointed out that it should be possible to determine the s–f exchange energy in several rare-earth metals directly by observing optical absorption corresponding to energy gaps at magnetic Brillouin zone boundaries. Experimental evidence for the observation of this absorption in Ho in the infra-red region was obtained by Schüler (1964). An extensive discussion of the physical properties of the rare-earth metals is given by Spedding *et al.* (1957).

The existence of a partially filled 4f shell in most of the rare-earth metals results in the existence of a ferromagnetic or antiferromagnetic structure for most of these metals at low temperatures. Consequently one would expect to be able to detect the existence of spin waves, either directly by inelastic neutron scattering (Møller and Houmann 1966, Møller *et al.* 1967, 1968, Woods *et al.* 1967, Nicklow *et al.* 1969, Stringfellow *et al.* 1969, Koehler *et al.* 1970) or indirectly from their contributions to such properties as the specific heat (Lounasmaa 1964 a, c, Janovec and Morrison 1965, Dempesy *et al.* 1966, Lounasmaa and Sundström 1966, 1967, Morrison and Newsham 1968), thermal conductivity (Arajs and Colvin 1964 a, b, Colvin and Arajs 1964, Aliev and Volkenshteĭn 1965 a, Gallo 1965, Nikol'skiĭ and Eremenko 1966, Rao 1967, Boys and Legvold 1968), electrical resistance (Coles 1958, Kondorskiĭ *et al.* 1958, Turov 1958, Colvin *et al.* 1960, Volkenshteĭn and Startsev 1964, Lodge and Taylor 1969).

The interpretation of the specific heat measurements on the rare-earth metals is complicated by the fact that, in addition to the usual lattice and electronic contributions which are proportional to T^3 and T respectively, the specific heats of most of the rare-earth metals at low temperatures have some extra contributions which are not present in most other metals. Apart from any specific heat anomalies arising from changes in the magnetic ordering there are two special contributions connected with the incomplete shell of 4f electrons. One of these contributions is associated with the motions of the magnetic moments of the ion cores, which are usually described in terms of spin waves, and the other contribution is associated with changes in the orientation of the nuclear magnetic moment in the intense magnetic field at the nucleus produced by the incomplete shell of 4f electrons in each ion core. In the very simple theories of spin

Table 8. Experimental values of γ, the temperature coefficient of the electronic contribution to the specific heat, for rare-earth metals

Element	γ (mJ mole^{-1} deg^{-2})	Reference
La	10·1	Gschneidner (1964)
	9·4 ± 0·1	Johnson and Finnemore (1967)
	11·5 ± 0·3	
Ce	21	Lounasmaa (1964 c)
	10·5	
Pr	19·0 ± 0·5	Dreyfus, Goodman, Lacaze and Trolliet (1961)
	24·4	Lounasmaa (1964 a)
	57·9 ± 1·0	Janovec and Morrison (1965)
	26·2 ± 0·2	Morrison and Newsham (1968)
Nd	8·92	Gschneidner (1964)
	22·5	Lounasmaa (1964 a)
	58 ± 2	Morrison and Newsham (1968)
Pm	10	estimated by Gschneidner (1964)
Sm	10·6 ± 1·5	Gschneidner (1964)
	12·4 ± 0·2	Morrison and Newsham (1968)
Eu	5·8 ± 1·0	Lounasmaa (1964 c)
	12·1 ± 0·5	Morrison and Newsham (1968)
Gd	11	Dreyfus *et al.* (1967)
Tb	9·05	Lounasmaa and Roach (1962)
	10·4 ± 0·4	Morrison and Newsham (1968)
Dy	10	Dash *et al.* (1960)
	9·0 ± 0·3	Dreyfus, Goodman, Trolliet and Weil (1961)
	9·5 ± 0·9	Lounasmaa and Guenther (1962)
	17·9 ± 0·4	Morrison and Newsham (1968)
Ho	26 ± 5	Dreyfus, Goodman, Lacaze and Trolliet (1961)
	10	Lounasmaa (1962 b)
	50 ± 2	Morrison and Newsham (1968)
Er	13 ± 1	Dreyfus, Goodman, Lacaze and Trolliet (1961)
Tm	19·7 ± 1·8	Gschneidner (1964)
	22·3 ± 0·2	Morrison and Newsham (1968)
Yb	2·92 ± 0·01	Lounasmaa (1963)
	2·90 ± 0·05	Morrison and Newsham (1968)
Lu	9·5	Jennings *et al.* (1960)
	11·27	Lounasmaa (1964 b)
	11·31 ± 0·13	Morrison and Newsham (1968)

waves the spin-wave contribution to the specific heat is proportional to $T^{3/2}$ for a ferromagnet and to T^3 for a simple antiferromagnet (Van Kranendonk and Van Vleck 1958). The presence of all these various contributions makes the interpretation of a set of experimental results quite complicated and it may sometimes be convenient to use graphical methods instead of the more conventional computer methods (Morrison and

Newsham 1968). Measured values of γ, the temperature coefficient of the electronic contribution to the specific heat, for the rare-earth metals are given in table 8. There are very few band structure calculations of γ; Dimmock and Freeman (1964) calculated a value of 4·2 mJ mole^{-1} deg^{-2} for Gd which is to be compared with the experimental value of 11 mJ mole^{-1} deg^{-2} given in table 8, and Andersen and Loucks (1968) calculated a band structure value of γ for Eu that was about one third of the experimental value of 5·8 mJ mole^{-1} deg^{-2} given in table 8. These results suggest that the phonon enhancement of γ in the rare-earth metals is of comparable magnitude to that in the transition metals. Since many of the rare-earth metals are magnetically ordered there is also the possibility of the enhancement of γ by electron–magnon interactions; this has been discussed in considerable detail by Nakajima (1967).

10.4. *The Actinides*

It is not very surprising that very little work has been done so far on the Fermi surfaces of the actinides because most of them do not occur naturally and all of them are radioactive.

The electronic structures of the free atoms of the actinides are very similar to those of the lanthanides except that it is now the 5f, 6d and 7s shells rather than the 4f, 5d and 6s shells which are of importance. By analogy with the lanthanides we may suppose that the conduction bands of the actinides are derived from the 6d and 7s electrons and that, in general, the partially filled 5f levels are sufficiently far away from the Fermi energy, either above it (for the lighter actinides) or below it (for the heavier actinides), so as not to affect the shape of the Fermi surface. The Fermi surface of any of these actinides can therefore be expected to be similar to that of the appropriate isoelectronic transition metal with the same crystal structure (if that exists). However, as is the case with the 4f levels in Ce, there will probably be at least one of the actinide metals in which the 5f levels are very near the Fermi energy so that a comparison with the transition metals would not be profitable (Gupta and Loucks 1969).

The group-theoretical analysis of the Brillouin zone and a study of the free-electron energy bands in α-U was performed by Jones (1960) and Suffczyński (1960); however, from our experience with the lanthanides and with the transition metals together with the fact that the atomic number of U is so high, it seems unlikely that the free-electron model will give even a first approximation to the shape of the Fermi surface of U. A band-structure calculation for Th was performed by Lehman (1959) considering only the d bands. Relativistic A.P.W. band structure calculations for Ac and Th, which have the f.c.c. structure, were performed by Keeton and Loucks (1966 a). According to Keeton and Loucks the 5f shell is empty in both Ac, which is analogous to La, and in Th, which is similar to Ce; there are therefore three conduction electrons per atom in Ac (. . . $6d^1 7s^2$) and four conduction electrons per atom in Th (. . . $6d^2 7s^2$).

Fig. 54

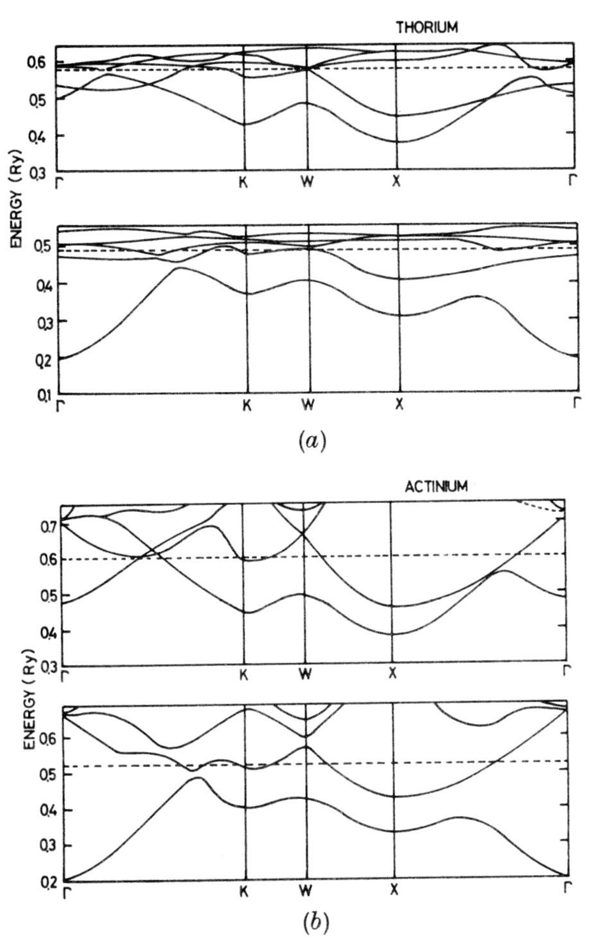

Non-relativistic (upper) and relativistic (lower) band structures for (*a*) Th and (*b*) Ac (Keeton and Loucks 1966 a).

The differences between the relativistic and non-relativistic band structures for these metals is very striking, see fig. 54 ; what is perhaps most unusual is that in the non-relativistic band structure the lowest level is not at Γ but at one of the other points of symmetry (X). The difference between the bandwidths in the non-relativistic calculations (about 0·2 Ry for Th) and in the relativistic calculations (about 0·3 Ry for Th) should be capable of being investigated experimentally by soft x-ray measurements. The first direct experimental work on the Fermi surface of Th was performed by Thorsen *et al.* (1967) using the de Haas–van Alphen effect. Their results were interpreted as indicating the existence of a nearly spherical piece of Fermi surface situated at the centre of the Brillouin zone and a set of six closed segments located along the ⟨100⟩ axes at X points. These

Fig. 55

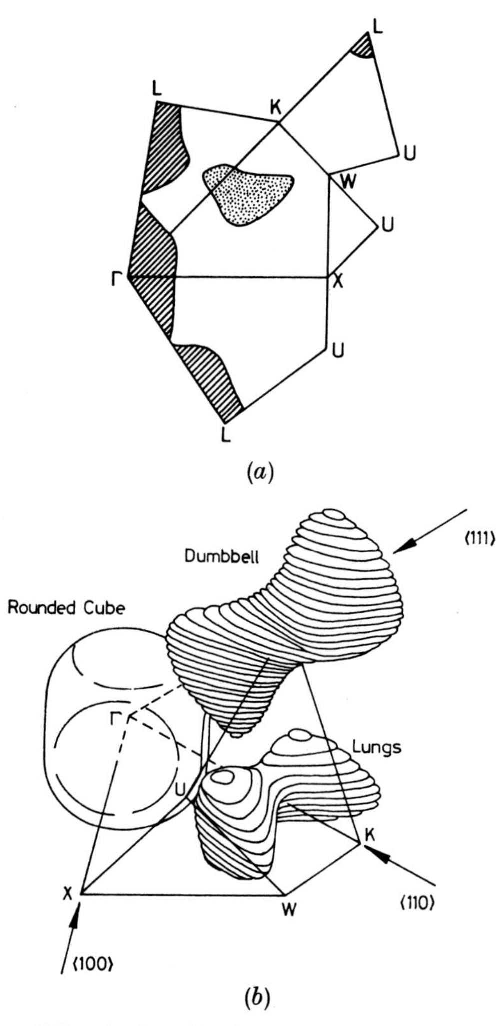

The Fermi surface of Th calculated by Gupta and Loucks (1969), (*a*) cross sections (holes shaded, electrons dotted), and (*b*) sketch.

results were not in agreement with the band structure calculations mentioned above. The reason for this is explained by Gupta and Loucks (1969) and is connected with the position of the 5f levels. The difficulty was overcome in a new relativistic A.P.W. band structure calculation for Th performed by Gupta and Loucks (1969) and, apart from the increased importance of relativistic effects, the calculated band structure showed a considerable resemblance to the band structures of the f.c.c. transition metals. The shape of the calculated Fermi surface is illustrated in fig. **55**; this consists of (i) a hole surface around Γ shaped like a rounded cube (**a** ' super-egg '), (ii) a dumb-bell-shaped hole surface with triangular ends centred at L and directed along the line ΓL, and (iii) regions of electrons on the line ΓK shaped like a pair of lungs or female breasts. The existence of the regions of holes around Γ is consistent with the de Haas–van Alphen results of Thorsen *et al.* (1967). The de Haas–van Alphen periods that were assigned by Thorsen *et al.* (1967) to an ellipsoidal pocket at X can readily be reassigned to the dumb-bell surface or the lung surface shown in fig. 55 (*b*). Further extensive de Haas–van Alphen measurements on Th by Boyle and Gold (1969) lend strong support to the essential correctness of the shape of the calculated Fermi surface for Th shown in fig. 55 (*b*). The experimental orbits on the lung surfaces were in fairly good agreement with those calculated from the Fermi surface shown in fig. 55 (*b*), but the experimental orbits assigned to the other two surfaces suggested that the calculated dumb-bell is slightly too small while the calculated super-egg at Γ is slightly too large.

This agreement between the theoretical and experimental work on the Fermi surface of Th is particularly important because it provides valuable support for the model of separating the magnetic f electrons from the conduction bands, which we have been using for most of the lanthanides and actinides. In spite of all that was said in § 10.2 about the Fermi surfaces of the lanthanides there was a slightly disturbing lack of convincing direct experimental Fermi surface measurements for those metals. This work on Th provides some indirect evidence that, at least for the non-magnetic phases of the rare-earth metals, the Fermi surfaces which were described in § 10.2, and which are quite similar to some of the transition metal Fermi surfaces, are not too far removed from the truth.

Experimental work indirectly related to the band structures and Fermi surfaces of the actinide metals include measurements of the Hall effect, electrical resistivity and magnetoresistance of Th and U (Berlincourt 1959), the electrical resistivity of Th (Peterson *et al.* 1967), positron annihilation in Th and U (Rozenfeld and Szuszkiewicz, 1966), and the thermoelectric power of Th and U (Vedernikov 1969). Several experimental values of γ, for Th, obtained by various workers and ranging from 4·7 mJ mole^{-1} deg^{-2} to 7·9 mJ mole^{-1} deg^{-2} are quoted by Wallace (1960).

§ 11. Conclusion

In general we have seen that the shapes of the Fermi surfaces of the d-block and f-block metals depart quite substantially from the predictions of the free-electron model. We have also seen that where magnetic ordering exists the effect on the band structure and Fermi surface of the metal is quite dramatic. We now indicate the areas in which it seems likely that further experimental or theoretical work would be useful.

Groups IB and IIB. The situation for these metals is very similar to that for the simple metals described in part I; no further geometrical investigations are really necessary and the shape of the Fermi surface of each of these metals is now known sufficiently accurately to be used in connection with describing many of the properties that are only indirectly related to the details of the band structure and Fermi surface of the metal (see, for example, Ziman (1959, 1961), Abarenkov and Vedernikov (1966), Hall (1967), and Hasegawa and Kasuya (1968, 1970)).

Group IIIA. Direct measurements of the shape of the Fermi surface of the h.c.p. phase of Sc are needed to distinguish between the two somewhat different calculated Fermi surfaces described in § 4.2 for this metal. We noted in § 4.2 that the cellular calculations for Sc and Y give very similar Fermi surfaces for the two metals and also that the A.P.W. calculations for Sc and Y give very similar results for the two metals ; however, the results of the cellular calculation differ from the results of the A.P.W. calculation. Therefore experimental work on Y as well as Sc would be valuable to distinguish between the two different theoretical Fermi surfaces for each of these metals. New band structure calculations for the f.c.c. phase of Sc and, if possible, experimental measurements as well, would be useful in connection with the investigation of the relationship between the band structures on the two sides of the phase transition in this metal.

Group IVA. Direct experimental measurements of the shape of the Fermi surface of Ti are needed to distinguish between the predicted Fermi surfaces described in § 5.2. Further experimental work on Zr is required in order to distinguish between the two somewhat different Fermi surfaces predicted by the cellular and A.P.W. methods. Since Hf has the same structure as Ti and Zr it is not unreasonable to suppose that the Fermi surface of Hf should be quite similar to those of Ti and Zr, although spin–orbit coupling effects can be expected to be more important in Hf than in Ti and Zr. There is clearly scope for both experimental and theoretical work on Hf.

Group VA. We have seen in § 6.2 how the rigid-band model has been used to give some theoretical predictions of the shapes of the Fermi surfaces of V, Nb and Ta. The Fermi surfaces of Nb and Ta have been investigated fairly thoroughly experimentally, but it still remains an open question as to whether the roughly octahedral region of holes around Γ in band 2, which is predicted by the A.P.W. calculations, actually exists or not. Although we can expect the Fermi surface of V to be quite similar to those of Nb and Ta it is probably still worthwhile investigating further the Fermi surface of V experimentally.

Group VIA. The general features of the Fermi surfaces of Mo, W and paramagnetic Cr are now fairly clearly established and have been described in §§ 7.2–7.4. Also the effect of the introduction of the sinusoidal antiferromagnetic ordering on the motions of the conduction electrons in Cr has been studied quite extensively.

Group VIIA. The very complicated structure of Mn with its very large unit cell and very small Brillouin zone presents such great difficulties that, in the past, hardly any work has been done either theoretically or experimentally on its band structure and Fermi surface. There is obviously scope for the enthusiast here. Most of the Fermi surface of Re has been mapped successfully except that there is still some doubt about the exact shape of the region of electrons in band 9. Nothing appears to have been done on the Fermi surface of Tc so far, presumably because of the instability and consequent rarity of this metal, and there is therefore considerable scope for both theoretical and experimental work on this metal ; however, it is unlikely that its Fermi surface will differ very much from that of Re, except that the splitting of the bands due to spin–orbit coupling (see fig. 35) is unlikely to be very significant in Tc.

Group VIII. As mentioned in § 9.2 the calculation of the band structure of a ferromagnetic metal including *ab initio* the effects of the magnetic ordering is far from easy and the problem has generally been simplified by only including the magnetic effects at the end in the form of an internal field $\mathbf{B}_{int}$ which causes a separation of $g\beta|\mathbf{B}_{int}|$ to appear between the spin-up and spin-down bands. There is scope for more sophisticated treatments of band structure calculations for ferromagnetic metals. Experimental work has been quite extensive on the Fermi surface of Ni but has been much more sparse on Fe and Co. The remaining metals in group VIII are not ferromagnetic and so should, in principle at least, be easier to study. Work on Pd and Pt has been quite extensive and the shapes of their Fermi surfaces are now quite well established (see § 9.5) while a considerable amount of work has also been done on Rh and Ir (see § 9.4). For Ru and Os, however, little work has been done so far and the shapes discussed in § 9.3 are only very speculative.

Lanthanides and actinides. In the absence of magnetic ordering the shapes of the Fermi surfaces of the lanthanides are very similar to those of the appropriate simple metals with the same structure and valence. The effects of the introduction of various forms of magnetic ordering at low temperatures have been described in § 10.2 and the general principles seem to be clearly established although for a number of the rare-earth metals many details remain to be clarified. Among the actinides the Fermi surfaces of Ac and Th, which have the f.c.c. structure, have been studied in some detail, but for various reasons mentioned in § 10.4 the Fermi surfaces of the remaining actinides have not so far received much attention.

For the quite large number of metals that we have discussed in both part I and part II and for which the detailed shape of the Fermi surface is well established, there is still a considerable amount of work to be done in parametrizing these Fermi surfaces by expressing their shapes in analytical form, as has already been done for certain Fermi surfaces (for example, Bi (see § 7.2 of part I), the noble metals (see §§ 2.2 and 2.3) and one sheet of the Fermi surface of Ni (see § 9.2)†).

The general principles involved in using the band structure and Fermi surface to try to explain any of the properties that depend only indirectly

---

† Also Pt, see note added in proof.

on the electronic structure of a metal have already been discussed in § 8 of part I ; that discussion is equally applicable to the d-block and f-block metals although, even more than in the case of the simple metals, the application of the knowledge of the detailed shapes of the Fermi surfaces is hampered by the present inadequate knowledge of the electrons' relaxation times, or mean free paths, for various types of collision process. However, there are further features which may arise for those metals in which spontaneous magnetic ordering exists with the consequent existence in the crystal of spin-wave excitations of which the quanta are magnons. We have already mentioned the spin-wave contribution to the specific heat in §§ 9.2 and 10.3. The spin waves can also influence the various transport properties such as the electrical conductivity or the thermal conductivity in a magnetically ordered crystal (for references see, for example, the book by Akhiezer *et al.* (1968)).

In the two parts of this article we have confined our attention to the consideration of the Fermi surfaces of metals which are chemical elements. There are, of course, other kinds of metals that are not elements but which may be alloys or simple chemical compounds. We have occasionally mentioned work related to the electronic properties of very dilute alloys in which it is conventional to use a rigid-band model and assume that the small percentage of impurity atoms does not significantly alter the shape of the energy bands from those of the host metal. If the valence of the impurity is different from the valence of the host metal the effect of the impurity atoms will be to change the average number of conduction electrons per atom and consequently to move the Fermi level. By using controlled amounts of an impurity of known valence and measuring one or other of the properties of the metal one can often extract some information about the band structure and Fermi surface of the pure host metal. This is a well-established technique and dates back at least to the very early work by Jones on the determination of the sizes of the pockets of carriers in Bi (see § 7.2 of part I). Once large concentrations of impurity are added to a host metal the situation generally becomes quite complicated both theoretically and experimentally. However, the special case of an ordered alloy, such as beta brass for example, in which the atoms of the different metallic elements are regularly arranged on a superlattice with a recognizable crystallographic structure, is not very different from that of a metallic element. The theoretical and experimental methods involved in studying the band structure and Fermi surface of an ordered alloy are the same as those used for a metallic element ; although we have not discussed it in this review a considerable amount of work has indeed been done on the band structures and Fermi surfaces of ordered alloys. Just to give one of many possible examples, the cross-sectional areas of the Fermi surface of ordered beta brass have been measured by de Haas–van Alphen experiments in just the same way as for metallic elements (Jan *et al.* 1964, Jan 1966) ; several other examples are cited by Ahn and Sellmyer (1970 a, b). In addition to ordered alloys or intermetallic compounds there are also metallic compounds which contain at least one non-metallic element; the study of the electronic properties of these materials owes much to a number of papers by Professor Sir Nevill Mott (for a recent review see, for example, Mott (1968)). Although in the past the band structures and Fermi surfaces of these metals may have been studied less than those of ordered alloys there is now a considerable interest in these materials. Again we have not discussed the Fermi surfaces of these materials in this review but the standard theoretical and experimental techniques that we have had occasion to mention in connection with metallic elements can also be applied to these materials; just to mention one or two examples selected almost at random, extremal areas of cross section of the Fermi surface of metallic ReO_3 have been measured by Graebner and Greiner (1969) using magnetothermal oscillations and Berko and Weger (1970) have performed positron annihilation experiments on V_3Si.

Note added in proof.—A number of miscellaneous new references related to the determination of the Fermi surfaces of d-block and f-block metals have come to hand since the manuscript of this article was prepared.

Group IB. The analysis discussed by Segall and Ham (for reference see *note added in proof* in part I) in which the partial-wave phase shifts η_l are regarded as adjustable parameters for use in band structure calculations has recently been applied to Cu using the A.P.W. method (M. J. G. Lee, *Phys. Rev.*, **187,** 901 (1969)) and the K.K.R. method (J. F. Cooke, H. L. Davis and R. F. Wood, *Phys. Rev. Lett.*, **25,** 28 (1970)).

A novel time-of-flight method for the measurement of the velocity of the group of electrons at the Fermi surface which are responsible for the radio-frequency size effect has been performed with Cu (G. A. Baraff and T. G. Phillips, *Phys. Rev. Lett.*, **24,** 1428 (1970)).

Group IIB. A new extensive set of measurements of caliper dimensions of the Fermi surface of Zn has recently been obtained using the radio-frequency size effect (O. L. Steenhaut and R. G. Goodrich, *Phys. Rev.* B, **1,** 4511 (1970)). The results were in good agreement (in many cases better than 1%) with caliper dimensions calculated on the basis of the non-local pseudopotential band structure calculation of Stark and Falicov

(1967). It is thought that the remaining small discrepancies would be removed by including spin–orbit coupling in the calculation.

Magnetoacoustic giant quantum oscillations have recently been observed in Hg (G. Bellessa, M. Ribault and R. Reich, *Physics Lett.* A, **31,** 556 (1970)).

Group VIA. Recent work on Mo includes magnetoacoustic attenuation measurements (A. A. Galkin, S. E. Zhevago, T. B. Butenko and E. P. Degtyar, *Ukr. fiz. Zh.*, **13,** 1106 (1968)) and radio-frequency size effect measurements (J. R. Cleveland and J. L. Stanford, *Phys. Rev. Lett.*, **24,** 1482 (1970)). From their measurements Cleveland and Stanford estimated the separation between the electron jack and each of the hole octahedra along the $\langle 100 \rangle$ directions to be $(7{\cdot}5 \pm 1)\%$ of ΓH. This quite large separation was taken to imply that the importance of spin–orbit coupling in Mo metal is considerably greater than was previously thought to be the case.

Galvanomagnetic measurements have been made on thin films of Cr and W (O. A. Panchenko, P. P. Lutsishin and Yu. G. Ptushinskiĭ, *Zh. éksp. teor. Fiz.*, **56,** 134 (1969). English translation : *Soviet Phys. JETP*, **29,** 76 (1969)).

Group VIII. A calculation of the band structure of paramagnetic h.c.p. Co has been performed using the Green's function (or K.K.R.) method (S. Wakoh and J. Yamashita, *J. phys. Soc. Japan*, **28,** 1151 (1970)) and the shape of the Fermi surface of ferromagnetic Co was predicted by assigning 10·56 and 7·44 electrons per unit cell to the majority- and minority-spin bands, respectively. This Fermi surface, which is illustrated in figs. 5–8 of the paper by Wakoh and Yamashita, consisted of :

(i) for the majority-spin electrons a large sphere, in the double-zone scheme, centred at Γ and elongated in the k_z direction. In the single-zone scheme this reduces to a smaller closed and roughly spherical Fermi surface around Γ together with a cylinder in the k_z direction and

(ii) for the minority-spin carriers there are several pieces of Fermi surface including cylinders of electrons directed parallel to the k_z direction, with their axes along ML lines and connected to one another at K points, one set of pockets centred at L, another set of smaller pockets near to L, and three closed pockets around Γ. The three concentric surfaces at Γ separate, respectively, bands 9 and 10, bands 10 and 9, and bands 9 and 8 as one moves out from Γ towards the surface of the Brillouin zone.

A comparison between this Fermi surface deduced for ferromagnetic Co and the rather meagre available experimental evidence on Co is given in the paper by Wakoh and Yamashita.

It was found (P. T. Coleridge, *J. low Temp. Phys.*, **1,** 577 (1969)), using the rigid-band model for a h.c.p. metal, that the newer de Haas–van Alphen data for Ru could be explained much better with a non-relativistic band structure than with the relativistic band structure calculated by Mattheiss that was mentioned in § 9.3. On the basis of such a non-relativistic band structure the Fermi surface proposed by Coleridge for Ru consisted of

(i) a small region of holes around Γ and another small region of holes on the line LM, and

(ii) two large closed electron surfaces centred at Γ and one large multiply-connected electron surface that touches the small pockets of holes around L.

The principal cross sections of all these sheets of the proposed Fermi surface of Ru are illustrated in fig. 7 of the paper by Coleridge.

Further band structure calculations have been performed for Pd (F. M. Mueller, A. J. Freeman, J. O. Dimmock and A. M. Furdyna, *Phys. Rev.* B, **1,** 4617 (1970)) using both the A.P.W. method and the combined interpolation scheme ; it was shown that, as expected, an improvement in the details of the agreement between theory and experiment can be obtained for Pd by the inclusion of relativistic effects in the calculations.

The parameters in a Fourier series expansion of the energy bands $E_{\mathbf{k}}$ of Pt near the Fermi surface have been determined for the large multiply-connected region of holes in band 5 and the closed region of electrons at Γ in band 6 (J. B. Ketterson, F. M. Mueller and L. R. Windmiller, *Phys. Rev.*, **186,** 656 (1969)).

Galvanomagnetic measurements have been made on thin films of Pd and Pt by Panchenko *et al.* (1969, loc. cit.).

Lanthanides and actinides. The positions of the 4f levels in Yb have been investigated by x-ray photoemission measurements (S. B. M. Hagström, P. O. Hedén and H. Löfgren, *Solid St. Commun.*, **8,** 1245 (1970)); two sets of 4f levels, separated as a result of spin–orbit coupling, were found at $(1{\cdot}4 \pm 0{\cdot}4)$ ev and $(2{\cdot}7 \pm 0{\cdot}4)$ ev below the Fermi energy.

Some further de Haas–van Alphen measurements have recently been made on Yb (S. Tanuma, W. R. Datars, H. Doi and A. Dunsworth, *Solid St. Commun.*, **8,** 1107 (1970)). The single crystals that were used were found to have the h.c.p. structure whereas earlier x-ray diffraction measurements had indicated that the f.c.c. structure of Yb was stable at room temperature and pressure. Indeed some of the single crystals prepared by Tanuma *et al.* did exhibit the f.c.c. structure initially at room temperature, but at low temperatures they transformed to the h.c.p. structure and on warming them to room temperature again the h.c.p. phase persisted. Thus it appears that it is the h.c.p. phase of Yb that is important. If the valence of Yb is considered to be equal to two the Fermi surface of h.c.p. Yb can be expected to bear some resemblance to those of the heavier divalent h.c.p. metals in group II. The de Haas–van Alphen frequencies obtained indicated the existence of hyperboloidal regions on the Fermi surface of Yb. It was claimed that the previous de Haas–van Alphen results of Tanuma *et al.* (1967) also apply to the h.c.p. rather than the f.c.c. phase of Yb. It was also suggested that it would be useful for a band structure calculation for h.c.p. Yb to be performed.

REFERENCES

ABARENKOV, I. V., and VEDERNIKOV, M. V., 1966, *Fizika tverd. Tela*, **8,** 236. English translation : *Soviet Phys. Solid St.*, **8,** 186.

ABATE, E., and ASDENTE, M., 1965, *Phys. Rev.*, **140,** A1303.

ABEL'SKIĬ, SH. SH., and IRKHIN, YU. P., 1968, *Fizika tverd. Tela*, **10,** 2245. English translation : *Soviet Phys. Solid St.*, **10,** 1768.

AFANAS'EVA, L. A., BOLOTIN, G. A., and NOSKOV, M. M., 1966, *Fizika Metall.*, **22,** 828. English translation : *Physics Metals Metallogr.*, **22,** (6), 25.

AHLERS, G., 1967, *J. Phys. Chem. Solids*, **28,** 525.

AHN, J., and SELLMYER, D. J., 1970 a, *Phys. Rev.* B, **1,** 1273 ; 1970 b, *Ibid.*, **1,** 1285.

AKHIEZER, A. I., BAR'YAKHTAR, V. G., and PELETMINSKII, S. V., 1968, *Spin waves* (Amsterdam : North-Holland).

AKSENOV, S. I., 1958, *Zh. éksp. teor. Fiz.*, **35,** 300. English translation : *Soviet Phys. JETP*, **8,** 207.

ALDERSON, J. E. A., FARRELL, T., and HURD, C. M., 1968, *Phys. Rev.*, **174,** 729.

ALEKSANDROV, B. N., LOMONOS, O. I., TSIVINSKIĬ, S. V., and ANTONOVA, N. N., 1967, *Zh. éksp. teor. Fiz.*, **53,** 79. English translation : *Soviet Phys. JETP*, **26,** 53.

ALEKSEEVSKIĬ, N. E., BERTEL', K. KH., and DUBROVIN, A. V., 1969, *Zh. éksp. teor. Fiz., Pis'ma*, **10,** 116. English translation : *Soviet Phys. JETP Lett.*, **10,** 74.

ALEKSEEVSKIĬ, N. E., BERTEL', K. KH., DUBROVIN, A. V., and KARSTENS, G. E., 1967, *Zh. éksp. teor. Fiz., Pis'ma*, **6,** 637. English translation : *Soviet Phys. JETP Lett.*, **6,** 132.

ALEKSEEVSKIĬ, N. E., EGOROV, V. S., KARSTENS, G. É., and KAZAK, B. N., 1962, *Zh. éksp. teor. Fiz.*, **43,** 731. English translation : *Soviet Phys. JETP*, **16,** 519.

ALEKSEEVSKIĬ, N. E., EGOROV, V. S., and KAZAK, B. N., 1963, *Zh. éksp. teor. Fiz.*, **44,** 1116. English translation : *Soviet Phys. JETP*, **17,** 752.

ALEKSEEVSKIĬ, N. E., and GAĬDUKOV, YU. P., 1958, *Zh. éksp. teor. Fiz.*, **35,** 554. English translation : *Soviet Phys. JETP*, **8,** 383 ; 1959 a, *Ibid.*, **36,** 447. English translation : *Soviet Phys. JETP*, **9,** 311 ; 1959 b, *Ibid.*, **37,** 672. English translation : *Soviet Phys. JETP*, **10,** 481 ; 1960, *Ibid.*, **38,** 1720. English translation : *Soviet Phys. JETP*, **11,** 1242 ; 1962 a, *Ibid.*, **42,** 69. English translation : *Soviet Phys. JETP*, **15,** 49 ; 1962 b, *Ibid.*, **43,** 2094. English translation : *Soviet Phys. JETP*, **16,** 1481. Erratum : *Ibid.*, **44,** 1421. English translation : *Soviet Phys. JETP*, **16,** 1488.

ALEKSEEVSKIĬ, N. E., KARSTENS, G. É., and MOZHAEV, V. V., 1964, *Zh. éksp. teor. Fiz.*, **46,** 1979. English translation : *Soviet Phys. JETP*, **19,** 1333.

ALERS, G. A., and SWIM, R. T., 1963, *Phys. Rev. Lett.*, **11,** 72.

ALIEV, N. G., and VOLKENSHTEĬN, N. V., 1965 a, *Fizika tverd. Tela*, **7,** 2560. English translation : *Soviet Phys. Solid St.*, **7,** 2068 ; 1965 b, *Zh. éksp. teor. Fiz.*, **49,** 1450. English translation : *Soviet Phys. JETP*, **22,** 997.

ALIG, R. C., and RODRIGUEZ, S., 1967, *Phys. Rev.*, **157,** 500.

ALLAN, G., LEMAN, G., and LENGLART, P., 1968, *J. Phys., Paris*, **29,** 885.

ALLEN, P. B., COHEN, M. L., FALICOV, L. M., and KASOWSKI, R. V., 1968, *Phys. Rev. Lett.*, **21,** 1794.

ALTMANN, S. L., 1958, *Proc. R. Soc.* A, **244,** 153.

ALTMANN, S. L., and BRADLEY, C. J., 1962, *Physics Lett.*, **1,** 336 ; 1964, *Phys. Rev.*, **135,** A1253 ; 1967, *Proc. phys. Soc.*, **92,** 764.

ALTMANN, S. L., DAVIES, B. L., and HARFORD, A. R., 1968, *J. Phys.* C (*Solid St. Phys.*), **1,** 1633.

AMITIN, E. B., and KOVALEVSKAYA, YU. A., 1968, *Fizika tverd. Tela*, **10,** 1884. English translation : *Soviet Phys. Solid St.*, **10,** 1483.

AMITIN, E. B., KOVALEVSKAYA, YU. A., and KOVDRYA, YU. Z., 1967, *Fizika tverd. Tela*, **9,** 905. English translation : *Soviet Phys. Solid St.*, **9,** 704.

ANDERSEN, O. K., and LOUCKS, T. L., 1968, *Phys. Rev.*, **167,** 551.

ANDERSEN, O. K., and MACKINTOSH, A. R., 1968, *Solid St. Commun.*, **6,** 285.

ANDERSON, G. S., LEGVOLD, S., and SPEDDING, F. H., 1958, *Phys. Rev.*, **111,** 1257.

ANDERSON, J. R., and GOLD, A. V., 1963, *Phys. Rev. Lett.*, **10,** 227.

ANDERSON, J. R., MCCAFFREY, J. W., and PAPACONSTANTOPOULOS, D. A., 1969, *Solid St. Commun.*, **7,** 1439.

ANDRES, K., and JENSEN, M. A., 1968, *Phys. Rev.*, **165,** 533.

ANDREW, E. R., 1949, *Proc. phys. Soc.* A, **62,** 77.

ANIMALU, A. O. E., and HEINE, V., 1965, *Phil. Mag.*, **12,** 1249.

ANTONIEWICZ, P. A., WOOD, L. T., and GAVENDA, J. D., 1968, *Phys. Rev. Lett.*, **21,** 998.

ARAJS, S., 1960 a, *Phys. Rev.*, **120,** 756 ; 1960 b, *J. Chem. Phys.*, **32,** 951.

ARAJS, S., and COLVIN, R. V., 1964 a, *J. appl. Phys.*, **35,** 1043 ; 1964 b, *Phys. Rev.*, **136,** A439.

ARAJS, S., and DUNMYRE, G. R., 1965 a, *J. appl. Phys.*, **36,** 3555 ; 1965 b, *Physica, 's Grav.*, **31,** 1466.

ARAJS, S., and MILLER, D. S., 1960, *J. appl. Phys.*, **31,** 325S.

ARKO, A. J., and MARCUS, J. A., 1964, *Proc. 9th Int. Conf. Low Temp. Phys.*, p. 748.

ARKO, A. J., MARCUS, J. A., and REED, W. A., 1966, *Physics Lett.*, **23,** 617 ; 1968, *Phys. Rev.*, **176,** 671 ; 1969, *Ibid.*, **185,** 901.

ARP, V., EDMONDS, D., and PETERSEN, R., 1959, *Phys. Rev. Lett.*, **3,** 212.

ARROTT, A., and COLES, B. R., 1961, *J. appl. Phys.*, **32,** 51S.

ASANO, S., and YAMASHITA, J., 1967, *J. phys. Soc. Japan*, **23,** 714.

ASDENTE, M., 1962, *Phys. Rev.*, **127,** 1949.

ASDENTE, M., and DELITALA, M., 1967, *Phys. Rev.*, **163,** 497.

ASDENTE, M., and FRIEDEL, J., 1961, *Phys. Rev.*, **124,** 384. Erratum : *Phys. Rev.*, **126,** 2262 (1962).

BABUSHKINA, N. A., 1964, *Dokl. Akad. Nauk SSSR*, **155,** 1290. English translation : *Soviet Phys. Dokl.*, **9,** 299 ; 1965 a, *Fizika tverd. Tela*, **7,** 2540. English translation : *Soviet Phys. Solid St.*, **7,** 2048 ; 1965 b, *Ibid.*, **7,** 3026. English translation : *Soviet Phys. Solid St.*, **7,** 2450.

BAGCHI, D. K., and CULLITY, B. D., 1967, *J. appl. Phys.*, **38,** 999.

BALAIN, K. S., GRENIER, C. G., and REYNOLDS, J. M., 1960, *Phys. Rev.*, **119,** 935.

BALLINGER, R. A., and MARSHALL, C. A. W., 1967, *Proc. phys. Soc.*, **91,** 203 ; 1969, *J. Phys.* C (*Solid St. Phys.*), **2,** 1822.

BAMBAKIDIS, G., 1970, *J. Phys. Chem. Solids*, **31,** 503.

BARDEEN, J., COOPER, L. N., and SCHRIEFFER, J. R., 1957, *Phys. Rev.*, **108,** 1175.

BARNES, R. G., BORSA, F., SEGEL, S. L., and TORGESON, D. R., 1965, *Phys. Rev.*, **137,** A1828.

BARRETT, C. S., 1957, *Acta crystallogr.*, **10,** 58.

BATTERMAN, B. W., CHIPMAN, D. R., and DEMARCO, J. J., 1961, *Phys. Rev.*, **122,** 68.

BAZAN, C., 1965, *Acta Phys. Polon*, **28,** 777.

BEAGLEHOLE, D., 1965, *Proceedings of the International Colloquium on Optical Properties and Electronic Structure of Metals and Alloys*, Paris (1965), edited by F. Abelès (Amsterdam : North-Holland), p. 154 ; 1966, *Proc. phys. Soc.*, **87,** 461.

BEATTIE, A. G., 1969, *Phys. Rev.*, **184,** 668.

BEHRENDT, D. R., LEGVOLD, S., and SPEDDING, F. H., 1958, *Phys. Rev.*, **109,** 1544.

BELLESSA, G., REICH, R., and MERCOUROFF, W., 1969, *J. Phys., Paris*, **30,** 823.

BELOV, K. P., BUROV, I. V., ERGIN, YU. V., PED'KO, A. V., and SAVITSKIĬ, E. M., 1964, *Zh. éksp. teor. Fiz.*, **47,** 860. English translation : *Soviet Phys. JETP*, **20,** 574.

BELOV, K. P., LEVITIN, R. Z., NIKITIN, S. A., and PED'KO, A. G., 1961, *Zh. éksp. teor. Fiz.*, **40,** 1562. English translation : *Soviet Phys. JETP*, **13,** 1096.

BELOV, K. P., and NIKITIN, S. A., 1962, *Fizika Metall.*, **13,** 43. English translation : *Physics Metals Metallogr.*, **13,** (1), 39.

BELOV, K. P., and PED'KO, A. V., 1962, *Zh. éksp. teor. Fiz.*, **42,** 87. English translation : *Soviet Phys. JETP*, **15,** 62.

BELSON, H. S., 1966, *J. appl. Phys.*, **37,** 1348.

BENNEMANN, K. H., 1967, *Physics Lett.* A, **25,** 233.

BENNETT, A. J., and FALICOV, L. M., 1964, *Phys. Rev.*, **136,** A998.

BENNETT, H. E., BENNETT, J. M., ASHLEY, E. J., and MOTYKA, R. J., 1968, *Phys. Rev.*, **165,** 755.

BERG, W. T., 1969, *J. Phys. Chem. Solids*, **30,** 69.

BERGERON, C. J., GRENIER, C. G., and REYNOLDS, J. M., 1960, *Phys. Rev.*, **119,** 925.

BERGLUND, C. N., and SPICER, W. E., 1964 a, *Phys. Rev.*, **136,** A1030 ; 1964 b, *Ibid.*, **136,** A1044.

BERGWALL, S., 1965, *Physics Lett.*, **19,** 539.

BERKO, S., CUSHNER, S., and ERSKINE, J. C., 1968, *Physics Lett.* A, **27,** 668.

BERKO, S., and PLASKETT, J. S., 1958, *Phys. Rev.*, **112,** 1877.

BERKO, S., and WEGER, M., 1970, *Phys. Rev. Lett.*, **24,** 55.

BERKO, S., and ZUCKERMAN, J., 1964, *Phys. Rev. Lett.*, **13,** 339a.

BERLINCOURT, T. G., 1954, *Phys. Rev.*, **94,** 1172 ; 1959, *Ibid.*, **114,** 969.

BERLINCOURT, T. G., and STEELE, M. C., 1954, *Phys. Rev.*, **95,** 1421.

BERTHEL, K. H., 1964, *Phys. Stat. Sol.*, **7,** 633.

BEZUGLYĬ, P. A., ZHEVAGO, S. E., and DENISENKO, V. I., 1965, *Zh. éksp. teor. Fiz.*, **49,** 1457. English translation : *Soviet Phys. JETP*, **22,** 1002.

BHATNAGAR, S., 1969, *Phys. Rev.*, **183,** 657.

BIONDI, M. A., and GUOBADIA, A. I., 1968, *Phys. Rev.*, **166,** 667.

BIRSS, R. R., 1960, *Proc. R. Soc.* A, **255,** 398 ; 1964, *Symmetry and Magnetism* (Amsterdam : North-Holland).

BLODGETT, A. J., and SPICER, W. E., 1965, *Phys. Rev. Lett.*, **15,** 29 ; 1966, *Phys. Rev.*, **146,** 390 ; 1967, *Ibid.*, **158,** 514.

BLODGETT, A. J., SPICER, W. E., and YU, A. Y. C., 1965, *Proceedings of the International Colloquium on Optical Properties and Electronic Structure of Metals and Alloys*, Paris (1965), edited by F. Abelès (Amsterdam : North-Holland), p. 246.

BLUMBERG, W. E., EISENGER, J., JACCARINO, V., and MATTHIAS, B. T., 1960, *Phys. Rev. Lett.*, **5,** 52.

BOERSTOEL, B. M., DU CHATENIER, F. J., and VAN DEN BERG, G. J., 1964, *Proc. 9th Int. Conf. Low Temp. Phys.*, p. 1071.

BOERSTOEL, B. M., VAN DISSEL, W. J. J., and JACOBS, M. B. M., 1968, *Physica, 's Grav.*, **38,** 287.

BOGLE, T. E., COON, J. B., and GRENIER, C. G., 1969, *Phys. Rev.*, **177,** 1122.

BOGOD, YU. A., and EREMENKO, V. V., 1965, *Phys. Stat. Sol.*, **11,** K51.

BOHM, H. V., and EASTERLING, V. J., 1962, *Phys. Rev.*, **128,** 1021.

BOHM, H. V., and MACKINNON, L., 1964, *Proc. 9th Int. Conf. Low Temp. Phys.*, p. 786.

BOĬKO, V. V., GASPAROV, V. A., and GVERDTSITELI, I. G., 1967, *Zh. éksp. teor. Fiz., Pis'ma*, **6,** 737. English translation : *Soviet Phys. JETP Lett.*, **6,** 212 ; 1969, *Zh. éksp. teor. Fiz.*, **56,** 489. English translation : *Soviet Phys. JETP*, **29,** 267.

BOLEF, D. I., DE CLERK, J., and BRANDT, C. B., 1961, *Bull. Am. phys. Soc.*, **7,** 236.
BOLOTIN, G. A., and CHUKINA, T. P., 1967, *Optika Spektrosk.*, **23,** 620. English translation : *Optics Spectrosc.*, **23,** 333.
BOLOTIN, G. A., KIRILLOVA, M. M., NOMEROVANNAYA, L. V., and NOSKOV, M. M., 1967, *Fizika Metall.*, **23,** 463. English translation : *Physics Metals Metallogr.*, **23,** (3), 72.
BOORSE, H. A., and HIRSHFELD, A., 1958, *Physica, 's Grav.*, **24,** S148.
BORISOV, N. D., and NEMOSHKALENKO, V. V., 1961, *Izv. Akad. Nauk SSSR, Ser. Fiz.*, **25,** 1002. English translation : *Bull. Acad. Sci. USSR, Phys. Ser.*, **25,** 1011.
BOROVIK, E. S., 1950, *Dokl. Akad. Nauk SSSR*, **70,** 601 ; 1956, *Zh. éksp. teor. Fiz.*, **30,** 262. English translation : *Soviet Phys. JETP*, **3,** 243.
BOROVIK, E. S., and VOLOTSKAYA, V. G., 1958, *Fizika Metall.*, **6,** 60. English translation : *Physics Metals Metallogr.*, **6,** (1), 56 ; 1959, *Zh. éksp. teor. Fiz.*, **36,** 1650. English translation : *Soviet Phys. JETP*, **9,** 1175.
BORSA, F., and BARNES, R. G., 1966, *J. Phys. Chem. Solids*, **27,** 567.
BOSNELL, J. R., and MYERS, A., 1964, *Physics Lett.*, **12,** 297.
BOYD, J. R., and GAVENDA, J. D., 1966, *Phys. Rev.*, **152,** 645.
BOYLE, D. J., and GOLD, A. V., 1969, *Phys. Rev. Lett.*, **22,** 461.
BOYS, D. W., and LEGVOLD, S., 1968, *Phys. Rev.*, **174,** 377.
BRADLEY, C. J., and CRACKNELL, A. P., 1970, *J. Phys.* C (*Solid St. Phys.*), **3,** 610.
BRADLEY, C. J., and DAVIES, B. L., 1968, *Rev. mod. Phys.*, **40,** 359.
BRANDT, G. B., and RAYNE, J. A., 1962, *Physics Lett.*, **3,** 148 ; 1963, *Phys. Rev.*, **132,** 1945 ; 1965, *Physics Lett.*, **15,** 18 ; 1966, *Phys. Rev.*, **148,** 644.
BREWER, L., 1963, *Electronic structure and alloy chemistry of the transition elements*, edited by P. A. Beck (New York : Interscience), p. 211 ; 1964, *High Strength Materials*, Proceedings of the Second Berkeley International Materials Conference, ' High-Strength Materials—Present status and anticipated developments ', held at the University of California, Berkeley, June 15–18, 1964, edited by V. F. Zackay (New York : Wiley), p. 12 ; 1967, *Acta metall.*, **15,** 553 ; 1968, *Science*, **161,** 115.
BROSS, H., and JUNGINGER, H. G., 1963, *Physics Lett.*, **8,** 240.
BRUN, T. O., and LANDER, G. H., 1969, *Phys. Rev. Lett.*, **23,** 1295.
BUCHER, E., CHU, C. W., MAITA, J. P., ANDRES, K., COOPER, A. S., BUEHLER, E., and NASSAU, K., 1969, *Phys. Rev. Lett.*, **22,** 1260.
BUDWORTH, D. W., HOARE, F. E., and PRESTON, J., 1960, *Proc. R. Soc.* A, **257,** 250.
BURDICK, G. A., 1961, *Phys. Rev. Lett.*, **7,** 156 ; 1963, *Phys. Rev.*, **129,** 138.
BURGER, I. P., and TAYLOR, M. A., 1961, *Phys. Rev. Lett.*, **6,** 185.
BUSEY, R. H., and GIAUQUE, W. F., 1952, *J. Am. chem. Soc.*, **74,** 3157.
BUTLER, F. A., BLOOM, F. K., and BROWN, E., 1969, *Phys. Rev.*, **180,** 744.
CABLE, J. W., MOON, R. M., KOEHLER, W. C., and WOLLAN, E. O., 1964, *Phys. Rev. Lett.*, **12,** 553.
CABLE, J. W., and WOLLAN, E. O., 1968, *Phys. Rev.*, **165,** 733.
CABLE, J. W., WOLLAN, E. O., KOEHLER, W. C., and WILKINSON, M. K., 1961, *J. appl. Phys.*, **32,** 49S.
CALLAWAY, J., 1955, *Phys. Rev.*, **99,** 500 ; 1959, *Ibid.*, **115,** 346 ; 1960, *Ibid.*, **120,** 731.
CALLAWAY, J., and ZHANG, H. M., 1970, *Phys. Rev.* B, **1,** 305.
CAPLIN, A. D., and SHOENBERG, D., 1965, *Physics Lett.*, **18,** 238.
CAROLINE, D., and SCHIRBER, J. E., 1963, *Phil. Mag.*, **8,** 71.
CHAMBERS, R. G., 1952, *Proc. R. Soc.* A, **215,** 481 ; 1956, *Ibid.*, **238,** 344.
CHARI, M. S. R., 1967, *Phys. Stat. Sol.*, **19,** 169.
CHATTERJEE, S., and CHAKRABORTY, D. K., 1967, *Indian J. Phys.*, **41,** 134.

CHATTERJEE, S., and SEN, S. K., 1964, *Indian J. pure appl. Phys.*, **2,** 390 ; 1966, *Proc. phys. Soc.*, **87,** 779 ; 1967, *Ibid.*, **91,** 749 ; 1968, *J. Phys.* C (*Solid St. Phys.*), **1,** 759.

CHECHERNIKOV, V. I., POP, I., TEREKHOVA, V. F., and KOLESNICHENKO, V. E., 1964, *Zh. éksp. teor. Fiz.*, **46,** 444. English translation : *Soviet Phys. JETP*, **19,** 298.

CHEN, T. W., MA, I. J., and CHIEN, J. P., 1967, *Chinese J. Phys., Taiwan*, **5,** 31.

CHEN WANG, A. L., 1966, *Chinese J. Phys., Taiwan*, **4,** 71.

CHENG, C. H., WEI, C. T., and BECK, P. A., 1960, *Phys. Rev.*, **120,** 426.

CHEREMUSHKINA, A. V., and VASIL'EVA, R. P., 1966, *Fizika tverd. Tela*, **8,** 822. English translation : *Soviet Phys. Solid St.*, **8,** 659.

CHOLLET, L. F., and TEMPLETON, I. M., 1968, *Phys. Rev.*, **170,** 656.

CHOPRA, K. L., and BAHL, S. K., 1967, *J. appl. Phys.*, **38,** 3607.

CHOU, C., WHITE, D., and JOHNSTON, H. L., 1958, *Phys. Rev.*, **109,** 788.

CHRISTENSEN, N. E., 1969, *Phys. Stat. Sol.*, **31,** 635.

CHRISTENSEN, N. E., and SERAPHIN, B. O., 1970, *Solid St. Commun.*, **8,** 1221.

CHRISTOPHER, J. E., ISIN, A., and COLEMAN, R. V., 1967, *J. appl. Phys.*, **38,** 1322.

ČÍŽEK, A., and ADAM, J., 1969, *Česk. Časopis Fis.* A, **19,** 139.

CLARK, A. F., and POWELL, R. L., 1968, *Phys. Rev. Lett.*, **21,** 802.

CLAUS, H., and ULMER, K., 1963, *Z. Phys.*, **173,** 462 ; 1964, *Physics Lett.*, **12,** 170 ; 1965, *Z. Phys.*, **185,** 139.

CLOGSTON, A. M., JACCARINO, V., and YAFET, V., 1964, *Phys. Rev.*, **134,** A650.

COHEN, M. H., and FALICOV, L. M., 1960, *Phys. Rev. Lett.*, **5,** 544 ; 1961, *Ibid.*, **7,** 231.

COHEN, M. H., and HEINE, V., 1958, *Adv. Phys.*, **7,** 395.

COLEMAN, R. V., and ISIN, A., 1966, *J. appl. Phys.*, **37,** 1028.

COLERIDGE, P. T., 1965, *Physics Lett.*, **15,** 223 ; 1966 a, *Ibid.*, **22,** 367 ; 1966 b, *Proc. R. Soc.* A, **295,** 458.

COLES, B. R., 1958, *Adv. Phys.*, **7,** 40.

COLLINS, J. G., 1966, *Ann. Acad. Sci. Fennicae, A VI, Physica*, **210,** 239.

COLOMBANI, A., and GOUREAUX, G., 1959, *C. r. hebd. Séanc. Acad. Sci., Paris*, **249,** 381.

COLQUITT, L., and GOODINGS, D. A., 1964, *Proc. Int. Conf. Magnetism, Nottingham*, p. 29.

COLVIN, R. V., and ARAJS, S., 1963, *J. appl. Phys.*, **34,** 286 ; 1964, *Phys. Rev.*, **133,** A1076.

COLVIN, R. V., ARAJS, S., and PECK, J. M., 1961, *Phys. Rev.*, **122,** 14.

COLVIN, R. V., LEGVOLD, S., and SPEDDING, F. H., 1960, *Phys. Rev.*, **120,** 741.

COMBLEY, F. H., 1968, *Acta crystallogr.* B, **24,** 142.

CONDON, J. H., 1966, *Phys. Rev.*, **145,** 526.

CONDON, J. H., and WALSTEDT, R. E., 1968, *Phys. Rev. Lett.*, **21,** 612.

CONNOLLY, J. W. D., 1967, *Phys. Rev.*, **159,** 415.

COOPER, B. R., 1965, *Phys. Rev.*, **139,** A1506 ; 1967, *Phys. Rev. Lett.*, **19,** 900.

COOPER, B. R., and EHRENREICH, H., 1964, *Solid St. Commun.*, **2,** 171.

COOPER, B. R., EHRENREICH, H., and HODGES, L., 1964, *Proc. Int. Conf. Magnetism*, Nottingham, p. 110.

CORNWELL, J. F., 1961, *Phil. Mag.*, **6,** 727.

CORNWELL, J. F., HUM, D. M., and WONG, K. C., 1968, *Physics Lett.* A, **26,** 365.

CORNWELL, J. F., and WOHLFARTH, E. P., 1962, *J. phys. Soc. Japan*, **17,** Suppl. B-I, 32.

COSTELLO, J. M., and WEYMOUTH, J. W., 1969, *Phys. Rev.*, **184,** 694.

CRACKNELL, A. P., 1964, Thesis, Oxford University ; 1967, *Contemp. Phys.*, **8,** 459 ; 1969 a, *J. Phys.* C (*Solid St. Phys.*), **2,** 1425 ; 1969 b, *Rep. Prog. Phys.*, **32,** 633 ; 1969 c, *Adv. Phys.*, **18,** 681 ; 1970, *Phys. Rev.* B, **1,** 1261.

Crangle, J., and Smith, T. F., 1962, *Phys. Rev. Lett.*, **9,** 86.
Cullen, J. R., and Callen, E., 1968, *Physics Lett.* A, **28,** 20.
Curry, M. A., Legvold, S., and Spedding, F. H., 1960, *Phys. Rev.*, **117,** 953.
Daniel, M. R., and Mackinnon, L., 1962, *Proc. 8th Int. Conf. Low Temp. Phys.*, p. 202 ; 1963, *Phil. Mag.*, **8,** 537.
Dash, J. G., Taylor, R. D., and Craig, P. P., 1960, *Proc. 7th Int. Conf. Low Temp. Phys.*, p. 705.
Datars, W. R., and Dixon, A. E., 1967, *Phys. Rev.*, **154,** 576.
Datars, W. R., and Tanuma, S., 1968, *Physics Lett.* A, **27,** 182.
Davis, H. L., 1968, *Physics Lett.* A, **28,** 85.
Davis, H. L., Faulkner, J. S., and Joy, H. W., 1968, *Phys. Rev.*, **167,** 601.
Deaton, B. C., 1962, Thesis, University of Texas ; 1963, *Physics Lett.*, **7,** 7 ; 1965, *Phys. Rev.*, **140,** A2051.
Deaton, B. C., and Gavenda, J. D., 1963, *Phys. Rev.*, **129,** 1990 ; 1964, *Ibid.*, **136,** A1096.
De Benedetti, S., Cowan, C. E., Konneker, W. R., and Primakoff, H., 1950, *Phys. Rev.*, **77,** 205.
De Cicco, P. D., and Kitz, A., 1967, *Phys. Rev.*, **162,** 486.
Deegan, R. A., 1968, *J. Phys.* C (*Solid St. Phys.*), **1,** 763.
Deegan, R. A., and Twose, W. D., 1967, *Phys. Rev.*, **164,** 993.
Dekhtyar, I. Ya., and Mikhalenkov, V. S., 1961, *Dokl. Akad. Nauk SSSR*, **140,** 1293. English translation : *Soviet Phys. Dokl.*, **6,** 917.
Dekhtyar, I. Ya., Mikhalenkov, V. S., and Sakharova, S. G., 1966, *Dokl. Akad. Nauk SSSR*, **168,** 785. English translation : *Soviet Phys. Dokl.*, **11,** 537.
De Launay, J., Dolecek, R. L., and Webber, R. T., 1959, *J. Phys. Chem. Solids*, **11,** 37.
Dempesy, C. W., Gordon, J. E., and Soller, T., 1966, *Ann. Acad. Sci. Fennicae, A VI, Physica*, **210,** 204.
Dheer, P. N., 1967, *Phys. Rev.*, **156,** 637.
Dhillon, J. S., and Shoenberg, D., 1955, *Phil. Trans. R. Soc.* A, **248,** 1.
Dimmock, J. O., and Freeman, A. J., 1964, *Phys. Rev. Lett.*, **13,** 750.
Dingle, R. B., and Shoenberg, D., 1950, *Nature, Lond.*, **166,** 652.
Dishman, J. M., and Rayne, J. A., 1966, *Physics Lett.*, **20,** 348 ; 1968, *Phys. Rev.*, **166,** 728.
Dixon, A. E., and Datars, W. R., 1965, *Solid St. Commun.*, **3,** 377 ; 1968, *Phys. Rev.*, **175,** 928.
Dixon, M., Hoare, F. E., Holden, T. M., and Moody, D. E., 1965, *Proc. R. Soc.* A, **285,** 561.
Dmitrenko, I. M., Verkin, B. I., and Lazarev, B. G., 1957, *Zh. éksp. teor. Fiz.*, **33,** 287. English translation : *Soviet Phys. JETP*, **6,** 223 ; 1958, *Ibid.*, **35,** 328. English translation : *Soviet Phys. JETP*, **8,** 229.
Dresselhaus, G., 1969, *Solid St. Commun.*, **7,** 419.
Dreyfus, B., Goodman, B. B., Lacaze, A., and Trolliet, G., 1961, *C. r. hebd. Séanc. Acad. Sci., Paris*, **253,** 1764.
Dreyfus, B., Goodman, B. B., Trolliet, G., and Weil, L., 1961, *C. r. hebd. Séanc. Acad. Sci., Paris*, **253,** 1085.
Dreyfus, B., Michel, J. C., and Thoulouze, D., 1967, *Physics Lett.* A, **24,** 457.
Drickamer, H. G., 1965, *Solid St. Phys.*, **17,** 1.
Dutta Roy, S. K., and Subrahmanyam, A. V., 1969, *Phys. Rev.*, **177,** 1133.
Easterling, V. J., and Bohm, H. V., 1962, *Phys. Rev.*, **125,** 812.
Eastman, D. E., 1969 a, *Solid St. Commun.*, **7,** 1697 ; 1969 b, *J. appl. Phys.*, **40,** 1387.
Eastman, D. E., and Krolikowski, W. F., 1968, *Phys. Rev. Lett.*, **21,** 623.
Edelmann, F., and Ulmer, K., 1967, *Z. Phys.*, **205,** 476.

EDWARDS, L. R., and LEGVOLD, S., 1968, *Phys. Rev.*, **176,** 753.
EGGS, H., and ULMER, K., 1968, *Physics Lett.* A, **26,** 246.
EHRENREICH, H., 1965, *Proceedings of the International Colloquium on Optical Properties and Electronic Structure of Metals and Alloys*, Paris (1965), edited by F. Abelès (Amsterdam : North-Holland), p. 109.
EHRENREICH, H., PHILIPP, H. R., and OLECHNA, D. J., 1963, *Phys. Rev.*, **131,** 2469.
EHRLICH, A. C., HUGUENIN, R., and RIVIER, D., 1967, *J. Phys. Chem. Solids*, **28,** 253.
EHRLICH, A. C., and RIVIER, D., 1968, *J. Phys. Chem. Solids*, **29,** 1293.
ELLIOTT, J. F., LEGVOLD, S., and SPEDDING, F. H., 1954, *Phys. Rev.*, **94,** 1143.
ELLIOTT, R. J., 1954, *Phys. Rev.*, **94,** 564.
ELLIOTT, R. J., and WEDGWOOD, F. A., 1963, *Proc. phys. Soc.*, **81,** 846 ; 1964, *Ibid.*, **84,** 63.
ELLIOTT, R. O., and MINER, W. N., 1967, *Trans. metall. Soc. AIME.*, **239,** 166.
ENGEL, N., 1967, *Acta metall.*, **15,** 557.
ERSHOV, O. A., and BRYTOV, I. A., 1967, *Optika Spektrosk.*, **22,** 305. English translation : *Optics Spectrosc.*, **22,** 165.
ESTERMANN, I., FRIEDBERG, S. A., and GOLDMAN, J. E., 1952, *Phys. Rev.*, **87,** 582.
EVENSON, W. E., and LIU, S. H., 1968, *Phys. Rev. Lett.*, **21,** 432 ; 1969, *Phys. Rev.*, **178,** 783.
FALICOV, L. M., PIPPARD, A. B., and SIEVERT, P. R., 1966, *Phys. Rev.*, **151,** 498.
FALICOV, L. M., and RUVALDS, J., 1968, *Phys. Rev.*, **172,** 498.
FALICOV, L. M., and ZUCKERMANN, M. J., 1967, *Phys. Rev.*, **160,** 372.
FARACI, G., and TURRISI, E., 1968, *Nuovo Cim.* B, **58,** 308.
FAULKNER, J. S., DAVIS, H. L., and JOY, H. W., 1967, *Phys. Rev.*, **161,** 656.
FAWCETT, E., 1956, *Phys. Rev.*, **103,** 1582 ; 1961 a, *Phys. Rev. Lett.*, **6,** 534 ; 1961 b, *J. Phys. Chem. Solids*, **18,** 320 ; 1961 c, *Phys. Rev. Lett.*, **7,** 370 ; 1962, *Phys. Rev.*, **128,** 154.
FAWCETT, E., and GRIFFITHS, D., 1962, *J. Phys. Chem. Solids*, **23,** 1631.
FAWCETT, E., and REED, W. A., 1962, *Phys. Rev. Lett.*, **9,** 336 ; 1963, *Phys. Rev.*, **131,** 2463 ; 1964 a, *Ibid.*, **134,** A723 ; 1964 b, *Proc. 9th Int. Conf. Low Temp. Phys.*, p. 782.
FAWCETT, E., REED, W. A., and SODEN, R. R., 1967, *Phys. Rev.*, **159,** 533.
FAWCETT, E., and WALSH, W. M., 1962, *Phys. Rev. Lett.*, **8,** 476.
FENTON, E. W., and WOODS, S. B., 1966, *Phys. Rev.*, **151,** 424.
FÉRON, J.-L., and PAUTHENET, R., 1969, *C. r. hebd. Séanc. Acad. Sci., Paris* B, **269,** 549.
FERREIRA DA SILVA, J., DUYKEREN, N. W. J., and DOKOUPIL, Z., 1966, *Physica, 's Grav.*, **32,** 1253.
FINK, H. J., 1964, *Physics Lett.*, **13,** 105.
FLEMING, G. S., LIU, S. H., and LOUCKS, T. L., 1968, *Phys. Rev. Lett.*, **21,** 1524 ; 1969, *J. appl. Phys.*, **40,** 1285.
FLEMING, G. S., and LOUCKS, T. L., 1968, *Phys. Rev.*, **173,** 685.
FLETCHER, G. C., 1952, *Proc. phys. Soc.* A, **65,** 192 ; 1969, *J. Phys.* C (*Solid St. Phys.*), **2,** 1440.
FLETCHER, R., MACKINNON, L., and WALLACE, W. D., 1967, *Physics Lett.* A, **25,** 395 ; 1969, *Phil. Mag.*, **20,** 245.
FLIPPEN, R. B., 1964, *J. appl. Phys.*, **35,** 1047.
FLOTOW, H. E., and OSBORNE, D. W., 1967, *Phys. Rev.*, **160,** 467.
FONG, C. Y., and COHEN, M. L., 1970, *Phys. Rev. Lett.*, **24,** 306.
FRANKEN, B., and VAN DEN BERG, G. J., 1960, *Physica, 's Grav.*, **26,** 1030.
FREEMAN, A. J., DIMMOCK, J. O., and WATSON, R. E., 1966 a, *Quantum theory of atoms, molecules and the solid state*, a tribute to John C. Slater, edited by P. O. Löwdin (New York : Academic Press), p. 361 ; 1966 b, *Phys. Rev. Lett.*, **16,** 94.

FREEMAN, A. J., FURDYNA, A. M., and DIMMOCK, J. O., 1966, *J. appl. Phys.*, **37,** 1256.

FRIEDBERG, S. A., ESTERMANN, I., and GOLDMAN, J. E., 1952, *Phys. Rev.*, **85,** 375.

FRIEDEL, J., 1969, *The physics of metals. I. Electrons*, edited by J. M. Ziman (Cambridge University Press), Chapter 8.

FRIEDEL, J., LENGLART, P., and LEMAN, G., 1964, *J. Phys. Chem. Solids*, **25,** 781.

FUCHS, K., 1936 a, *Proc. R. Soc.* A, **151,** 585 ; 1936 b, *Ibid.*, **153,** 622.

FUNES, A. J., and COLEMAN, R. V., 1963, *Phys. Rev.*, **131,** 2084.

FUJIWARA, K., 1965, *J. phys. Soc. Japan*, **20,** 1533.

FUJIWARA, K., and SUEOKA, O., 1966, *J. phys. Soc. Japan*, **21,** 1947 ; 1967, *Ibid.*, **23,** 1242.

GAĬDUKOV, YU. P., 1959, *Zh. éksp. teor. Fiz.*, **37,** 1281. English translation : *Soviet Phys. JETP*, **10,** 913.

GAĬDUKOV, YU. P., and ITSKEVICH, E. S., 1963, *Zh. éksp. teor. Fiz.*, **45,** 71. English translation : *Soviet Phys. JETP*, **18,** 51.

GAĬDUKOV, YU. P., and KRECHETOVA, I. P., 1965 a, *Zh. éksp. teor. Fiz., Pis'ma*, **1,** (3), 25. English translation : *Soviet Phys. JETP Lett.*, **1,** 88 ; 1965 b, *Zh. éksp. teor. Fiz.*, **49,** 1411. English translation : *Soviet Phys. JETP*, **22,** 971 ; 1966, *Zh. éksp. teor. Fiz., Pis'ma*, **3,** 227. English translation : *Soviet Phys. JETP Lett.*, **3,** 144.

GALKIN, A. A., and KOROLYUK, A. P., 1960, *Zh. éksp. teor. Fiz.*, **38,** 1688. English translation : *Soviet Phys. JETP*, **11,** 1218.

GALLO, C. F., 1965, *J. appl. Phys.*, **36,** 3410.

GALT, J. K., and MERRITT, F. R., 1960, *The Fermi Surface.* Proceedings of a Conference held at Cooperstown, New York, on 22–24 August 1960, edited by W. A. Harrison and M. B. Webb (New York : Wiley), p. 159.

GALT, J. K., MERRITT, F. R., and KLAUDER, J. R., 1965, *Phys. Rev.*, **139,** A823.

GALT, J. K., MERRITT, F. R., and SCHMIDT, P. H., 1961, *Phys. Rev. Lett.*, **6,** 458.

GALT, J. K., MERRITT, F. R., YAGER, W. A., and DAIL, H. W., 1959, *Phys. Rev. Lett.*, **2,** 292.

GARCÍA-MOLINER, F., 1958, *Phil. Mag.*, **3,** 207.

GARDNER, W. E., and PENFOLD, J., 1965, *Phil. Mag.*, **11,** 549 ; 1968, *Physics Lett.* A, **26,** 204.

GARDNER, W. E., PENFOLD, J., and TAYLOR, M. A., 1965, *Proc. phys. Soc.*, **85,** 963.

GARLAND, J. W., BENNEMANN, K. H., and MUELLER, F. M., 1968, *Phys. Rev. Lett.*, **21,** 1315.

GAVENDA, J. D., and CHANG, F. H. S., 1969, *Phys. Rev.*, **186,** 630.

GAVENDA, J. D., and DEATON, B. C., 1962, *Phys. Rev. Lett.*, **8,** 208.

GERHARDT, U., 1968, *Phys. Rev.*, **172,** 651.

GERHARDT, U., BEAGLEHOLE, D., and SANDROCK, R., 1967, *Phys. Rev. Lett.*, **19,** 309.

GERSDORF, R., 1962, *J. Phys., Paris*, **23,** 726.

GERSTEIN, B. C., JELINEK, F. J., MULLALY, J. R., SHICKELL, W. D., and SPEDDING, F. H., 1967, *J. chem. Phys.*, **47,** 5194.

GIBBONS, D. F., 1961, *Phil. Mag.*, **6,** 445.

GIBBONS, D. F., and FALICOV, L. M., 1963, *Phil. Mag.*, **8,** 177.

GILAT, G., and RAUBENHEIMER, L. P., 1966, *Phys. Rev.*, **144,** 390.

GIRVAN, R. F., GOLD, A. V., and PHILLIPS, R. A., 1968, *J. Phys. Chem. Solids*, **29,** 1485.

GLASSER, M. L., 1962, *Revta mex. Fís.*, **11,** 31.

GMELIN, E., and GOBRECHT, K. H., 1967, *Z. Angew. Phys.*, **24,** 21.

GOLD, A. V., 1964, *Proc. Int. Conf. Magnetism*, Nottingham, p. 124 ; 1968, *J. appl. Phys.*, **39,** 768.

GOLOVASHKIN, A. I., LEKSINA, I. E., MOTULEVICH, G. P., and SHUBIN, A. A., 1969, *Zh. éksp. teor. Fiz.*, **56,** 51. English translation : *Soviet Phys. JETP*, **29,** 27.

GOLUBKOV, A. V., DEVYATKOVA, E. D., ZHUZE, V. P., SERGEEVA, V. M., and SMIRNOV, I. A., 1966, *Fizika tverd. Tela*, **8,** 1761. English translation : *Soviet Phys. Solid St.*, **8,** 1403.

GOODRICH, R. G., and JONES, R. C., 1967, *Phys. Rev.*, **156,** 745.

GOUREAUX, G., and COLOMBANI, A., 1959, *C. r. hebd. Séanc. Acad. Sci., Paris*, **248,** 543 ; 1960, *Ibid.*, **250,** 4310.

GOUREAUX, G., HUET, P., and COLOMBANI, A., 1958, *C. r. hebd. Séanc. Acad. Sci., Paris*, **247,** 189.

GRAEBNER, J. E., and GREINER, E. S., 1969, *Phys. Rev.*, **185,** 992.

GRAEBNER, J. E., and MARCUS, J. A., 1966, *J. appl. Phys.*, **37,** 1262 ; 1968, *Phys. Rev.*, **175,** 659.

GRASSIE, A. D. C., 1964, *Phil. Mag.*, **9,** 847.

GRAVES, R. H. W., and LENHAM, A. P., 1968 a, *J. Opt. Soc. Am.*, **58,** 126 ; 1968 b, *Ibid.*, **58,** 884.

GREEN, B. A., and CULBERT, H. V., 1965, *Phys. Rev.*, **137,** A1168.

GREEN, R. W., LEGVOLD, S., and SPEDDING, F. H., 1961, *Phys. Rev.*, **122,** 827.

GREENE, J. B., and MANNING, M. F., 1943, *Phys. Rev.*, **63,** 203.

GREISEN, F. C., 1968, *Phys. Stat. Sol.*, **25,** 753.

GRIFFEL, M., SKOCHDOPOLE, R. E., and SPEDDING, F. H., 1956, *J. chem. Phys.*, **25,** 75.

GRODSKI, J. J., and DIXON, A. E., 1969, *Solid St. Commun.*, **7,** 735.

GSCHNEIDNER, K. A., 1964, *Solid St. Phys.*, **16,** 275.

GUBANOV, A. I., and NIKULIN, V. K., 1966, *Phys. Stat. Sol.*, **17,** 815.

GUPTA, K. P., CHENG, C. H., and BECK, P. A., 1964, *J. Phys. Chem. Solids*, **25,** 73.

GUPTA, R. P., and LOUCKS, T. L., 1968, *Phys. Rev.*, **176,** 848 ; 1969, *Phys. Rev. Lett.*, **22,** 458.

GUSTAFSON, D. R., and MACKINTOSH, A. R., 1964, *J. Phys. Chem. Solids*, **25,** 389.

GUSTAFSON, D. R., MACKINTOSH, A. R., and ZAFFARANO, D. J., 1963, *Phys. Rev.*, **130,** 1455.

HAGSTRUM, H. D., and BECKER, G. E., 1967, *Phys. Rev.*, **159,** 572.

HALL, L. H., 1967, *Phys. Rev.*, **153,** 779.

HALL, P. M., LEGVOLD, S., and SPEDDING, F. H., 1960, *Phys. Rev.*, **117,** 971.

HALLORAN, M. H., CONDON, J. H., GRAEBNER, J. E., KUNZLER, J. E., and HSU, F. S. L., 1970, *Phys. Rev.* B, **1,** 366.

HALSE, M. R., 1969, *Phil. Trans. R. Soc.* A, **265,** 507.

HAMBOURGER, P. D., MARCUS, J. A., and MUNARIN, J. A., 1967, *Physics Lett.* A, **25,** 461.

HANNA, S. S., and PRESTON, R. S., 1958, *Phys. Rev.*, **109,** 716.

HANUS, J., 1962, *Massachusetts Institute of Technology, Solid State Theory and Molecular Theory Group, Quarterly Progress Report no. 44* p. 62.

HANUS, J., FEINLEIB, J., and SCOULER, W. J., 1967, *Phys. Rev. Lett.*, **19,** 16 ; 1968, *J. appl. Phys.*, **39,** 1272.

HARPER, A. F. A., KEMP, W. R. G., KLEMENS, P. G., TAINSH, R. J., and WHITE, G. K., 1957, *Phil. Mag.*, **2,** 577.

HARRIS, R., 1969, *Physics Lett.* A, **30,** 473.

HARRISON, W. A., 1960, *Phys. Rev.*, **118,** 1190 ; 1962, *Ibid.*, **126,** 497 ; 1963, *Ibid.*, **129,** 2512 ; 1966, *Ibid.*, **147,** 467.

HASEGAWA, A., and KASUYA, T., 1968, *J. phys. Soc. Japan*, **25,** 141 ; 1970, *Ibid.*, **28,** 75.

HASEGAWA, A., WAKOH, S., and YAMASHITA, J., 1965, *J. phys. Soc. Japan*, **20,** 1865.

HÄUSSLER, P., and WELLES, S. J., 1966, *Phys. Rev.*, **152,** 675.

HEDGECOCK, F. T., and MUIR, W. B., 1963, *Phys. Rev.*, **129,** 2045.

HEER, C. V., and ERICKSON, R. A., 1957, *Phys. Rev.*, **108,** 896 ; 1958, *Physica, 's Grav.*, **24,** S155.

HEGLAND, D., LEGVOLD, S., and SPEDDING, F. H., 1963, *Phys. Rev.*, **131,** 158.

HEINIGER, F., BUCHER, E., and MULLER, J., 1966, *Phys. Kondens. Materie*, **5,** 243.

HENNING, W., KAINDL, G., KIENLE, P., KÖRNER, H. J., KULZER, H., REHM, K. E., and EDELSTEIN, N., 1968, *Physics Lett.* A, **28,** 209.

HENNINGSEN, J. O., 1967, *Phys. Stat. Sol.*, **22,** 441 ; 1969 a, *Solid St. Commun.*, **7,** 763 ; 1969 b, *Phys. Stat. Sol.*, **32,** 239.

HENRY, J. F., 1962, *J. geophys. Res.*, **67,** 4843.

HERRING, C., 1960, *J. appl. Phys.*, **31,** 3S ; 1966, *Magnetism*, edited by G. T. Rado and H. Suhl (New York : Academic Press), Volume 4, ' Exchange interactions among itinerant electrons '.

HERRMANN, R., 1967, *Physics Lett.* A, **25,** 607 ; 1968, *Phys. Stat. Sol.*, **25,** 661.

HIGGINS, R. J., and MARCUS, J. A., 1966, *Phys. Rev.*, **141,** 553 ; 1967, *Ibid.*, **161,** 589.

HIGGINS, R. J., MARCUS, J. A., and WHITMORE, D. H., 1964, *Proc. 9th Int. Conf. Low Temp. Phys.*, p. 859 ; 1965, *Phys. Rev.*, **137,** A1172.

HIRSHFELD, A. T., LEUPOLD, H. A., and BOORSE, H. A., 1962, *Phys. Rev.*, **127,** 1501.

HOARE, F. E., and MATTHEWS, J. C., 1952, *Proc. R. Soc.* A, **212,** 137.

HODGES, L., and EHRENREICH, H., 1965, *Physics Lett.*, **16,** 203 ; 1968, *J. appl. Phys.*, **39,** 1280.

HODGES, L., EHRENREICH, H., and LANG, N. D., 1966, *Phys. Rev.*, **152,** 505.

HODGES, L., LANG, N. D., EHRENREICH, H., and FREEMAN, A. J., 1966, *J. appl. Phys.*, **37,** 1449.

HODGES, L., STONE, D. R., and GOLD, A. V., 1967, *Phys. Rev. Lett.*, **19,** 655.

HODGSON, J. N., and CLEYET, B., 1969, *J. Phys.* C (*Solid St. Phys.*), **2,** 97.

HOHENEMSER, C., WEINGART, J. M., and BERKO, S., 1968, *Physics Lett.* A, **28,** 41.

HÖRNFELDT, S., 1970, *Solid St. Commun.*, **8,** 673.

HOWARD, D. G., 1965, *Phys. Rev.*, **140,** A1705.

HOWARTH, D. J., 1953, *Proc. R. Soc.* A, **220,** 513 ; 1955, *Phys. Rev.*, **99,** 469.

HUBBARD, J., 1964, *Proc. phys. Soc.*, **84,** 455.

HUBBARD, J., and DALTON, N. W., 1968, *J. Phys.* C (*Solid St. Phys.*), **1,** 1637.

HUDGINS, A. C., and PAVLOVIC, A. S., 1965, *J. appl. Phys.*, **36,** 3628.

HUGHES, R. S., and LAWSON, A. W., 1967, *Physics Lett.* A, **25,** 473.

HUI, S. W., 1969, *Phys. Rev.*, **185,** 988.

HUM, D. M., and WONG, K. C., 1969, *J. Phys.* C (*Solid St. Phys.*), **2,** 833.

HUME-ROTHERY, W., 1963, *Electronic structure and alloy chemistry of the transition elements*, edited by P. A. Beck (New York : Wiley), p. 83.

HUME-ROTHERY, W., and COLES, B. R., 1954, *Adv. Phys.*, **3,** 149.

HUME-ROTHERY, W., and ROAF, D. J., 1961, *Phil. Mag.*, **6,** 55.

HYGH, E. H., and WELCH, R. M., 1970, *Phys. Rev.* B, **1,** 2424.

ISAACS, L. L., 1965, *J. chem. Phys.*, **43,** 307.

ISIN, A., and COLEMAN, R. V., 1965, *Phys. Rev.*, **137,** A1609 ; 1966, *Ibid.*, **142,** 372.

ITSKEVICH, E. S., 1962, *Zh. éksp. teor. Fiz.*, **42,** 1173. English translation : *Soviet Phys. JETP*, **15,** 811.

ITSKEVICH, E. S., and VORONOVSKIĬ, A. N., 1966, *Zh. éksp. teor. Fiz., Pis'ma*, **4,** 226. English translation : *Soviet Phys. JETP Lett.*, **4,** 154.

JACKSON, C., 1969, *Phys. Rev.*, **178,** 949.

JACKSON, C., and DONIACH, S., 1969, *Physics Lett.* A, **30,** 328.

JACOBS, R. L., 1968 a, *J. Phys.* C (*Solid St. Phys.*), **1,** 1296; 1968 b, *Ibid.*, **1,** 1307.
JAMES, N. R., LEGVOLD, S., and SPEDDING, F. H., 1952, *Phys. Rev.*, **88,** 1092.
JAN, J. P., 1966, *Can. J. Phys.*, **44,** 1787 ; 1968, *J. Phys. Chem. Solids*, **29,** 561.
JAN, J. P., PEARSON, W. B., and SPRINGFORD, M., 1964, *Proc. 9th Int. Conf. Low Temp. Phys.*, p. 776.
JAN, J. P., and TEMPLETON, I. M., 1967, *Phys. Rev.*, **161,** 556.
JANAK, J. F., 1969, *Physics Lett.* A, **28,** 570.
JANOVEC, V., and MORRISON, J. A., 1965, *Physics Lett.*, **17,** 226.
JAYARAMAN, A., 1964, *Phys. Rev.*, **135,** A1056.
JENNINGS, L. D., MILLER, R. E., and SPEDDING, F. H., 1960, *J. Chem. Phys.*, **33,** 1849.
JENNINGS, L. D., STANTON, R. M., and SPEDDING, F. H., 1957, *J. Chem. Phys.*, **27,** 909.
JERICHO, M. H., and SIMPSON, A. M., 1968, *Phil. Mag.*, **17,** 267.
JEROME, D., and RIEUX, M., 1969, *Solid St. Commun.*, **7,** 957.
JOHANSEN, G., 1969, *Solid St. Commun.*, **7,** 731.
JOHANSEN, G., and MACKINTOSH, A. R., 1970, *Solid St. Commun.*, **8,** 121.
JOHNSON, D. L., and FINNEMORE, D. K., 1967, *Phys. Rev.*, **158,** 376.
JOHNSON, E. W., and JOHNSON, H. H., 1965, *J. appl. Phys.*, **36,** 1286.
JONES, C. K., and RAYNE, J. A., 1964 a, *Physics Lett.*, **8,** 155 ; 1964 b, *Ibid.*, **13,** 282 ; 1964 c, *Proc. 9th Int. Conf. Low Temp. Phys.*, p. 790 ; 1965 a, *Physics Lett.*, **14,** 13 ; 1965 b, *Phys. Rev.*, **139,** A1876.
JONES, H., 1937, *Proc. phys. Soc.*, **49,** 250 ; 1955, *Ibid.* A, **68,** 1191 ; 1960, *The Theory of Brillouin Zones and Electronic States in Crystals* (Amsterdam : North-Holland).
JONES, R. C., GOODRICH, R. G., and FALICOV, L. M., 1968, *Phys. Rev.*, **174,** 672.
JONES, W. H., and MILFORD, F. J., 1962, *Phys. Rev.*, **125,** 1259.
JONGENBURGER, P., 1961, *Acta metall.*, **9,** 985.
JOSEPH, A. S., and GORDON, W. L., 1962, *Phys. Rev.*, **126,** 489.
JOSEPH, A. S., GORDON, W. L., REITZ, J. R., and ECK, T. G., 1961, *Phys. Rev. Lett.*, **7,** 334.
JOSEPH, A. S., and THORSEN, A. C., 1963 a, *Phys. Rev. Lett.*, **11,** 67 ; 1963 b, *Ibid.*, **11,** 554 ; 1964 a, *Phys. Rev.*, **134,** A979 ; 1964 b, *Phys. Rev. Lett.*, **13,** 9 ; 1964 c, *Phys. Rev.*, **133,** A1546 ; 1964 d, *Proc. Int. Conf. Magnetism*, Nottingham, p. 117 ; 1965, *Phys. Rev.*, **138,** A1159.
JOSEPH, A. S., THORSEN, A. C., and BLUM, F. A., 1965, *Phys. Rev.*, **140,** A2046.
JOSEPH, A. S., THORSEN, A. C., GERTNER, E., and VALBY, L. E., 1966, *Phys. Rev.*, **148,** 569.
JUENKER, D. W., LEBLANC, L. J., and MARTIN, C. R., 1968, *J. opt. Soc. Am.*, **58,** 164.
KAMM, G. N., 1970, *Phys. Rev.* B, **1,** 554.
KARAGEZYAN, A. G., and RAO, K. V., 1968, *Zh. éksp. teor. Fiz.*, **55,** 1168. English translation : *Soviet Phys. JETP*, **28.** 609.
KASPER, J. S., and ROBERTS, B. W., 1956, *Phys. Rev.*, **101,** 537.
KASUYA, T., 1956, *Prog. theor. Phys.*, **16,** 58.
KATSUKI, S., and TSUJI, M., 1965, *J. phys. Soc. Japan*, **20,** 1136.
KATYAL, O. P., and GERRITSEN, A. N., 1969 a, *Phys. Rev.*, **178,** 1037 ; 1969 b, *Ibid.*, **185,** 1017.
KEESOM, W. H., and CLARK, C. W., 1939, *Physica, 's Grav.*, **6,** 513.
KEETON, S. C., and LOUCKS, T. L., 1966 a, *Phys. Rev.*, **146,** 429 ; 1966 b, *Ibid.*, **152,** 548 ; 1968, *Ibid.*, **168,** 672.
KEMP, W. R. G., KLEMENS, P. G., SREEDHAR, A. K., and WHITE, G. K., 1955, *Phil. Mag.*, **46,** 811.
KEMP, W. R. G., KLEMENS, P. G., and WHITE, G. K., 1956, *Aust. J. Phys.*, **9,** 180.

KETTERSON, J. B., PRIESTLEY, M. G., and VUILLEMIN, J. J., 1966, *Physics Lett.*, **20,** 452.

KETTERSON, J. B., and WINDMILLER, L. R., 1968, *Phys. Rev. Lett.*, **20,** 321.

KETTERSON, J. B., WINDMILLER, L. R., and HÖRNFELDT, S., 1968, *Physics Lett.* A, **26,** 115.

KEVANE, C. J., LEGVOLD, S., and SPEDDING, F. H., 1953, *Phys. Rev.*, **91,** 1372.

KIP, A. F., LANGENBERG, D. N., and MOORE, T. W., 1961, *Phys. Rev.*, **124,** 359.

KITTEL, C., 1963, *Phys. Rev. Lett.*, **10,** 339.

KLAUDER, J. R., REED, W. A., BRENNERT, G. F., and KUNZLER, J. E., 1966, *Phys. Rev.*, **141,** 592.

KLEMENS, P. G., 1954, *Aust. J. Phys.*, **7,** 70.

KNEIP, G. D., BETTERTON, J. O., and SCARBROUGH, J. O., 1963, *Phys. Rev.*, **130,** 1687.

KNIGHT, W. D., BERGER, A. G., and HEINE, V., 1959, *Ann. Phys.*, **8,** 173.

KOBAYASHI, H., 1966, *J. phys. Soc. Japan*, **21,** 201.

KOBAYASI, S., and MATUMOTO, S., 1967, *J. phys. Soc. Japan*, **22,** 933.

KOCH, J. F., STRADLING, R. A., and KIP, A. F., 1964, *Phys. Rev.*, **133,** A240.

KOEHLER, W. C., 1965, *J. appl. Phys.*, **36,** 1078.

KOEHLER, W. C., CABLE, J. W., WILKINSON, M. K., and WOLLAN, E. O., 1966, *Phys. Rev.*, **151,** 414.

KOEHLER, W. C., CABLE, J. W., WOLLAN, E. O., and WILKINSON, M. K., 1962 a, *Phys. Rev.*, **126,** 1672 ; 1962 b, *J. phys. Soc. Japan*, **17,** Suppl. B-III, 32.

KOEHLER, W. C., CHILD, H. R., NICKLOW, R. M., SMITH, H. G., MOON, R. M., and CABLE, J. W., 1970, *Phys. Rev. Lett.*, **24,** 16.

KOEHLER, W. C., CHILD, H. R., WOLLAN, E. O., and CABLE, J. W., 1963, *J. appl. Phys.*, **34,** 1335.

KOEHLER, W. C., MOON, R. M., TREGO, A. L., and MACKINTOSH, A. R., 1966, *Phys. Rev.*, **151,** 405.

KOJIME, H., TEBBLE, R. S., and WILLIAMS, D. E. G., 1961, *Proc. R. Soc.* A, **260,** 237.

KOLOUCH, R. J., and MCCARTHY, K. A., 1965, *Phys. Rev.*, **139,** A700.

KONDORSKIĬ, E. I., GALKINA, O. S., and CHERNIKOVA, L. A., 1958, *Zh. éksp. teor. Fiz.*, **34,** 1070. English translation : *Soviet Phys. JETP*, **7,** 741.

KOROLYUK, A. P., and PRUSHCHAK, T. A., 1961, *Zh. éksp. teor. Fiz.*, **41,** 1689. English translation : *Soviet Phys. JETP*, **14,** 1201.

KOSTINA, T. L., KOSLOVA, T. N., and KONDORSKIĬ, E. I., 1963, *Zh. éksp. teor. Fiz.*, **45,** 1352. English translation : *Soviet Phys. JETP*, **18,** 931.

KREBS, K., and HÖLZL, K., 1967, *Solid St. Commun.*, **5,** 159.

KRINCHIK, G. S., 1964 a, *J. appl. Phys.*, **35,** 1089 ; 1964 b, *Proc. Int. Conf. Magnetism*, Nottingham, p. 114.

KRINCHIK, G. S., and ARTEM'EV, V. A., 1967, *Zh. éksp. teor. Fiz.*, **53,** 1901. English translation : *Soviet Phys. JETP*, **26,** 1080 ; 1968, *Physics Lett.* A, **27,** 127.

KRINCHIK, G. S., and GAN'SHINA, E. A., 1966, *Physics Lett.*, **23,** 294.

KRINCHIK, G. S., and GORBACHER, A. A., 1961, *Fizika Metall.*, **11,** 203. English translation : *Physics Metals Metallogr.*, **11,** (2), 49.

KRINCHIK, G. S., GUSHCHIN, V. S., and GAN'SHINA, E. A., 1968, *Zh. éksp. teor. Fiz., Pis'ma*, **8,** 53. English translation : *Soviet Phys. JETP Lett.*, **8,** 31.

KRINCHIK, G. S., and NURALIEVA, R. D., 1959, *Zh. éksp. teor. Fiz.*, **36,** 1022. English translation : *Soviet Phys. JETP*, **9,** 724.

KRINCHIK, G. S., and NURMUKHAMEDOV, G. M., 1965, *Zh. éksp. teor. Fiz.*, **48,** 34. English translation : *Soviet Phys. JETP*, **21,** 22.

KRIVITSKIĬ, V. V., and NIKIFOROV, I. YA., 1967, *Izv. Akad. Nauk SSSR, Ser. Fiz.*, **31,** 970. English translation : *Bull. Acad. Sci. USSR, Phys. Ser.*, **31,** 985.

KROLIKOWSKI, W. F., and SPICER, W. E., 1969, *Phys. Rev.*, **185,** 882.

KUNZLER, J. E., and KLAUDER, J. R., 1961, *Phil. Mag.*, **6,** 1045.

KUPRATAKULN, S., and FLETCHER, G. C., 1969, *J. Phys.* C (*Solid St. Phys.*), **2,** 1886.

KURTI, N., and SAFRATA, R. S., 1958, *Phil. Mag.*, **3,** 780.

KUSHIDA, T., and RIMAI, L., 1966, *Phys. Rev.*, **148,** 593.

KUSMISS, J. H., and SWANSON, J. W., 1968, *Physics Lett.* A, **27,** 517.

LANG, G., and DE BENEDETTI, S., 1957, *Phys. Rev.*, **108,** 914.

LANG, G., DE BENEDETTI, S., and SMOLUCHOWSKI, R., 1955, *Phys. Rev.*, **99,** 596.

LANGENBERG, D. N., and MARCUS, S. M., 1964, *Phys. Rev.*, **136,** A1383.

LAPEYRE, G. J., and KRESS, K. A., 1968, *Phys. Rev.*, **166,** 589.

LARSON, D. C., COLEMAN, R. V., and BOIKO, B. T., 1965, *Appl. Phys. Lett.*, **6,** 30.

LAUTZ, G., and TITTES, E., 1958, *Z. Naturforsch.*, **13a,** 866.

LAWSON, A. W., and TANG, T. Y., 1949, *Phys. Rev.*, **76,** 301.

LAWSON, J. R., and GORDON, W. L., 1964, *Proc. 9th Int. Conf. Low Temp. Phys.*, p. 854.

LAZAREV, B. G., NAKHIMOVICH, N. M., and PARFENOVA, E. A., 1939, *Zh. éksp. teor. Fiz.*, **9,** 1169.

LEAVER, G., and MYERS, A., 1969, *Phil. Mag.*, **19,** 465.

LEBÉGUE, G., 1963, *J. Phys.*, *Paris*, **24,** 709.

LEE, E. W., and ASGAR, M. A., 1969, *Phys. Rev. Lett.*, **22,** 1436.

LEE, R. S., and LEGVOLD, S., 1967, *Phys. Rev.*, **162,** 431.

LEHMAN, G. W., 1959, *Phys. Rev.*, **116,** 846.

LENGLART, P., 1967, *J. Phys. Chem. Solids*, **28,** 2011 ; 1968, *Ann. Phys.* (*France*), **3,** 27.

LENGLART, P., LEMAN, G., and LELIEUR, J. P., 1966, *J. Phys. Chem. Solids*, **27,** 377.

LENHAM, A. P., 1967, *J. opt. Soc. Am.*, **57,** 473.

LENHAM, A. P., and TREHERNE, D. M., 1966, *J. opt. Soc. Am.*, **56,** 683 ; 1967, *Ibid.*, **57,** 476.

LETTINGTON, A. H., 1964, *Physics Lett.*, **9,** 98 ; 1965, *Proceedings of the International Colloquium on Optical Properties and Electronic Structure of Metals and Alloys*, Paris (1965), edited by F. Abelès (Amsterdam : North-Holland), p. 147.

LEUPOLD, H. A., and BOORSE, H. A., 1964, *Phys. Rev.*, **134,** A1322.

LEWIS, P. E., and LEE, P. M., 1968, *Phys. Rev.*, **175,** 795.

LIDIARD, A. B., 1954, *Proc. R. Soc.* A, **224,** 161.

LIKHTER, A. I., and VENTTSEL', V. A., 1962, *Fizika tverd. Tela*, **4,** 485. English translation : *Soviet Phys. Solid St.*, **4,** 352.

LIPSON, S. G., 1964, *Proc. 9th Int. Conf. Low Temp. Phys.*, p. 814.

LOCK, J. M., 1957, *Proc. phys. Soc.* B, **70,** 566.

LODGE, F. M. K., and TAYLOR, K. N. R., 1969, *Physics Lett.* A, **30,** 359.

LOMER, W. M., 1962, *Proc. phys. Soc.*, **80,** 489 ; 1964 a, *Ibid.*, **84,** 327 ; 1964 b, *Proc. Int. Conf. Magnetism*, Nottingham, p. 127.

LOMER, W. M., and MARSHALL, W., 1958, *Phil. Mag.*, **3,** 185.

LOUCKS, T. L., 1965 a, *Phys. Rev.*, **139,** A1181 ; 1965 b, *Phys. Rev. Lett.*, **14,** 693 ; 1966 a, *Phys. Rev.*, **143,** 506 ; 1966 b, *Ibid.*, **144,** 504 ; 1967, *Ibid.*, **159,** 544.

LOUNASMAA, O. V., 1962, *Phys. Rev.*, **128,** 1136 ; 1963, *Ibid.*, **129,** 2460 ; 1964 a, *Ibid.*, **133,** A211 ; 1964 b, *Ibid.*, **133,** A219 ; 1964 c, *Ibid.*, **133,** A502.

LOUNASMAA, O. V., and GUENTHER, R. A., 1962, *Phys. Rev.*, **126,** 1357.

LOUNASMAA, O. V., and ROACH, P. R., 1962, *Phys. Rev.*, **128,** 622.

LOUNASMAA, O. V., and SUNDSTRÖM, L. J., 1966, *Phys. Rev.*, **150,** 399 ; 1967, *Ibid.*, **158,** 591.

LÜCK, R., and SAEGER, K. E., 1967, *Phys. Stat. Sol.*, **21,** 671.

LUKHVICH, A. A., 1969, *Fizika tverd. Tela*, **11,** 1051. English translation : *Soviet Phys. Solid St.*, **11,** 857.

LÜTHI, B., 1959, *Phys. Rev. Lett.*, **2,** 503 ; 1960, *Helv. phys. Acta*, **33,** 161.

LYNAM, P., SCURLOCK, R. G., and WRAY, E. M., 1964, *Proc. 9th Int. Conf. Low Temp. Phys.*, p. 905.

LYNCH, R. W., and DRICKAMER, H. G., 1965, *J. Phys. Chem. Solids*, **26,** 63.

MCALISTER, A. J., STERN, E. A., and MCGRODDY, J. C., 1965, *Phys. Rev.*, **140,** A2105.

MACFARLANE, R. E., and RAYNE, J. A., 1967, *Phys. Rev.*, **162,** 532.

MACFARLANE, R. E., RAYNE, J. A., and JONES, C. K., 1965, *Physics Lett.*, **19,** 354 ; 1967, *Ibid.* A, **24,** 197.

MCGRODDY, J. C., MCALISTER, A. J., and STERN, E. A., 1965, *Phys. Rev.*, **139,** A1844.

MCHARGUE, C. J., and YAKEL, H. L., 1960, *Acta metall.*, **8,** 637.

MACKEY, H. J., SYBERT, J. R., and FIELDER, J. T., 1967, *Phys. Rev.*, **157,** 578.

MACKINNON, L., and DANIEL, M. R., 1962 a, *Physics Lett.*, **1,** 157 ; 1962 b, *Proc. 8th Int. Conf. Low Temp. Phys.*, p. 203.

MACKINNON, L., TAYLOR, M. T., and DANIEL, M. R., 1962, *Phil. Mag.*, **7,** 523.

MACKINTOSH, A. R., 1962, *Phys. Rev. Lett.*, **9,** 90 ; 1968 *Physics Lett.* A, **28,** 217 (1968).

MACKINTOSH, A. R., WILLIAMS, R. W., and LOUCKS, T. L., 1965, *Positron annihilation.* Proceedings of a conference held at Wayne State University on July 27–29, 1965, edited by A. T. Stewart and L. O. Roellig (New York : Academic Press), p. 287.

MCMILLAN, W. L., 1968, *Phys. Rev.*, **167,** 331.

MCWHAN, D. B., RICE, T. M., and SCHMIDT, P. H., 1969, *Phys. Rev.*, **177,** 1063.

MAJOR, R. W., and LEINHARDT, T. E., 1967, *J. Phys. Chem. Solids*, **28,** 1669.

MAMIYA, T., FUKUROI, T., and TANUMA, S., 1965, *J. phys. Soc. Japan*, **20,** 1559.

MANNING, M. F., 1943, *Phys. Rev.*, **63,** 190.

MANNING, M. F., and CHODOROW, M. I., 1939, *Phys. Rev.*, **56,** 787.

MARCUS, J. A., 1947, *Phys. Rev.*, **71,** 559 ; 1949, *Ibid.*, **76,** 621 ; 1950, *Ibid.*, **77,** 750.

MARCUS, S. M., and LANGENBERG, D. N., 1963, *J. appl. Phys.*, **34,** 1367.

MARSOCCI, V. A., 1965, *Phys. Rev.*, **137,** A1842.

MARTIN, D. H., 1967, *Magnetism in Solids* (London : Iliffe).

MARTIN, D. H., DONIACH, S., and NEAL, K. J., 1964, *Physics Lett.*, **9,** 224.

MARTIN, D. H., NEAL, K. J., and DEAN, T. J., 1965, *Proc. phys. Soc.*, **86,** 505.

MARTIN, D. L., 1966, *Phys. Rev.*, **141,** 576 ; 1968 a, *Ibid.*, **167,** 640 ; 1968 b, *Ibid.*, **170,** 650 ; 1969, *Ibid.*, **186,** 642.

MAR'YAKHIN, A. A., and SVECHKAREV, I. V., 1967, *Phys. Stat. Sol.*, **23,** K133 ; 1969, *Ibid.*, **33,** K37.

MASUMOTO, H., SAITÔ, H., and KIKUCHI, M., 1966, *Sci. Rep. Res. Insts Tôhoku Univ.* A, **18,** Suppl., 84.

MATSUMOTO, T., SAMBONGI, T., and MITSUI, T., 1969, *J. phys. Soc. Japan*, **26,** 209.

MATTHEISS, L. F., 1964, *Phys. Rev.*, **134,** A970 ; 1965, *Ibid.*, **139,** A1893 ; 1966, *Ibid.*, **151,** 450 ; 1970, *Ibid.* B, **1,** 373.

MATTHEISS, L. F., and WATSON, R. F., 1964, *Phys. Rev. Lett.*, **13,** 526.

MATZKANIN, G. A., and SCOTT, T. A., 1966, *Phys. Rev.*, **151,** 360.

MEADEN, G. T., RAO, K. V., LOO, H. Y., and SZE, N. H., 1969, *J. phys. Soc. Japan*, **27,** 1073.

MERZ, H., and ULMER, K., 1966, *Physics Lett.*, **22,** 251.

MIHALISIN, T. W., and PARKS, R. D., 1966, *Physics Lett.*, **21,** 610 ; 1967, *Phys. Rev. Lett.*, **18,** 210 ; 1969, *Solid St. Commun.*, **7,** 33.
MIJNARENDS, P. E., 1969, *Phys. Rev.*, **178,** 622.
MIJNARENDS, P. E., and HAMBRO, L., 1964, *Physics Lett.*, **10,** 272.
MITCHELL, O. M. M., and YATES, G. H., 1967, *Phys. Rev. Lett.*, **18,** 603.
MIWA, H., 1963, *Prog. theor. Phys.*, **29,** 477.
MØLLER, H. B., and HOUMANN, J. C. G., 1966, *Phys. Rev. Lett.*, **16,** 737.
MØLLER, H. B., HOUMANN, J. C. G., and MACKINTOSH, A. R., 1967, *Phys. Rev. Lett.*, **19,** 312 ; 1968, *J. appl. Phys.*, **39,** 807.
MONTALVO, R. A., and MARCUS, J. A., 1964, *Physics Lett.*, **8,** 151.
MOOK, H., and SHULL, C. G., 1966, *J. appl. Phys.*, **37,** 1034.
MOON, R. M., CABLE, J. W., and KOEHLER, W. C., 1964, *J. appl. Phys.*, **35,** 1041.
MORIARTY, J. A., 1970, *Phys. Rev.* B, **1,** 1363.
MORIN, F. J., and MAITA, J. P., 1963, *Phys. Rev.*, **129,** 115.
MORRISON, J. A., and NEWSHAM, D. M. T., 1968, *J. Phys.* C (*Solid St. Phys*), **1,** 370.
MORSE, R. W., 1960, *The Fermi Surface.* Proceedings of a Conference held at Cooperstown, New York, on 22–24 August 1960, edited by W. A. Harrison and M. B. Webb (New York : Wiley), p. 214.
MORSE, R. W., MYERS, A., and WALKER, C. T., 1960, *Phys. Rev. Lett.*, **4,** 605 ; 1961, *J. acoust. Soc. Am.*, **33,** 699.
MORTON, V. M., 1960, Thesis, Cambridge University.
MOSS, J. S., and DATARS, W. R., 1967, *Physics Lett.* A, **24,** 630.
MOTT, N. F., 1935, *Proc. phys. Soc.*, **47,** 571 ; 1964, *Adv. Phys.*, **13,** 325 ; 1966, *Phil. Mag.*, **13,** 989 ; 1968, *Rev. mod. Phys.*, **40,** 677.
MOTT, N. F., and JONES, H., 1936, *The Theory of the Properties of Metals and Alloys* (Oxford University Press).
MOTT, N. F., and STEVENS, K. W. H., 1957, *Phil. Mag.*, **2,** 1364.
MOTULEVICH, G. P., and SHUBIN, A. A., 1964, *Zh. éksp. teor. Fiz.*, **47,** 840. English translation : *Soviet Phys. JETP*, **20,** 560 ; 1969, *Ibid.*, **56,** 45. English translation : *Soviet Phys. JETP*, **29,** 24.
MUELLER, F. M., 1966, *Phys. Rev.*, **148,** 636 ; 1967, *Bull. Am. phys. Soc.*, **12,** 287.
MUELLER, F. M., and PHILLIPS, J. C., 1967, *Phys. Rev.*, **157,** 600.
MUELLER, F. M., and PRIESTLEY, M. G., 1966, *Phys. Rev.*, **148,** 638.
MÜLLER, W. E., 1965, *Physics Lett.*, **17,** 82 ; 1966, *Solid St. Commun.*, **4,** 581 ; 1967, *Phys. Kondens. Materie*, **6,** 243.
MÜLLER, W. E., and THOMPSON, J. C., 1969, *Phys. Rev. Lett.*, **23,** 1037.
MUKHOPADHYAY, G., and MAJUMDAR, C. K., 1969, *J. Phys.* C (*Solid St. Phys.*), **2,** 924.
MYERS, A., and BOSNELL, J. R., 1965, *Physics Lett.*, **17,** 9 ; 1966, *Phil. Mag.*, **13,** 1273.
MYRON, H. W., and LIU, S. H., 1970, *Phys. Rev.* B, **1,** 2414.
NABEREZHNȲKH, V. P., and DAN'SHIN, N. K., 1969, *Zh. éksp. teor. Fiz.*, *Pis'ma*, **10,** 22. English translation : *Soviet Phys. JETP Lett.*, **10,** 14.
NABEREZHNȲKH, V. P., and MAR'YAKHIN, A. A., 1967, *Phys. Stat. Sol.*, **20,** 737.
NABEREZHNȲKH, V. P., MAR'YAKHIN, A. A., and MEL'NIK, V. L., 1967, *Zh. éksp. teor. Fiz.*, **52,** 617. English translation : *Soviet Phys. JETP*, **25,** 403.
NABEREZHNȲKH, V. P., and MEL'NIK, V. L., 1965 a, *Zh. éksp. teor. Fiz.*, **47,** 873. English translation : *Soviet Phys. JETP*, **20,** 583 ; 1965 b, *Fizika tverd. Tela*, **7,** 258. English translation : *Soviet Phys. Solid St.*, **7,** 197.
NABEREZHNȲKH, V. P., and TSYMBAL, D. T., 1967, *Zh. éksp. teor. Fiz.*, *Pis'ma*, **5,** 319. English translation : *Soviet Phys. JETP Lett.*, **5,** 263.

NAGASAWA, H., and SUGAWARA, T., 1967, *J. phys. Soc. Japan*, **23,** 701.
NAKAJIMA, S., 1967, *Prog. theor. Phys.*, **38,** 23.
NARATH, A., 1967, *Phys. Rev.*, **163,** 232.
NATHANS, R., SHULL, C. G., SHIRANE, G., and ANDRESEN, A., 1959, *J. Phys. Chem. Solids*, **10,** 138.
NELLIS, W. J., and LEGVOLD, S., 1969, *J. appl. Phys.*, **40,** 2267.
NELSON, K. S., STANFORD, J. L., and SCHMIDT, F. A., 1968, *Physics Lett.* A, **28,** 402.
NEMNONOV, S. A., and SOROKINA, M. F., 1967, *Fizika metall.*, **23,** 732. English translation : *Physics Metals Metallogr.*, **23,** (4), 162.
NEMNONOV, S. A., TRAPEZNIKOV, V. A., KOLOBOVA, K. M., and TROFIMOVA, V. A., 1966, *Fizika Metall.*, **22,** 470. English translation : *Physics Metals Metallogr.*, **22,** (3), 154.
NERESON, N. G., OLSEN, C. E., and ARNOLD, G. P., 1964, *Phys. Rev.*, **135,** A176.
NEUBERT, W., 1967, *Z. Naturf.* A, **22,** 1639 ; 1969, *Ibid.*, **24,** 922.
NICKLOW, R. M., MOOK, H. A., SMITH, H. G., REED, R. E., and WILKINSON, M. K., 1969, *J. appl. Phys.*, **40,** 1452.
NIGH, H. E., LEGVOLD, S., and SPEDDING, F. H., 1963, *Phys. Rev.*, **132,** 1092.
NIGH, H. E., LEGVOLD, S., SPEDDING, F. H., and BEAUDRY, B. J., 1964, *J. chem. Phys.*, **41,** 3799.
NIKIFOROV, I. YA., 1961, *Fizika Metall.*, **11,** 927. English translation : *Physics Metals Metallogr.*, **11,** (6), 100.
NIKIFOROV, I. YA., and BLOKHIN, M. A., 1964, *Izv. Akad. Nauk SSSR, Ser. Fiz.*, **28,** 786. English translation : *Bull. Acad. Sci. USSR, Phys. Ser.*, **28,** 695.
NIKIFOROV, I. YA., and SACHENIKO, V. P., 1963, *Izv. Akad. Nauk SSSR, Ser. Fiz.*, **27,** 310. English translation : *Bull. Acad. Sci. USSR, Phys. Ser.*, **27,** 319.
NIKOL'SKIĬ, G. S., and EREMENKO, V. V., 1966, *Phys. Stat. Sol.*, **18,** K123.
NIKOL'SKIĬ, G. S., ZVYAGINA, N. M., and EREMENKO, V. V., 1969, *Fizika tverd. Tela*, **11,** 2489. English translation : *Soviet Phys. Solid St.*, **11,** 2009.
NIKULIN, V. K., 1966 a, *Phys. Stat. Sol.*, **16,** K125 ; 1966 b, *Fizika tverd. Tela*, **8,** 619. English translation : *Soviet Phys. Solid St.*, **8,** 499.
NIKULIN, V. K., and TRZHASKOVSKAYA, M. B., 1968, *Phys. Stat. Sol.*, **28,** 801.
NILSSON, P. O., NORRIS, C., and WALLDEN, L., 1969, *Solid St. Commun.*, **7,** 1705.
NILSSON, P. O., PERSSON, A., and NORDÉN, H., 1967, *Ark. Fys.*, **35,** 165.
O'BRIEN, E. J., and FRANKLIN, J., 1966, *J. appl. Phys.*, **37,** 2809.
OLSEN, C. E., NERESON, N. G., and ARNOLD, G. P., 1962, *J. appl. Phys.*, **33,** 1135.
OPECHOWSKI, W., and GUCCIONE, R., 1965, *Magnetism*, edited by G. T. Rado and H. Suhl (New York : Academic Press), volume 2A, p. 105.
OSBORNE, D. W., FLOTOW, H. E., and SCHREINER, F., 1967, *Rev. scient. Instrum.*, **38,** 159.
O'SULLIVAN, W. J., and SCHIRBER, J. E., 1965, *Phys. Lett.*, **18,** 212 ; 1966 a, *Phys. Rev.*, **151,** 484 ; 1966 b, *Phys. Rev. Lett.*, **16,** 691 ; 1967 a, *Cryogenics*, **7,** 118 ; 1967 b, *Phys. Rev.*, **162,** 519 ; 1968, *Ibid.*, **170,** 667 ; 1969, *Ibid.*, **181,** 1367.
O'SULLIVAN, W. J., SWITENDICK, A. C., and SCHIRBER, J. E., 1970, *Phys. Rev.* B, **1,** 1443.
OVERHAUSER, A. W., 1962, *Phys. Rev.*, **128,** 1437.
PADALKA, V. G., and SHKLIAREVSKIĬ, I. N., 1961, *Optika Spektrosk.*, **11,** 527. English translation : *Optics Spectrosc.*, **11,** 285 ; 1962, *Ibid.*, **12,** 291. English translation : *Optics Spectrosc.*, **12,** 158.
PANT, M. M., and JOSHI, S. K., 1969, *Phys. Rev.*, **184,** 639.

PARKINSON, D. H., and ROBERTS, L. M., 1957, *Proc. phys. Soc.* B, **70,** 471.
PARKINSON, D. H., SIMON, F. E., and SPEDDING, F. H., 1951, *Proc. R. Soc.* A, **207,** 137.
PAULING, L., 1953, *Proc. natn. Acad. Sci., U.S.A.*, **39,** 551.
PAWEL, R. E., and STANSBURY, E. E., 1965, *J. Phys. Chem. Solids*, **26,** 757.
PERRIN, B., WEISBUCH, G., and LIBCHABER, A., 1970, *Phys. Rev.* B, **1,** 1501.
PETERSON, D. T., PAGE, D. F., RUMP, R. B., and FINNEMORE, D. K., 1967, *Phys. Rev.*, **153,** 701.
PETTIFOR, D. G., 1970, *J. Phys.* C (*Solid St. Phys.*), **3,** 367.
PHILLIPS, J. C., 1964, *Phys. Rev.*, **133,** A1020.
PHILLIPS, J. C., and MATTHEISS, L. F., 1963, *Phys. Rev. Lett.*, **11,** 556.
PHILLIPS, J. C., and MUELLER, F. M., 1967, *Phys. Rev.*, **155,** 594.
PHILLIPS, N. E., 1964, *Phys. Rev.*, **134,** A385.
PICKART, S. J., ALPERIN, H. A., MINKIEWICZ, V. J., NATHANS, R., SHIRANE, G., and STEINSVOLL, O., 1967, *Phys. Rev.*, **156,** 623.
PIPER, W. W., 1961, *Phys. Rev.*, **123,** 1281.
PIPPARD, A. B., 1947, *Proc. R. Soc.* A, **191,** 385 ; 1957, *Phil. Trans. R. Soc.* A, **250,** 325 ; 1960, *The Fermi Surface.* Proceedings of a Conference held at Cooperstown, New York, on 22–24 August 1960, edited by W. A. Harrison and M. B. Webb (New York : Wiley), p. 330 ; 1964, *Phil. Trans. R. Soc.* A, **256,** 317.
POWELL, R. L., 1964, *Proc. 9th Int. Conf. Low Temp. Phys.*, p. 732.
POWELL, R. L., CLARK, A. F., and FICKETT, F. R., 1969, *Phys. Kondens. Materie*, **9,** 104.
POWELL, R. W., 1966, *J. appl. Phys.*, **37,** 2518.
POWELL, R. W., and JOLIFFE, B. W., 1965, *Physics Lett.*, **14,** 171.
PRIESTLEY, M. G., 1960, *Phil. Mag.*, **5,** 111.
PRIESTLEY, M. G., and MONDINO, M. A., 1964, *Bull. Am. phys. Soc.*, **9,** 551.
QUADROS, C., and OLIVEIRA, N. F., 1968, *Physics Lett.* A, **27,** 227.
RAMCHANDANI, M. G., 1970, *J. Phys.* C (*Solid St. Phys.*), **3,** Metal Phys. Suppl., S1.
RAO, K. V., 1967, *Physics Lett.* A, **24,** 39.
RAYNE, J. A., 1964, *Phys. Rev.*, **133,** A1104.
RAYNE, J. A., and KEMP, W. R. G., 1956, *Phil. Mag.*, **1,** 918.
RAYNE, J. A., and SELL, H., 1962, *Phys. Rev. Lett.*, **8,** 199.
REED, W. A., and BRENNERT, G. F., 1963, *Phys. Rev.*, **130,** 565.
REED, W. A., and FAWCETT, E., 1964 a, *Phys. Rev.*, **136,** A422 ; 1964 b, *J. appl. Phys.*, **35,** 754 ; 1964 c, *Proc. Int. Conf. Magnetism*, Nottingham, p. 120.
REED, W. A., FAWCETT, E., and SODEN, R. R., 1965, *Phys. Rev.*, **139,** A1557.
REED, W. A., and SODEN, R. R., 1968, *Phys. Rev.*, **173,** 677.
RENTON, C. A., 1960, *Proc. 7th Int. Conf. Low Temp. Phys.*, p. 216.
RHYNE, J. J., 1968, *Phys. Rev.*, **172,** 523.
ROAF, D. J., 1962, *Phil. Trans. R. Soc.* A, **255,** 135.
RODDA, J. L., and STEWART, M. G., 1963, *Phys. Rev.*, **131,** 255.
RORER, D. C., ONN, D. G., and MEYER, H., 1965, *Phys. Rev.*, **138,** A1661.
ROSENBERG, H. M., 1955, *Phil. Trans. R. Soc.* A, **247,** 441.
ROTH, L. M., ZEIGER, H. J., and KAPLAN, T. A., 1966, *Phys. Rev.*, **149,** 519.
ROZENFELD, B., and SZUSZKIEWICZ, M., 1966, *Nukleonika* (*Poland*), **11,** 693.
RUDERMAN, M. A., and KITTEL, C., 1954, *Phys. Rev.*, **96,** 99.
RUVALDS, J., and FALICOV, L. M., 1968, *Phys. Rev.*, **172,** 508.
SABO, J. J., 1970, *Phys. Rev.* B, **1,** 1479.
SAEGER, K. E., 1968, *Phys. Stat. Sol.*, **28,** 589.
SAMBONGI, T., and MITSUI, T., 1968, *J. phys. Soc. Japan*, **24,** 1168.
SAMSONOV, G. V., 1968, *Handbook of the Physicochemical Properties of the Elements* (London : Oldbourne).

SCHIRBER, J. E., 1964, *Proc. 9th Int. Conf. Low Temp. Phys.*, p. 863 ; 1965 a, *Phys. Rev. Lett.*, **14,** 66 ; 1965 b, *Phys. Rev.*, **140,** A2065.

SCHONE, H. E., 1964, *Phys. Rev. Lett.*, **13,** 12.

SCHÜLER, C. C., 1964, *Physics Lett.*, **12,** 84 ; 1965, *Proceedings of the International Colloquium on Optical Properties and Electronic Structure of Metals and Alloys*, Paris (1965), edited by F. Abelès (Amsterdam : North-Holland), p. 221.

SCHLOSSER, H., 1970, *Phys. Rev.* B, **1,** 491.

SCHRIEMPF, J. T., 1967, *J. Phys. Chem. Solids*, **28,** 2581 ; 1968, *Phys. Rev. Lett.*, **20,** 1034.

SCHRÖDER, K., and GIANNUZZI, A., 1969, *Phys. Stat. Sol.*, **34,** K133.

SCHWEISS, P., FURRER, A., and BÜHRER, W., 1967, *Helv. phys. Acta*, **40,** 378.

SCOTT, G. B., and SPRINGFORD, M., 1970, *Proc. R. Soc.* A, **320,** 115.

SCOTT, G. B., SPRINGFORD, M., and STOCKTON, J. R., 1968 a, *Physics Lett.* A, **27,** 655 ; 1968 b, *Proc. 11th Int. Conf. Low Temp. Phys.*, p. 1129.

SCOVIL, G. W., 1966, *Appl. Phys. Lett.*, **9,** 247.

SEDOV, V. L., 1963, *Zh. éksp. teor. Fiz.*, **45,** 2070. English translation : *Soviet Phys. JETP*, **18,** 1419 ; 1965, *Ibid.*, **48,** 1200. English translation : *Soviet Phys. JETP*, **21,** 800.

SEGALL, B., 1961, *Phys. Rev. Lett.*, **7,** 154 ; 1962, *Phys. Rev.*, **125,** 109.

SEITCHIK, J. A., GOSSARD, A. C., and JACCARINO, V., 1964, *Phys. Rev.*, **136,** A1119.

SEITCHIK, J. A., JACCARINO, V., and WERNICK, J. H., 1965, *Phys. Rev.*, **138,** A148.

SEMENENKO, E. E., SUDOVTSOV, A. I., and SHVETS, A. D., 1962, *Zh. éksp. teor. Fiz.*, **42,** 1488. English translation : *Soviet Phys. JETP*, **15,** 1033.

SEMENENKO, E. E., SUDOVTSOV, A. I., and VOLKENSHTEĬN, N. V., 1963, *Zh. éksp. teor. Fiz.*, **45,** 1387. English translation : *Soviet Phys. JETP*, **18,** 957.

SHARMA, S. N., and WILLIAMS, D. L., 1967, *Physics Lett.* A, **25,** 738.

SHARP, R. I., 1969 a, *J. Phys.* C (*Solid St. Phys.*), **2,** 421 ; 1969 b, *Ibid.*, **2,** 432.

SHAW, M. P., and ECK, T. G., 1964, *Proc. 9th Int. Conf. Low Temp. Phys.*, p. 761.

SHAW, M. P., ECK, T. G., and ZYCH, D. A., 1966, *Phys. Rev.*, **142,** 406.

SHAW, M. P., SAMPATH, P. I., and ECK, T. G., 1966, *Phys. Rev.*, **142,** 399.

SHEN, L. Y. L., SENOZAN, N. M., and PHILLIPS, N. E., 1965, *Phys. Rev. Lett.*, **14,** 1025.

SHIBATANI, A., MOTIZUKI, K., and NAGAMIYA, T., 1969, *Phys. Rev.*, **177,** 984.

SHIMIZU, M., KATO, T., and TSUIKI, T., 1969, *Physics Lett.* A, **28,** 656.

SHIMIZU, M., and KATSUKI, A., 1964, *J. phys. Soc. Japan*, **19,** 1856.

SHIMIZU, M., KATSUKI, A., and OHMORI, K., 1966, *J. phys. Soc. Japan*, **21,** 1922.

SHIMIZU, M., TAKAHASHI, T., and KATSUKI, A., 1962, *J. phys. Soc. Japan*, **17,** 1740 ; 1963, *Ibid.*, **18,** 240.

SHIRANE, G., MINKIEWICZ, V. J., and NATHANS, R., 1968, *J. appl. Phys.*, **39,** 383.

SHIROKOVSKIĬ, V. P., and KULAKOVA, Z. V., 1968, *Fizika metall.*, **25,** 404. English translation : *Physics Metals Metallogr.*, **25,** (3), 20.

SHKLIAREVSKIĬ, I. N., and PADALKA, V. G., 1959 a, *Optika Spektrosk.*, **6,** 78. English translation : *Optics Spectrosc.*, **6,** 45 ; 1959 b, *Ibid.*, **6,** 776. English translation : *Optics Spectrosc.*, **6,** 505.

SHKLIAREVSKIĬ, I. N., and YAROVAYA, R. G., 1966, *Optika Spektrosk.*, **21,** 197. English translation : *Optics Spectrosc.*, **21,** 115.

SHOEMAKE, G. E., and RAYNE, J. A., 1968, *Physics Lett.* A, **26,** 222.

SHOENBERG, D., 1952, *Phil. Trans. R. Soc.* A, **245,** 1 ; 1959, *Nature, Lond.*, **183,** 171 ; 1960 a, *Phil. Mag.*, **5,** 105 ; 1960 b, *The Fermi Surface.* Proceedings of a Conference held at Cooperstown, New York, on 22–24 August 1960, edited by W. A. Harrison and M. B. Webb (New York : Wiley), p. 74 ; 1962, *Phil. Trans. R. Soc.* A, **255,** 85 ; 1968, *Can. J. Phys.*, **46,** 1915.

SHOENBERG, D., and TEMPLETON, I. M., 1968, *Can. J. Phys.*, **46,** 1925.
SHOENBERG, D., and WATTS, B. R., 1964, *Proc. 9th Int. Conf. Low Temp. Phys.*, p. 831 ; 1967, *Phil. Mag.*, **15,** 1275.
SHULL, C. G., 1963, *Electronic Structure and Alloy Chemistry of the Transition Elements*, edited by P. A. Beck (New York : Wiley), p. 69.
SHULL, C. G., and WILKINSON, M. K., 1953, *Rev. mod. Phys.*, **25,** 100.
SIDGWICK, N. V., 1950, *The Chemical Elements and their Compounds* (Oxford University Press).
SILHOUETTE, D., 1968, *C. r. hebd. Séanc. Acad. Sci., Paris*, **267,** 1451.
SINHA, S. K., BRUN, T. O., MUHLESTEIN, L. D., and SAKURAI, J., 1970, *Phys. Rev.* B, **1,** 2430.
SIROTA, N. N., and OLEKHNOVICH, N. M., 1961, *Dokl. Akad. Nauk SSSR*, **139,** 844. English translation : *Soviet Phys. Dokl.*, **6,** 704.
SLATER, J. C., 1936, *Phys. Rev.*, **49,** 537.
SLATER, J. C., and KOSTER, G. F., 1954, *Phys. Rev.*, **94,** 1498.
SMIRNOV, YU. N., and FINKEL, V. A., 1965, *Zh. éksp. teor. Fiz.*, **49,** 1077. English translation : *Soviet Phys. JETP*, **22,** 750.
SMITH, D. A., 1967, *Proc. R. Soc.* A, **296,** 476.
SMITHELLS, C. J., 1967, *Metals Reference Book* (London : Butterworths).
SNOW, E. C., 1968 a, *Phys. Rev.*, **171,** 785 ; 1968 b, *Ibid.*, **172,** 708.
SNOW, E. C., and WABER, J. T., 1967, *Phys. Rev.*, **157,** 570 ; 1969, *Acta metall.*, **17,** 623.
SNOW, E. C., WABER, J. T., and SWITENDICK, A. C., 1966, *J. appl. Phys.*, **37,** 1342.
SOFFER, S. B., 1968, *Phys. Rev.*, **176,** 861.
SOMMERS, C. B., and AMAR, H., 1970, *Phys. Rev.*, **188,** 1117.
SOUERS, P. C., and JURA, G., 1963, *Science, N.Y.*, **140,** 481.
SOULE, D. E., and ABELE, J. C., 1969, *Phys. Rev. Lett.*, **23,** 1287.
SPARLIN, D. M., and MARCUS, J. A., 1966, *Phys. Rev.*, **144,** 484.
SPECHT, F., 1967, *Phys. Rev.*, **162,** 389.
SPEDDING, F. H., LEGVOLD, S., DAANE, A. H., and JENNINGS, L. D., 1957, *Prog. low. Temp. Phys.*, **2,** 368.
SPICER, W. E., 1965, *Proceedings of the International Colloquium on Optical Properties and Electronic Structure of Metals and Alloys*, Paris (1965), edited by F. Abelès (Amsterdam : North-Holland), p. 296 ; 1966, *J. appl. Phys.*, **37,** 947.
SPICER, W. E., and BERGLUND, C. N., 1964, *Phys. Rev. Lett.*, **12,** 9.
ŚREDNIAWA, B., 1960, *Helv. phys. Acta*, **33,** 131.
STAFLEU, M. D., and DE VROOMEN, A. R., 1965, *Physics Lett.*, **19,** 81.
STARK, R. W., 1962, *Phys. Rev. Lett.*, **9,** 482 ; 1964, *Phys. Rev.*, **135,** A1698.
STARK, R. W., and FALICOV, L. M., 1967, *Phys. Rev. Lett.*, **19,** 795.
STARK, R. W., and TSUI, D. C., 1968, *J. appl. Phys.*, **39,** 1056.
STERN, F., 1959, *Phys. Rev.*, **116,** 1399 ; 1961, *Phys. Rev. Lett.*, **6,** 675.
STEWART, A. T., 1957, *Can. J. Phys.*, **35,** 168.
STOLL, M.-PH., 1970, *Solid St. Commun.*, **8,** 1207.
STRINGER, J., HILL, J., and HUGLIN, A. S., 1970, *Phil. Mag.*, **21,** 53.
STRINGFELLOW, M. W., HOLDEN, T. M., POWELL, B. M., and WOODS, A. D. B., 1969, *J. appl. Phys.*, **40,** 1443.
STROM-OLSEN, J. O., 1967, *Proc. R. Soc.* A, **302,** 83.
SUEOKA, O., 1967, *J. phys. Soc. Japan*, **23,** 1246 ; 1969, *Ibid.*, **26,** 863.
SUFFCZYŃSKI, M., 1957, *Acta Physiol. Pol.*, **16,** 161 ; 1959, *Proc. phys. Soc.*, **73,** 671 ; 1960, *J. Phys. Chem. Solids*, **16,** 174.
SUZUKI, H., MINOMURA, S., and MIYAHARA, S., 1966, *J. phys. Soc. Japan*, **21,** 2089.
SUZUKI, H., and MIYAHARA, S., 1965, *J. phys. Soc. Japan*, **20,** 2102.

SWITENDICK, A. C., 1966, *J. appl. Phys.*, **37,** 1022.

TAMARIN, P. V., CHUPRIKOV, G. E., and SHALYT, S. S., 1968, *Zh. éksp. teor. Fiz.*, **55,** 1595. English translation : *Soviet Phys. JETP*, **28,** 836.

TAN, D. S., 1965, *J. Phys. Chem. Solids*, **26,** 1623.

TANUMA, S., ISHIZAWA, Y., NAGASAWA, H., and SUGAWARA, T., 1967, *Physics Lett.* A, **25,** 669.

TAYLOR, A., and KAGLE, B. J., 1963, *Crystallographic Data on Metal and Alloy Structures* (New York : Dover).

TAYLOR, M. A., and LLEWELLYN SMITH, C. H., 1962, *Physica, 's Grav.*, **28,** 453.

TELESNIN, R. V., AL'MEN'EVA, D. V., and POGOSHEV, V. A., 1962, *Fizika tverd. Tela*, **4,** 357. English translation : *Soviet Phys. Solid St.*, **4,** 256.

TEMPLETON, I. M., 1966, *Proc. R. Soc.* A, **292,** 413.

TESTARDI, L. R., and SODEN, R. R., 1967, *Phys. Rev.*, **158,** 581.

THEYE, M. L., 1967, *Physics Lett.* A, **25,** 764.

THORSEN, A. C., and BERLINCOURT, T. G., 1961, *Phys. Rev. Lett.*, **7,** 244.

THORSEN, A. C., and JOSEPH, A. S., 1963, *Phys. Rev.*, **131,** 2078.

THORSEN, A. C., JOSEPH, A. S., and VALBY, L. E., 1966, *Phys. Rev.*, **150,** 523 ; 1967, *Ibid.*, **162,** 574.

THORSEN, A. C., VALBY, L. E., and JOSEPH, A. S., 1964, *Proc. 9th Int. Conf. Low Temp. Phys.*, p. 867.

TOMBOULIAN, D. H., and BEDO, D. E., 1961, *Phys. Rev.*, **121,** 146.

TOMPA, K., 1966, *Rep. Cen. Res. Inst. Phys., Hung.*, **14,** 227.

TORGESON, D. R., and BARNES, R. G., 1964, *Phys. Rev.*, **136,** A738.

TROMBE, F., 1945, *C. r. hebd. Séanc. Acad. Sci., Paris*, **221,** 19 ; 1951, *J. Phys., Paris*, **12,** 222 ; 1953, *C. r. hebd. Séanc. Acad. Sci., Paris*, **236,** 591.

TSUI, D. C., 1967, *Phys. Rev.*, **164,** 669.

TSUI, D. C., and STARK, R. W., 1966 a, *Phys. Rev. Lett.*, **16,** 19 ; 1966 b, *Ibid.*, **17,** 871 ; 1967, *Ibid.*, **19,** 1317.

TSUJI, M., and KUNIMUNE, M., 1963, *J. phys. Soc. Japan*, **18,** 1569.

TUROV, E. A., 1958, *Fizika Metall.*, **6,** 203. English translation : *Physics Metals Metallogr.*, **6,** (2), 13.

UMEBAYASHI, H., SHIRANE, G., FRAZER, B. C., and DANIELS, W. B., 1968, *Phys. Rev.*, **165,** 688.

VAN BAARLE, C., ROEST, G. J., ROEST-YOUNG, M. K., and GORTER, F. W.,1966, *Physica, 's Grav.*, **32,** 1700.

VAN DER HOEVEN, B. J. C., and KEESOM, P. H., 1964 a, *Phys. Rev.*, **134,** A1320 ; 1964 b, *Ibid.*, **135,** A631.

VAN DYKE, J. P., MCCLURE, J. W., and DOAR, J. F., 1970, *Phys. Rev.* B, **1,** 2511.

VAN KRANENDONK, J., and VAN VLECK, J. H., 1958, *Rev. mod. Phys.*, **30,** 1.

VAN OSTENBURG, D. O., LAM, D. J., SHIMIZU, M., and KATSUKI, A., 1963, *J. phys. Soc. Japan*, **18,** 1744.

VAN WITZENBURG, W., and LAUBITZ, M. J., 1968, *Can. J. Phys.*, **46,** 1887.

VASVARI, B., ANIMALU, A. O. E., and HEINE, V., 1967, *Phys. Rev.*, **154,** 535.

VAUTIER, C., and COLOMBANI, A., 1966, *C. r. hebd. Séanc. Acad. Sci. Paris*, B, **263,** 997.

VEDERNIKOV, M. V., 1969, *Adv. Phys.*, **18,** 337.

VEHSE, R. C., ARAKAWA, E. T., and WILLIAMS, M. W., 1970, *Phys. Rev.* B, **1,** 517.

VENTTSEL', V. A., 1968, *Zh. éksp. teor. Fiz.*, **55,** 1191. English translation : *Soviet Phys. JETP*, **28,** 622.

VENTTSEL', V. A., LIKHTER, A. I., and RUDNEV, A. V., 1966, *Zh. éksp. teor. Fiz., Pis'ma*, **4,** 216. English translation : *Soviet Phys. JETP Lett.*, **4,** 148 ; 1967, *Zh. éksp. teor. Fiz.*, **53,** 108. English translation : *Soviet Phys. JETP*, **26,** 73.

VERKIN, B. I., and DMITRENKO, I. M., 1955, *Izv. Akad. Nauk SSSR, Ser. Fiz.*, **19,** 409. English translation : *Bull. Acad. Sci. USSR, Phys. Ser.*, **19,** 365 ; 1958, *Zh. éksp. teor. Fiz.*, **35,** 201. English translation : *Soviet Phys. JETP*, **8,** 200 ; 1959, *Dokl. Akad. Nauk SSSR*, **124,** 557. English translation : *Soviet Phys. Dokl.*, **4,** 118.

VERKIN, B. I., KUZMICHEVA, L. B., and SVECHKARYOV, I. V., 1968, *Zh. éksp. teor. Fiz.*, **54,** 74. English translation : *Soviet Phys. JETP*, **27,** 41.

VERKIN, B. I., LAZAREV, B. G., and RUDENKO, N. S., 1950 a, *Zh. éksp. teor. Fiz.*, **20,** 93 ; 1950 b, *Ibid.*, **20,** 995 ; 1951, *Dokl. Akad. Nauk SSSR*, **80,** 45.

VERKIN, B. I., PELIKH, L. N., and EREMENKO, V. V., 1964, *Dokl. Akad. Nauk SSSR*, **159,** 771. English translation : *Soviet Phys. Dokl.*, **9,** 1076.

VOLKENSHTEĬN, N. V., and DYAKINA, V. P., 1966, *Zh. éksp. teor. Fiz., Pis'ma*, **4,** 396. English translation : *Soviet Phys. JETP Lett.*, **4,** 268.

VOLKENSHTEĬN, N. V., DYAKINA, V. P., NOVOSELOV, V. A., and STARTSEV, V. E., 1966, *Fizika metall.*, **21,** 674. English translation : *Physics Metals Metallogr.*, **21,** (5), 31.

VOLKENSHTEĬN, N. V., and FEDOROV, G. V., 1963, *Zh. éksp. teor. Fiz.*, **44,** 825. English translation : *Soviet Phys. JETP*, **17,** 560 ; 1965, *Fizika tverd. Tela*, **7,** 3213. English translation : *Soviet Phys. Solid St.*, **7,** 2599 ; 1966, *Ibid.*, **8,** 1895. English translation : *Soviet Phys. Solid St.*, **8,** 1500.

VOLKENSHTEĬN, N. V., and GALOSHINA, É. V., 1963, *Fizika metall.*, **16,** 298. English translation : *Physics Metals Metallogr.*, **16,** (2), 117.

VOLKENSHTEĬN, N. V., GALOSHINA, É. V., and SHCHEGOLIKHINA, N. I., 1969, *Zh. éksp. teor. Fiz.*, **56,** 139. English translation : *Soviet Phys. JETP*, **29,** 79.

VOLKENSHTEĬN, N. V., GRIGOROVA, I. K., and FEDOROV, G. V., 1966, *Zh. éksp. teor. Fiz.*, **51,** 780. English translation : *Soviet Phys. JETP*, **24,** 519.

VOLKENSHTEĬN, N. V., KACHINSKIĬ, V. N., and STAROSTINA, L. S., 1963, *Zh. éksp. teor. Fiz.*, **45,** 43. English translation : *Soviet Phys. JETP*, **18,** 32.

VOLKENSHTEĬN, N. V., and STARTSEV, V. E., 1964, *Zh. éksp. teor. Fiz.*, **46,** 457. English translation : *Soviet Phys. JETP.*, **19,** 308.

VOLKOV, D. I., KOZLOVA, T. M., PRUDNIKOV, V. N., and KOZIS, E. V., 1968, *Zh. éksp. teor. Fiz.*, **55,** 2103. English translation : *Soviet Phys. JETP*, **28,** 1113.

VUILLEMIN, J. J., 1966, *Phys. Rev.*, **144,** 396.

VUILLEMIN, J. J., and BRYANT, H. J., 1969, *Phys. Rev. Lett.*, **23,** 914.

VUILLEMIN, J. J., and PRIESTLEY, M. G., 1965, *Phys. Rev. Lett.*, **15,** 307.

WAKOH, S., 1965, *J. phys. Soc. Japan*, **20,** 1894.

WAKOH, S., and YAMASHITA, J., 1964, *J. phys. Soc. Japan*, **19,** 1342 ; 1966, *Ibid.*, **21,** 1712.

WALKER, C. B., and EGELSTAFF, P. A., 1969, *Phys. Rev.*, **117,** 1111.

WALLACE, D. C., 1960, *Phys. Rev.*, **120,** 84.

WALLACE, W. D., and BOHM, H. V., 1968, *J. Phys. Chem. Solids*, **29,** 721.

WALLACE, W. D., TEPLEY, N., BOHM, H. V., and SHAPIRA, Y., 1965, *Physics Lett.*, **17,** 184.

WALLING, J. C., and BUNN, P. B., 1959, *Proc. phys. Soc.*, **74,** 417.

WALMSLEY, R. H., 1962, *Phys. Rev. Lett.*, **8,** 242.

WALSH, W. M., 1964, *Phys. Rev. Lett.*, **12,** 161.

WALSH, W. M., and GRIMES, C. C., 1964, *Phys. Rev. Lett.*, **13,** 523.

WATABE, A., and KASUYA, T., 1969, *J. phys. Soc. Japan*, **26,** 64.

WATSON, R. E., FREEMAN, A. J., and DIMMOCK, J. O., 1968, *Phys. Rev.*, **167,** 497.

WATTS, B. R., 1964 a, *Physics Lett.*, **10,** 275 ; 1964 b, *Proc. 9th Int. Conf. Low Temp. Phys.*, p. 779.

WEINERT, R. W., and SCHUMACHER, R. T., 1968, *Phys. Rev.*, **172,** 711.
WEISS, R. J., 1970, *Phys. Rev. Lett.*, **24,** 883.
WEISS, R. J., and DE MARCO, J. J., 1958, *Rev. mod. Phys.*, **30,** 59.
WEISS, R. J., and FREEMAN, A. J., 1959, *J. Phys. Chem. Solids*, **10,** 147.
WEISS, W. D., and KOHLHAAS, R., 1967, *Z. angew. Phys.*, **23,** 175.
WERNER, S. A., ARROTT, A., and KENDRICK, H., 1966, *J. appl. Phys.*, **37,** 1260.
WESOŁOWSKI, J., ROZENFELD, B., and SZUSZKIEWICZ, M., 1963, *Acta phys. pol.*, **24,** 729.
WHITE, G. K., 1956, *Can. J. Phys.*, **34,** 1328.
WHITE, G. K., and WOODS, S. B., 1957 a, *Can. J. Phys.*, **35,** 248 ; 1957 b, *Ibid.*, **35,** 346 ; 1957 c, *Ibid.*, **35,** 656 ; 1957 d, *Ibid.*, **35,** 892 ; 1958, *Ibid.*, **36,** 875.
WHITE, G. K., and TAINSH, R. J., 1967, *Phys. Rev. Lett.*, **19,** 165.
WHITTEN, W. B., and PICCINI, A., 1966, *Physics Lett.*, **20,** 248.
WIGNER, E., and SEITZ, F., 1933, *Phys. Rev.*, **43,** 804 ; 1934, *Ibid.*, **46,** 509.
WILDING, M. D., 1967, *Proc. phys. Soc.*, **90,** 801.
WILDING, M. D., and LEE, E. W., 1965, *Proc. phys. Soc.*, **85,** 955.
WILKINSON, M. K., CHILD, H. R., MCHARGUE, C. J., KOEHLER, W. C., and WOLLAN, E. O., 1961, *Phys. Rev.*, **122,** 1409.
WILKINSON, M. K., KOEHLER, W. C., WOLLAN, E. O., and CABLE, J. W., 1961. *J. appl. Phys.*, **32,** 48S.
WILL, G., 1968, *Z. angew. Phys.*, **26,** 67.
WILL, T. A., and GREEN, B. A., 1966, *Phys. Rev.*, **150,** 519.
WILLIAMS, D. L., BECKER, E. H., PETIJEVICH, P., and JONES, G., 1968, *Phys. Rev. Lett.*, **20,** 448.
WILLIAMS, R. W., and DAVIS, H. L., 1968, *Physics Lett.* A, **28,** 412.
WILLIAMS, R. W., LOUCKS, T. L., and MACKINTOSH, A. R., 1966, *Phys. Rev. Lett.*, **16,** 168.
WILLIAMS, R. W., and MACKINTOSH, A. R., 1968, *Phys. Rev.*, **168,** 679.
WINDMILLER, L. R., and KETTERSON, J. B., 1968 a, *Phys. Rev. Lett.*, **30,** 324 ; 1968 b, *Ibid.*, **21,** 1076.
WINDMILLER, L. R., KETTERSON, J. B., and HORNFELDT, S., 1969, *J. appl. Phys.*, **40,** 1291.
WOHLFARTH, E. P., and CORNWELL, J. F., 1961, *Phys. Rev. Lett.*, **7,** 342.
WONG, K. C., 1970, *J. Phys.* C (*Solid St. Phys.*), **3,** 378.
WONG, K. C., WOHLFARTH, E. P., and HUM, D. M., 1969, *Physics Lett.* A, **29,** 452.
WOOD, J. H., 1960, *Phys. Rev.*, **117,** 714 ; 1962, *Ibid.*, **126,** 517.
WOODS, A. D. B., HOLDEN, T. M., and POWELL, B. M., 1967, *Phys. Rev. Lett.*, **19,** 908.
WUNSCH, K. M., WEISS, W. D., and KOHLHAAS, R., 1968, *Z. Naturf.* A, **23,** 1402.
YAMASHITA, J., FUKUCHI, M., and WAKOH, S., 1963, *J. phys. Soc. Japan*, **18,** 999.
YAROVAYA, R. G., and SHKLIAREVSKIĬ, I. N., 1965, *Optika Spektrosk.*, **18,** 832. English translation : *Optics Spectrosc.*, **18,** 465.
YAROVAYA, R. G., and TIMCHENKO, L. I., 1968, *Fizika tverd. Tela*, **10,** 1237. English translation : *Soviet Phys. Solid St.*, **10,** 983.
YOSIDA, K., 1957, *Phys. Rev.*, **106,** 893 ; 1964, *Prog. low Temp. Phys.*, **4,** 265.
YOSIDA, K., and WATABE, A., 1962, *Prog. Theor. Phys.*, **28,** 361.
YU, A. Y. C., DONOVAN, T. M., and SPICER, W. E., 1968, *Phys. Rev.*, **167,** 670.
YU, A. Y. C., and SPICER, W. E., 1966, *Phys. Rev. Lett.*, **17,** 1171 ; 1968 a, *Phys. Rev.*, **167,** 674 ; 1968 b, *Ibid.*, **169,** 497.
ZAĬTSEV, G. A., STEPANOVA, S. V., and KHOTKEVICH, V. I., 1965, *Zh. éksp. teor. Fiz.*, **48,** 760. English translation : *Soviet Phys. JETP*, **21,** 502.

ZAMIR, D., and SCHREIBER, D. S., 1964, *Phys. Rev.*, **136,** A1087.
ZENER, C., 1951 a, *Phys. Rev.*, **81,** 440 ; 1951 b, *Ibid.*, **82,** 403 ; 1951 c, *Ibid.*, **83,** 299 ; 1952, *Ibid.*, **85,** 324.
ZIMAN, J. M., 1959, *Proc. R. Soc.* A, **252,** 63 ; 1961, *Adv. Phys.*, **10,** 1.
ZORNBERG, E. I., 1968, *Solid St. Commun.*, **6,** 729 ; 1970, *Phys. Rev.* B, **1,** 244.
ZORNBERG, E. I., and MUELLER, F. M., 1966, *Phys. Rev.*, **151,** 557.

Subject Index